RENEWALS 458-4574

APPLIED METALLOGRAPHY

APPLIED METALLOGRAPHY

Edited by

George F. Vander Voort

VNR VAN NOSTRAND REINHOLD COMPANY
———————— New York

Copyright © 1986 by Van Nostrand Reinhold Company Inc.

Library of Congress Catalog Card Number 85-22635
ISBN 0-442-28836-0

All rights reserved. No part of this work covered by the copyright hereon may be reproduced or used in any form or by any means—graphic, electronic, or mechanical, including photocopying, recording, taping, or information storage and retrieval systems—without written permission of the publisher.

Printed in the United States of America

Van Nostrand Reinhold Company, Inc.
115 Fifth Avenue
New York, New York 10003

Van Nostrand Reinhold Company Limited
Molly Millars Lane
Wokingham, Berkshire RG11 2PY, England

Van Nostrand Reinhold
480 La Trobe Street
Melbourne, Victoria 3000, Australia

Macmillan of Canada
Division of Canada Publishing Corporation
164 Commander Boulevard
Agincourt, Ontario M1S 3C7, Canada

16 15 14 13 12 11 10 9 8 7 6 5 4 3 2 1

Library of Congress Cataloging-in-Publication Data

Applied metallography.

  Bibliography: p.
  Includes index.
  1. Metallography—Technique. 2. Physical metallurgy
—Technique. I. Vander Voort, George F.
TN690.A66 1986   669'.95   85-22635
ISBN 0-442-28836-0

PREFACE

This book is a collection of articles written to provide the student, technician, engineer, or researcher with detailed information on the nature and application of a number of important metallographic techniques and subjects critical to physical metallurgy studies, materials development, and failure analyses. Consequently, this text should be a valuable extension to basic text books on light and electron microscopy because it emphasizes the application of these procedures as well as describes the basic methodology of their use.

The first five chapters describe important advanced metallographic techniques for revealing the microstructure of materials, usually selectively. For example, Vander Voort's chapter reviews the more traditional selective etching methods, whereas the other four are devoted to more specialized techniques. Stansbury reviews the technique of potentiostatic etching, which represents the ultimate in control of the etching process. Bühler and Aydin describe the interference layer method, perhaps the only universally applicable etching procedure. Gray describes the magnetic etching used to study domain structure and for revealing ferromagnetic phases in paramagnetic materials. Ondracek's chapter details the newly developed gas-contrasting technique, another procedure that employs interference effects.

Three other chapters deal with the use of metallographic techniques with specific materials—stainless steels, low-carbon free-machining steels, and titanium alloys—all very interesting subjects for the metallographer. Crouse and Leslie have reviewed procedures for revealing the structure of stainless steels, which, by the nature of their inherent corrosion resistance, are difficult to etch. Watson has prepared a comprehensive review that shows how microscopy is fostering the development of better free-machining steels, whereas Rhodes shows how vital microscopy has been in the development of titanium alloys.

Quantitative metallography, or stereology, has grown in importance substantially in the past three decades. De-Hoff has prepared a review of basic measurement procedures and has shown how they are applied to generate meaningful data. Underwood's paper provides a comprehensive, up-to-date review of procedures for quantifying fracture features. Application of these techniques should produce further advances in our ability to control fractures.

Blau has written a review of microindentation hardness testing and shows that much more can be done with this technique than merely determining the hardness of materials.

Two chapters deal with electron metallographic techniques. In a mere two decades, the scanning electron microscope has become one of the most important metallographic tools for examining fractures and microstructures. A second chapter by Vander Voort demonstrates how natural and artificially produced contrast is used to reveal microstructures. Williams provides a detailed review of the scanning transmission electron microscope, perhaps the most powerful new tool for fine structure analysis.

Cross et al. have surveyed the use of metallography for the control of welding processes. Metallography has been the cornerstone of the development of strong, safe weldments.

Three chapters deal with the use of metallography in failure analysis. LeMay provides an overview of this important topic. Glaeser and White provide excellent detailed descriptions of the use of metallography for diagnosis of wear and corrosion failures, respectively.

GEORGE F. VANDER VOORT
Editor

CONTENTS

Preface   v

1. Phase Identification by Selective Etching   *G. F. Vander Voort*   1
Phase Identification Methods   1
Analytical Methods   3
Chemical Etchants   6
Tint Etchants   7
Electrolytic Etching   14
Anodizing   16
Heat Tinting   16
Summary   18

2. Potentiostatic Etching   *E. E. Stansbury*   21
Electrochemical Background   21
Etching As a Corrosion Process   23
Relationship of Half-Cell Polarization to Corrosion and Etching   24
Potentiostatic Polarization   25
Interpretation of Polarization Curves for Application to Etching   25
Principles of Potentiostatic Etching   27
Comments on Empirical Procedure   33
Survey of Etchants Used for Potentiostatic Etching   34
Examples of Potentiostatically Etched Microstructures   34

3. Applications of the Interference Layer Method   *H. E. H Bühler and I. Hydin*   41
Objectives of Metallography   41
Optical Characteristics of Metallic and Nonmetallic Phases   41
Basics of the Interference Layer Method (Contrast Enhancement)   42
Preparation of Specimens (Grinding and Polishing)   43
Application of Interference Layers   45
Characteristics of Layer Materials and Layers   46
Applications of the Interference Layer Method   49
Applications Involving Steels   49
Applications Involving Nonmetallic Inclusions (Sulfides)   50
Applications Involving Aluminum Alloys   51
Summary   51

4. Magnetic Etching   *R. J. Gray*   53
Results   53
Conclusions   60

5. The Gas-Contrasting Method   *G. Ondracek*   63
The Gas-Contrasting Apparatus   64
The Principle of Gas Contrasting   65
Gas-Contrasted Microstructures   66
The Interpretation of the Gas-Contrasting Effect   67
Summary   69

6. Techniques for Stainless Steel Microscopy   *R. S. Crouse and B. C. Leslie*   71
Historical Background   71
The Nature of Stainless Steels   71
Types of Stainless Steels   71
General Characteristics of Metallographic Preparation   75
Sensitization   79
Stress-Corrosion Cracking   80
Sigma Phase, Ferrite and Carbides in Austenitic Stainless Steel   85
Preparation for Transmission Electron Microscopy   86
Summary   87

7. Problem Using Quantitative Stereology   *R. T. DeHoff*   89
Making Measurements   89
What Is Measured   91
What the Measurements Mean   92
Instrumentation for Image Analysis   93
Planning an Experiment   94
Examples   96
Summary   98

8. Quantitative Fractography
E. E. Underwood — 101
- Background — 101
- Experimental Techniques — 102
- Analytical Procedures — 110
- Applications — 117
- Summary — 121

9. Methods and Applications of Microindentation Hardness Testing
P. J. Blau — 123
- Hardness, Microhardness, and Microindentation Hardness — 123
- Selecting a Microindentation Hardness Test Method — 123
- Interconversion of Microindentation Hardness Scales — 128
- Selecting a Microindentation Hardness Tester — 129
- Specimen Preparation and Fixing — 132
- Summary — 137

10. The SEM As a Metallographic Tool
G. F. Vander Voort — 139
- SEM Image Contrast Formation — 139
- Resolution — 150
- Microscope Variables — 150
- Specimen Preparation — 154
- SEM Quantitative Metallography — 165
- Summary — 167

11. Metallography in the Scanning Transmission Electron Microscope
D. B. Williams — 171
- The STEM Instrument — 171
- Specimen Preparation — 172
- Spatial Resolution in the STEM — 174
- Imaging in the STEM — 176
- Microdiffraction — 181
- X-Ray Microanalysis — 184
- Electron Energy-Loss Spectrometry (EELS) — 189
- Specimen Thickness — 192
- Summary — 194

12. Metallography and Welding Process Control
C. E. Cross, O. Grong, S. Liv, and J. F. Capes — 197
- Characteristic Features of Welds — 197
- Solidification Microstructure — 198
- Solid-State Transformation in Weld Metal — 203
- Heat-Affected Zone — 207
- Partly Molten Zone — 207
- Welding of Dissimilar Metals — 208

13. Microscopy and the Development of Free-Machining Steels
J. D. Watson — 211
- Specimen Preparation Methods — 212
- Macro Evidence of Microstructure — 213
- Qualitative Metallography — 213
- Quantitative Metallography — 224
- Electrolytic Inclusion Extraction (EIE) — 224
- Sterological Methods — 228
- Metallographic Techniques Applied to the Metal Cutting Process — 230
- Conclusion — 232

14. Microscopy and Titanium Alloy Development
C. G. Rhodes — 237
- Metallographic Techniques — 237
- Microstructure Development — 239
- Alpha and Near-Alpha Phase Alloys — 241
- Alpha+Beta Alloys — 243
- Beta Alloys — 245
- Summary — 247

15. Use of Microscopy in Failure Analysis
I. LeMay — 251
- Failure Mechanisms — 251
- Failure Analysis Methodology — 259
- Conclusion — 259

16. Microscopy and the Study of Wear
W. A. Glaeser — 261
- Wear Modes — 261
- Microscopy — 264
- Preparation of Specimens — 267
- Interpretation — 268
- Metallography — 271
- Stero-Pair Analysis — 274
- Transmission Electron Microscopy — 275
- Wear Debris — 275
- Summary — 277
- Appendix — 278

17. Microscopy and the Study of Corrosion
W. E. White — 281
- Forms and Mechanisms of Corrosion — 281
- Investigative Methods in Corrosion Science and Engineering Practice — 284
- Applied Microscopy—Case Studies — 285
- Pitting Corrosion—Laboratory Studies — 285
- Pitting Corrosion—Field Studies — 287
- Corrosion–Erosion — 289
- Stress Corrosion, Embrittlement, and Fracture — 289
- Summary — 295

Index — 297

APPLIED METALLOGRAPHY

1
PHASE IDENTIFICATION BY SELECTIVE ETCHING

George F. Vander Voort

R&D Laboratory, Carpenter Technology Corp.

The identification of phases or constituents in metals and alloys is a common metallographic problem. Metallographers can often predict which phases are likely to be present in a given alloy from a knowledge of its sample chemical composition, processing history, previously published information on its characteristics, or by reference to phase diagrams. Guided by such information, the metallographer examines the microstructure to identify the phases present. Even when one is possessed of such information, however, the task is often far from trivial, even for well-documented alloy systems. Only after the phases have been identified should a quantification of the amount and morphological characteristics of the phases be attempted.

Many techniques can be used to obtain a qualitative description of the microstructure of materials. Some methods require expensive X-ray or electron metallographic devices, but light microscopy remains the cornerstone of the analytical procedure. Conventional general-purpose etchants that outline the phases are often adequate, but errors are possible. Etchants selective to a particular phase provide positive, unambiguous phase identification and are ideal for image-analysis measurements that depend on gray-level image differences for phase detection. These etchants are also valuable for manual measurements. This chapter will present information on selective etching solutions and their implementation for a variety of commercial materials.

PHASE IDENTIFICATION METHODS

Phases are identified by means of known differences in composition, crystal structure, optical reflectance, morphology, or corrosion behavior. This information can be obtained for most systems in the literature. When one is dealing with unfamiliar systems, compilations of phase diagrams make a good starting point. Basic aids to phase identification will be described in the following sections.

Quantity or Morphology

In some instances, phase identification may be aided by differences in quantity of constituents or morphology. In simple two-phase systems with essentially equilibrium microstructures, the matrix (major) phase can usually be distinguished from isolated, dispersed, second-phase constituents of lower volume fraction. When approximately equal amounts of two phases are present with no obvious continuous-matrix phase, however, identification becomes more difficult.

Reflectance

Some phases of alloys in the as-polished condition, can be detected reliably by means of reflectance differences. Well-known examples include graphite in cast iron, inclu-

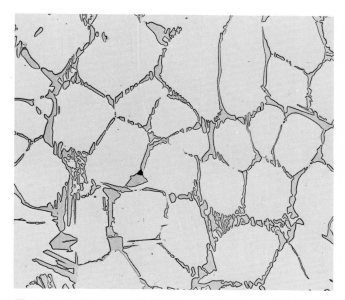

Fig. 1-1. Cu_2Sb in as-cast Sb-10%Cu revealed in the as-polished condition by reflectivity differences (at 100×).

sions, and certain nitrides and intermetallic phases. Figure 1-1 presents a less common example: η phase (Cu_2Sb) in as-cast Sb–10%Cu in the as-polished condition revealed solely by reflectance differences. Beta phase in alpha-beta brass can be detected in the as-polished condition because it appears as a dark yellow compared to the light yellow of the alpha phase. Reflectance differences are important because they are the basis of phase detection by image analyzers.[1]

Polarized Light Response

Phases with noncubic crystallographic structure can be detected by their positive response to crossed polarized light. Alpha phase in titanium alloys responds to polarized light, whereas beta phase does not. Sigma phase in stainless steels also responds to polarized light. Cu_2O in tough-pitch copper can be discriminated from copper sulfide in the as-polished condition by the use of polarized light;

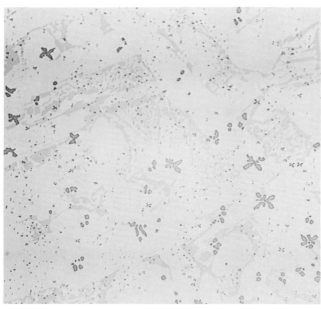

As-Polished

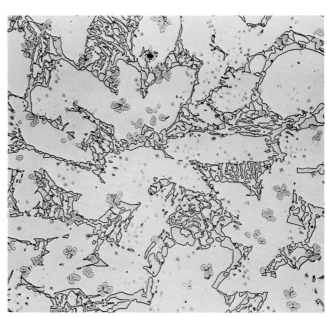

10% Ammonium Persulfate

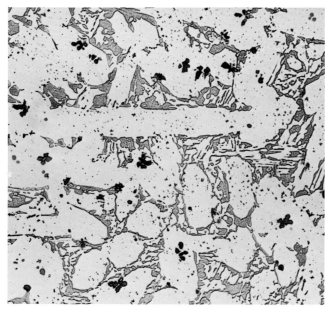

10% Sodium Thiosulfate
(etch 2a)

Beraha's Lead Sulfide Tint
Etch (etch 3h, no $NaNO_2$)

Fig. 1-2. In the as-polished condition, the γ_2 phase in aluminum bronze is faintly visible, whereas the iron particles are somewhat darker. Relief cannot be developed between these phases and the matrix because of inadequate hardness differences and etching must be employed. (All at 500×.)

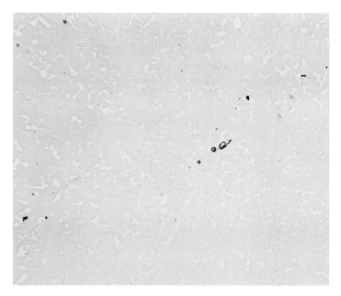 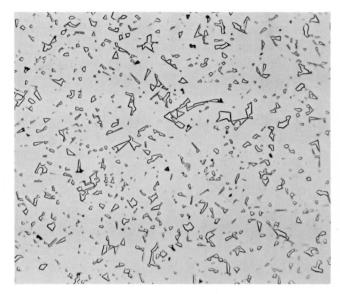

Fig. 1-3. In the as-polished condition, porosity and graphite can be observed on WC–Co samples (left). Relief polishing outlines the cobalt binder phase (right). (Both at 1000×.)

it appears ruby red under such illumination whereas the sulfide is dark.

Relief

In some systems, a minor phase may exhibit little or no difference in reflectivity from the matrix, but the difference in hardness, if appreciable, can be used to reveal the phase by introducing relief during final polishing. Although the relief effect can be observed with bright-field illumination, it can be enhanced by using oblique lighting or differential-interference contrast illumination.

The following two figures illustrate the use of reflectance and relief effects for phase discrimination. Figure 1-2 shows the microstructure of aluminum bronze (Cu–10.5%Al–3.8%Fe–0.3%Ni) heat-treated to produce γ_2 (Al_4Cu_9) in an alpha matrix with insoluble iron particles. In the as-polished condition, the iron particles appear light blue whereas the γ_2 particles are only faintly visible. Relief cannot be developed between these particles, and the matrix and etching must be utilized as shown to reveal them clearly. Figure 1-3 shows an as-polished WC-Co sintered-carbide cutting tool. In this condition, its porosity and graphite can be detected. After relief polishing, the cobalt binder phase appears to be outlined as a result of the very high difference in hardness between Co and WC.

ANALYTICAL METHODS

If these simple methods are inadequate for positive identification, other procedures must be utilized. In general, the preceding methods have limited, if important, applicability. The availability of more sophisticated tools simplifies phase identification for the metallographer. For example, X-ray diffraction can be employed to identify phases that are present in amounts greater than 1 to 2 percent. Although this technique tells the metallographer what phases are present, each must be identified on a polished cross section of the sample.

Because phases usually exhibit compositional differences, energy-dispersive or wavelength-dispersive detectors in scanning electron microscopes (SEM) or in electron microprobe analyzers (EMPA) can provide means for phase identification. Since energy-dispersive detectors cannot detect lightweight elements such as carbon, oxygen, and nitrogen, their value is somewhat limited. Neither instrument is suitable for identification of particles smaller than about 2 μ in diameter. For such particles, it is necessary to use electron diffraction of extracted particles or other advanced techniques. Identification procedures for fine particles exceed the scope of this chapter (but see Chap. 11).

Selective Etching

Many metallographers who do not have access to X-ray and electron metallographic instruments for routine work can utilize the etching response to specifically chosen reagents for phase identification, except in the case of very fine particles. Prior to the development of the SEM and the EMPA, the response to selected etchants was widely used to identify inclusions in steels and intermetallic constituents in aluminum alloys, but such procedures are rarely employed for these purposes today.

The use of selective etchants is invaluable for quantitative metallography, a field of growing importance. To detect features by image analysis, the desired phase must be darker or lighter than all other phases present.[1] Unfortunately, etchant selection is empirical in nature and often requires considerable trial-and-error. Selection of potential etchants is aided by referring to published compilations of standard and specialized etchants.[2-4] A list of typical etchants sufficiently specific in action to be suitable for phase identification appears in Table 1-1.

A wide variety of etching methods may be used for phase identification: chemical solutions (for immersion or swabbing), tint etchants, anodizing and electrolytic etchants, and potentiostatic etching. Additionally, related methods—e.g., heat tinting, gas contrasting, magnetic colloids, and vapor-deposited interference films—are very effective. The simpler methods will be discussed in this chapter. Some of the more specialized methods will be described elsewhere.

Table 1-1. Etchants for Selective Phase Identification.

METAL	PHASE DETECTED	ETCH COMPOSITION	COMMENTS
1. Aluminum alloys	(a) $FeAl_3$	0.1–10 ml HF 90–100 ml water*	Attacks $FeAl_3$; other constituents outlined.
	(b) Al_3Mg_2	1–2 ml HNO_3 98–99 ml ethanol	Colors Al_3Mg_2 brown (Al–Mg alloys). Immerse for 15 min for 1-percent solution.
	(c) $CuAl_2$	20 ml water 20 ml HNO_3 3 g ammonium molybdate	Mix one part of reagent with four parts ethanol. Immerse until surface is colored. $CuAl_2$ blue.
	(d) $CuAl_2$	200 ml water 1 g ammonium molybdate 6 g NH_4Cl	Immerse for 2 min. $CuAl_2$ violet. (Lienard and Pacque.)
2. Copper	(a) γ_2	100 ml water 10 g sodium thiosulfate	Colors γ_2 in aluminum bronze. Immerse 90 to 150 sec. (Aldridge.)
	(b) Alpha	One part water One part H_2O_2(3%) One part NH_4OH	Swab etch. Colors alpha in alpha-beta brass.
	(c) Beta	50 ml saturated aq. sodium thiosulfate 1 g potassium metabisulfite	Colors beta in alpha-beta brass. Immerse about 3 min. (Klemm's I reagent.)
	(d) Beta	10 g cupric ammonium chloride 100 ml water	Add NH_4OH until solution is neutral or slightly alkaline. Colors beta in alpha-beta brass.
3. Iron and Steel	(a) Ferrite	50 ml saturated aq. sodium thiosulfate 1 g potassium metabisulfite	Colors ferrite (blue and red), martensite (brown), austenite and carbide (not affected). Immerse 60 to 120 sec after light pre-etch with nital or picral. (Klemm's I reagent.)
	(b) Martensite	8–15 g sodium metabisulfite 100 ml water	Pre-etch lightly with nital. Immerse for about 20 sec. Darkens as-quenched martensite; carbides unaffected.
	(c) Ferrite	(1) 8 g sodium thiosulfate 6 g ammonium nitrate 100 ml water (2) 1 ml HNO_3 2.5 ml H_3PO_4	To 100 ml of (1) heated to 70°–75°C, add 0.5 ml of (2). Pre-etch lightly with picral. Immerse in solution for about 1 min. Etch only good for 15 min. Colors ferrite; cementite unaffected. (Beraha.)
	(d) Fe_3C	100 ml water 1 g sodium molybdate 0.1–0.5 g ammonium bifluoride	Add HNO_3 to decrease the pH to 2.5–3.5. Pre-etch lightly with picral. Immerse in solution about 15 sec. Colors Fe_3C yellow-orange; ferrite unaffected. (Beraha.)
	(e) Fe_3C Ferrite Fe_3P	1000 ml Water 240 g sodium thiosulfate 20–25 g cadmium chloride 30 g citric acid	Dissolve in order given. Allow each to dissolve before adding next. Age 24 hr at 20°C in dark bottle. Filter 100 ml of solution; good for 4 hr. Pre-etch with picral. Immerse 20 to 40 sec; colors ferrite. After 60 to 90 sec, ferrite is yellow or light blue, Fe_3P brown, carbides violet or blue. (Beraha.)

Table 1-1 (continued)

METAL	PHASE DETECTED	ETCH COMPOSITION	COMMENTS
	(f) Ferrite	1 g potassium metabisulfite 1 ml HCl 100 ml water	Pre-etch lightly with picral. Immerse in solution until surface is colored. Colors ferrite. (Beraha.)
	(g) M_3C M_6C	2 g picric acid 25 g NaOH 100 ml water	Immerse in boiling solution 1 to 15 min, or use electrolytically at 6 V dc, 20°C, 0.5–2 A/in², 30 to 120 sec. Colors M_3C and M_6C brown to black.
	(h) Ferrite Fe_3C MnS	1000 ml water 240 g sodium thiosulfate 24 g lead acetate 30 g citric acid	Mix and age in same manner as etch No. 3e. Add 0.2 g sodium nitrite to 100 ml of solution; good for 30 minutes only. Pre-etch with picral. Immerse in solution until surface is colored. Cementite will be darker than ferrite; MnS white. (Beraha.)
	(i) Oxygen	145 ml water 16 g CrO_3 80 g NaOH	Alkaline chromate etch to detect oxygen enrichment during or before hot working. Heat to 118° to 120°C; immerse for 7 to 20 min. Enriched areas are white. (Fine.)
	(j) Fe_3P Fe_3C Alloy carbides Sigma Delta ferrite	100 ml water 10 g $K_3Fe(CN)_6$ 10 g KOH	Immerse at 20°C; M_7C_3 is attacked; Mo_2C turns brown to black; Fe_3C is unaffected. Bring to boil. Fe_3P darkens in 10 sec; Fe_3C is lightly colored after 2 min. For stainless steels, immerse at 20°C to reveal carbides, sigma after 3 min. At 80°C to boiling for 2 to 60 min, carbides turn dark; sigma, blue; delta, yellow to brown; austenite is unaffected. (Murakami's reagent.)
	(k) Sigma Carbides Delta ferrite	100 ml water 20 g $K_3Fe(CN)_6$ 20 g KOH	For stainless steels, use cold to boiling: carbides turn dark; delta, yellow; sigma, blue; austenite is unaffected. (Le-May and White.)
	(l) Sigma Carbides Delta ferrite	100 ml water 30 g $K_3Fe(CN)_6$ 30 g KOH	For stainless steels, use at 95°C for 15 sec: sigma turns reddish brown; delta, dark gray; carbide, black; austenite is unaffected. (Kegley.)
	(m) Austenite	Heat tint in air at 500 to 700°C, up to 20 min.	For stainless steels: austenite colors before delta ferrite.
	(n) Sigma Delta ferrite	10 N KOH (56 g KOH in 100 ml water)	For stainless steels, use electrolytically at 2.5 V dc, 20°C for 10 sec to color sigma orange and delta light blue and brown; austenite is unaffected.
	(o) Sigma Delta ferrite	100 ml water 20 g NaOH	For stainless steels, use electrolytically at 5 V dc, 20°C for 20 sec to color sigma orange and delta tan; austenite is unaffected.
	(p) Carbide	NH_4OH (Conc)	For stainless steels, use electrolytically at 1.5 V dc; carbide completely revealed in 40 sec and sigma unaffected after 180 sec. At 6 V dc, sigma is etched after 40 sec.
4. Magnesium alloys	(a) $Mg_{17}Al_{12}$	90 ml water 10 ml HF	Immerse for 3 to 30 sec. Darkens $Mg_{17}Al_{12}$; $Mg_2Al_2Zn_3$ unetched. (George.)
	(b) MgZn	1000 ml water 50 g CrO_3 4 g Na_2SO_4	For Mg–Zn alloys, a 2-sec immersion severely attacks MgZn; Mg_7Zn_3 is slightly attacked. (Clarke and Rhines.)
5. Nickel alloys	(a) γ'	50 ml HCl 1 to 2 ml H_2O_2(30%)	For superalloys, immerse 10 to 15 sec to attack gamma prime.
	(b) γ' Carbides	100 ml ethanol 1 to 3 ml selenic acid 20 to 30 ml HCl	Tint etch for superalloys. Immerse 1 to 4 min at 20°C. Colors carbides and gamma prime, matrix unaffected. (Beraha.)
6. Titanium alloys	(a) Retained beta	10 ml 40% Aq. KOH 5 ml H_2O_2 (30%) 20 ml water	Swab up to 20 sec or immerse for 30 to 60 sec at 70° to 80°C. Stains alpha and transformed beta; retained beta remains white.

Table 1-1 (continued)

METAL	PHASE DETECTED	ETCH COMPOSITION	COMMENTS
	(b) Retained beta Alpha Martensite Carbides	Heat tint in air at 400° to 700°C	Tint until surface is colored red-violet. After 60 sec at 600°C, retained beta turns deep violet to bright blue; alpha, dull to golden yellow; martensite, yellow with violet coloring from fine retained beta; carbides, brilliant bright blue or yellow.
7. Sintered carbide	(a) Cobalt binder	HCl saturated with $FeCl_3$	Immerse 1 to 5 min at 20°C to blacken the cobalt binder phase. (Chaporova.)
	(b) Eta	100 ml water 10 g $K_3Fe(CN)_6$ 10 g KOH	Immerse 2 to 10 sec at 20°C to darken eta; longer times attack eta. TiC attacked; WC unaffected (outlined). (Murakami's reagent.)
	(c) Cobalt binder	Heat tint at 316° to 593°C in air	Tint in air at 316°C for 5 min to color cobalt brown; carbides are unaffected. At higher temperatures, carbides are colored, TiC before WC.

* Use distilled water in all reagents where water is required.

CHEMICAL ETCHANTS

Successful use of etching solutions requires development of controlled corrosive action between areas of different electrochemical potential. Etching is a result of electrolytic action at structural constituents caused by chemical or physical differences on the sample surface. Under the imposed conditions, these variations cause certain areas to be anodic and others cathodic.

Contrast between similar constituents may be developed as a result of variations in etch rate produced by differences in crystallographic orientation. Such etchants produce reflectivity differences and, hence, grain contrast. When more than one phase is present, an etch may attack one phase preferentially, thereby producing elevation differences between the constituents. If this attack is uniform, little reflectance difference will be noted between the constituents, but the phase boundaries will be visible. Such behavior is encountered when annealed steels containing ferrite and carbide are etched with either picral or nital. Both these etchants attack the ferrite phase, leaving the carbide in relief.

In some instances, the phase being attacked is roughened during etching, and this roughness produces reflectance differences. Figure 1-4, for example, shows an as-cast Zn–15%Sn sample whose zinc-rich proeutectic phase has been preferentially attacked by etching with a solution containing 82 ml water, 1 ml HF, and 15 ml H_2SO_4 at about 38°C for 30 sec. At low magnifications, the zinc-rich primary dendrites appear dark and vividly revealed in contrast to the tin-rich, unetched continuous-eutectic constituent. High magnification reveals the roughening of the zinc-rich phase. Such an etching response is unsuitable for high magnification work.

Examples

Two examples of two-phase alloys in which either phase can be selectively attacked by simple chemical etchants will be given here. The first, an unfamiliar example, is an as-cast Cd–37%Cu alloy that consists of two intermetallic compounds, $\delta(Cd_8Cu_5)$ and $\gamma(Cd_3Cu_4)$. Figure 1-5 shows the microstructure of this alloy—primary δ containing precipitated excess γ in a matrix of eutectic δ plus γ. Etching with a solution consisting of equal parts of 10-percent aqueous $FeCl_3$ and HCl attacked and darkened the γ phase, whereas etching with 10-percent aqueous ammonium persulfate attacked and darkened the δ phase.

A more common example of such dual etching behavior is shown in Fig. 1-6 where α/β brass (Cu–40%Zn) (β phase is ordered) has been etched with four different reagents. Etching with 10-percent aqueous ammonium persulfate attacked the β phase, but it was not roughened, and only the phase boundary is visible, producing an outlining of each β grain. Kehl and Metlay[5] have shown that nearly all chemical etchants are specific to the β phase. Only one etchant—equal parts of water, H_2O_2 (3 percent), and NH_4OH (swab)—will attack the α phase preferentially. This etch produces vivid grain contrast.

Several etchants have been commonly employed to darken the β phase preferentially. A commonly used etchant is 10-percent aqueous cupric ammonium chloride made neutral or slightly alkaline by adding NH_4OH (the solution has a pH of 2.5; addition of 6 ml NH_4OH to 100 ml of etch increased the pH to 7.25). This etch (No. 2d in Table 1-1) roughens the β phase. Tint etchants, such as Klemm's I reagent (No. 2c in Table 1-1), will color the β phase after a short immersion. After about an hour in this solution, the alpha matrix is slightly revealed.

For pure metals and single-phase alloys, etching response is due to potential differences between differently oriented grains or between grain boundaries and grain interiors or is due to concentration gradients within solid-solution phases. For two-phase and multi-phase alloys, there are also potential differences between phases of dif-

PHASE IDENTIFICATION BY SELECTIVE ETCHING

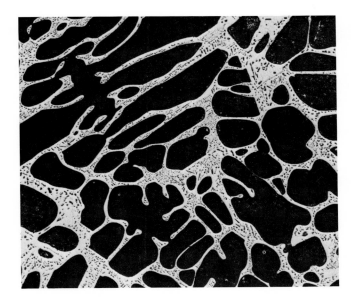

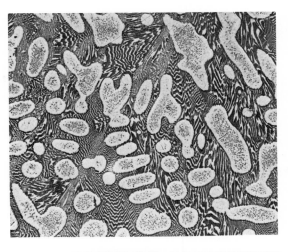

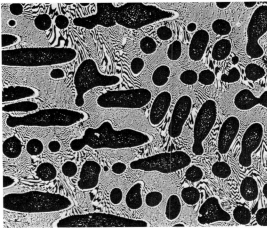

Fig. 1-5. Simple immersion etchants can reveal either phase in as-cast Cd–37%Cu. Top: etching with equal parts 10-percent aqueous $FeCl_3$ and HCl colors the γ phase (Cd_3Cu_4). Bottom: etching with 10-percent aqueous ammonium persulfate colors the δ phase (Cd_8Cu_5). (Both at 200×.)

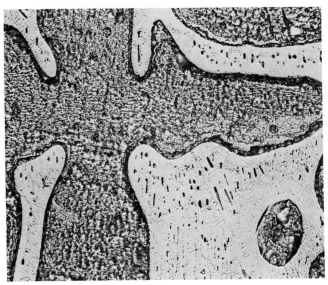

Fig. 1-4. Bottom: the primary zinc-rich dendrites in an as-cast Zn-15%Sn alloy and the zinc-rich portion of the Zn-Sn continuous eutectic were attacked and roughened by etching in a solution containing 82 ml water, 1 ml HF, and 15 ml H_2SO_4 (at 500X). Top: this attack produced good contrast for low magnification examination (at 100×).

ferent composition. Impurity segregation to grain boundaries also promotes selective grain-boundary etching. Such potential differences control the nature of etch dissolution, thus producing differences in the rate of attack that reveal the microstructure.

In a two-phase alloy, one phase will be higher in potential than the other phase; i.e., it is anodic to the other (cathodic) phase in a particular etchant. The more electropositive (anodic) phase is attacked, whereas the electronegative (cathodic) phase is attacked superficially or not at all. Duplex alloys etch more readily than pure metals or single-phase alloys because of the greater potential difference between the two phases.

The unetched cathodic phase will stand in relief and appear bright, whereas the anodic phase is recessed below the plane-of-polish and will appear either light or dark depending on whether or not its surface was roughened. Simple chemical etchants work in this manner. For most two-phase alloys, nearly all etchants will attack one phase—the anodic phase—producing either phase-boundary etching only or darkening of the anodic constituent. For most two-phase alloys, it is rather difficult to find a simple etchant that will attack the other phase; i.e., for most chemical immersion etchants, the same phase will be anodic.

TINT ETCHANTS

Considerable progress has been made in the development of simple immersion solutions that produce selective color contrast by film formation, substantially because of the work of Beraha.[6] Such reagents, generally referred to as tint etchants, are usually acidic solutions with either water or alcohol as solvents. They are chemically balanced to deposit a thin (40 to 500 nm) film of oxide, sulfide, complex molybdate, elemental selenium, or chromate, on the sample surface. Coloration is developed by interference between light rays reflected at the inner and outer film surfaces. Recombination of two light waves with a phase difference of 180 degrees causes light of a certain wavelength to disappear so that the reflected light is of a complementary color. Crystallographic orientation, film thickness, and refractive index of the film control the colors produced.

Tint etchants are either anodic reagents that precipitate a thin film on the anodic constituents to color only these areas, cathodic reagents that precipitate a thin film on the cathodic constituents to color only these areas, or complex reagents where the film is precipitated from a complex reaction.[6]

Some of the most common ingredients of tint etchants are sodium metabisulfite ($Na_2S_2O_5$), potassium metabisulfite ($K_2S_2O_5$), and sodium thiosulfate ($Na_2S_2O_3 \cdot 5H_2O$). When dissolved in water, the metabisulfite salt decomposes during etching, producing SO_2, H_2S, and H_2 (producing a characteristic odor). When etching passivated

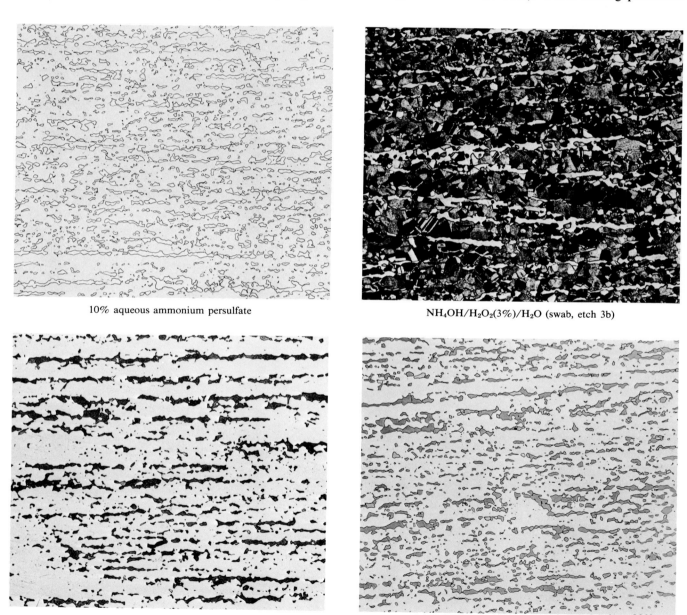

Fig. 1-6. Influence of different etchants on the microstructure of alpha–beta brass (Cu–40%Zn). Top left: beta phase outlined but not roughened. Top right: alpha phase contrast etched; beta not affected. Bottom left: beta phase attacked and darkened. Bottom right: beta phase colored by tint etch. (All at 100×.)

surfaces—e.g., stainless steel—SO_2 depassivates the surface, and the H_2S provides sulfur ions that combine with metallic ions from the sample to produce the sulfide staining film. The addition of HCl to tint etchants extends their use to the corrosion-resistant alloys.

In tint etchants based on these compounds, anodic constituents—e.g., ferrite, martensite, and austenite—are colored, whereas the cathodic constituents—e.g., carbides and nitrides—are unaffected and appear bright. A variety of hues are produced in the anodic constituents as a result of differences in crystallographic orientation. In most cases, coloration can be further enhanced by observation with polarized light.

Tint etchants based on selenic acid (a dangerous chemical) produce film deposition on cathodic areas by reduction of the acid to elemental selenium, which deposits on the cathodic features. Tint etchants containing sodium molybdate also color cathodic constituents by precipitation of a complex molybdate film on these features.

Beraha[6] also developed a complex thiosulfate tint etchant containing lead acetate and citric acid. This etchant (No. 3h in Table 1-1) is useful for tint etching copper-based alloys (see Fig. 1-2) and for staining cementite or phosphide in cast-iron or sulfide inclusions in steels. The film produced is due to the deposition of lead sulfide.

Mixing of these etchants should be done strictly according to the instructions provided by the developer (see References 4 and 6 and also Table 1-1). Some of these reagents must be aged in a dark bottle before use. Although selenic acid is highly corrosive and toxic and should be handled with extreme care, most of them are relatively safe to use and some can be stored as stock solutions. In most work, a light pre-etch with a general-purpose etch that produces uniform attack without roughening of the constituents improves image sharpness after tinting.

The sample is immersed in a suitable quantity of the tint etchant for the recommended time (not necessarily correct for all samples), usually long enough to produce a red-violet macroscopic color. After tinting, do not touch the sample surface. Polishing must be of a very high quality to obtain best results since tint etchants will reveal vividly even the most superficial surface scratches, even when none appear to be present before tinting. Final polishing with colloidal silica—aided in some cases by attack-polishing solutions[4]—is usually very effective for such work. When successfully performed, tint-etch results can be remarkably beautiful as well as technically valuable.

Examples: Nonferrous Alloys

Figure 1-6 illustrated the use of Klemm's I tint etch (etch 2c in Table 1-1), a very versatile reagent. Klemm's I will tint etch all-alpha, copper-based alloys, but the rate of film formation varies with sample composition and processing. For example, single-phased cartridge brass (Cu–30%Zn) may require up to an hour or more to tint etch, whereas single-phase phosphor bronze (Cu–5%Sn–0.2%P) etches in a few minutes. Cold-rolled cartridge brass etches much more quickly than annealed cartridge brass.

Figure 1-7 shows the microstructure of an annealed (790°C; cool 25°C/hr to 590°C; air cool), as-cast Cu–10%Sn–10%Ni alloy. Etching with aqueous 3-percent ammonium persulfate, 1-percent ammonium hydroxide outlined and slightly colored the γ phase $(Cu, Ni)_3Sn$ in the α plus γ lamellar structure. Tinting with Klemm's I reagent colored the α phase. Figure 1-8 shows the microstructure of an as-cast Ag–28%Cu eutectic alloy, a mixture of essentially pure copper and pure silver. Etching

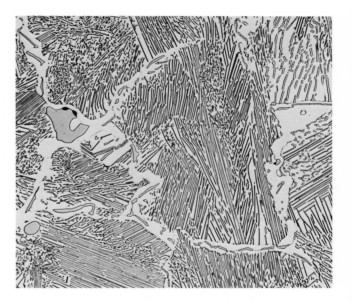

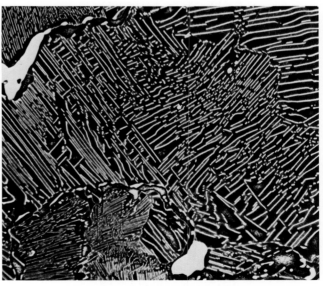

Fig. 1-7. Microstructure of annealed, as-cast Cu–10%Sn–10%Ni revealed by two etchants. Top: γ phase $(Cu, Ni)_3Sn$ outlined with aqueous 3-percent ammonium persulfate, 1-percent NH_4OH. Bottom: alpha matrix phase colored by Klemm's I reagent. (Both at 500×.)

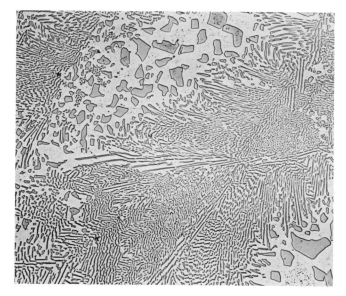

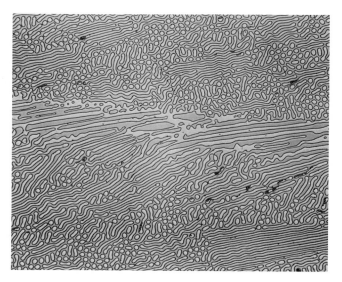

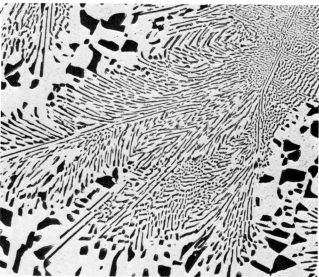

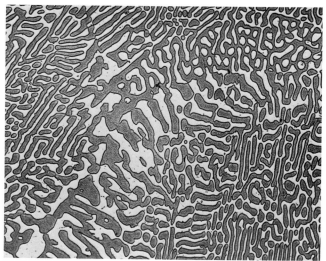

Fig. 1-8. Eutectic microstructure of Ag–28%Cu revealed by two etchants. Top: copper phase outlined with 500 ml H$_2$O, 7.5 g CrO$_3$, 4 ml H$_2$SO$_4$. Bottom: copper phase colored with Klemm's I reagent. (Both at 500×.)

with a solution containing 500 ml water, 7.5 g CrO$_3$, and 4 ml H$_2$SO$_4$ outlined the copper phase, which is visible because of the reflectance difference between copper and silver. Tint etching with Klemm's I reagent darkened the copper grains producing strong color contrast.

Several etchants will selectively color θ phase (CuAl$_2$) in Al–Cu alloys. Figure 1-9 illustrates the use of etchants to outline or color CuAl$_2$ in an as-cast eutectic Al–33%Cu alloy. Keller's reagent (95 ml water, 1.5 ml HCl, 1 ml HF, 2.5 ml HNO$_3$) outlines the CuAl$_2$ phase, which appears darker than the matrix as a result of reflectance differences. Two reagents (etchants Nos. 1c and 1d in Table 1-1)—both based on ammonium molybdate—were used. Etch No. 1c, from Anderson's compilation,[2] colored

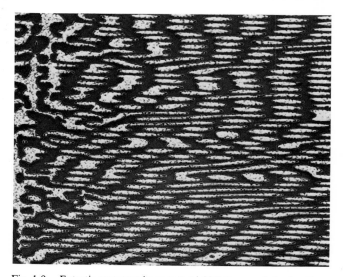

Fig. 1-9. Eutectic structure in as-cast Al-33%Cu revealed using three reagents. Top: Keller's reagent outlines CuAl$_2$. Middle: CuAl$_2$ colored blue with etch No. 1c. Bottom: CuAl$_2$ colored violet with etch No. 1d. (All at 500×.)

PHASE IDENTIFICATION BY SELECTIVE ETCHING 11

CuAl$_2$ blue, whereas etch No. 1d[7] colored CuAl$_2$ violet.

Examples: Ferrous Alloys

Figure 1-10 shows eight examples of how spheroidized cementite in a ferritic matrix (AISI W2 tool steel, Fe–1%C–0.3%Mn–0.1%V) can be revealed. Nital or picral are commonly used to etch such a structure. Both attack the anodic ferrite phase, leaving the carbide standing in relief. Nital, however, is sensitive to crystallographic orientation, and the carbides within some grains are poorly delineated. Nital also reveals the ferrite grain boundaries, an undesirable result if the amount of cementite is to be measured by image analysis. Strong contrast between these phases can be obtained by using selective etchants. Klemm's I reagent (etch No. 3a) colors the ferrite red and blue, leaving the cementite white. Boiling or electrolytic alkaline sodium picrate (etch No. 3g) colors cementite preferentially. Several of Beraha's tint etchants can be used to color either ferrite or cementite. Beraha's sodium thiosulfate-ammonium nitrate solution (etch No. 3c) and his potassium metabisulfite-HCl reagent (etch No. 3f) both color the ferrite, although the former produced pitting. Beraha's sodium-molybdate reagent (etch No. 3d) and his cadmium sulfide-reagent (etch No. 3e)

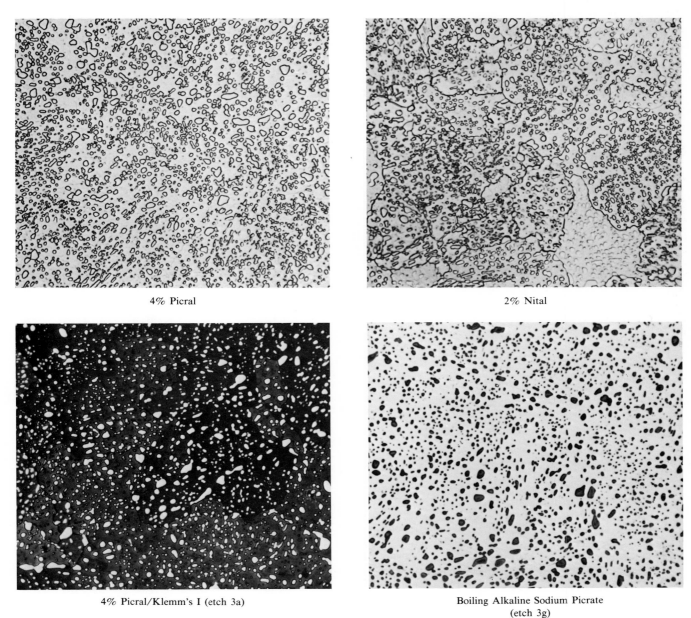

4% Picral

2% Nital

4% Picral/Klemm's I (etch 3a)

Boiling Alkaline Sodium Picrate (etch 3g)

Fig. 1-10. Examples of selective etching of ferrite or cementite in spheroidized annealed AISI W2 tool steel compared to etching with standard nital and picral immersion etchants. (All at 1000×.) Tint etchants provide selectivity required for image analysis. See Table 1-1 for etch compositions.

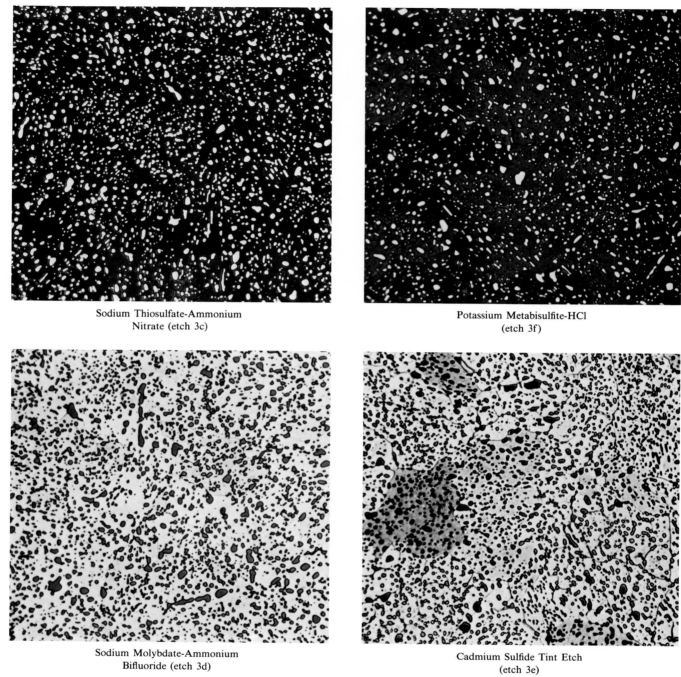

Sodium Thiosulfate-Ammonium Nitrate (etch 3c)

Potassium Metabisulfite-HCl (etch 3f)

Sodium Molybdate-Ammonium Bifluoride (etch 3d)

Cadmium Sulfide Tint Etch (etch 3e)

Fig. 1-10 (continued)

both preferentially color the cathodic cementite phase. Tint etch No. 3d colored the ferrite yellow and white and the cementite orange-brown. Longer etch times (30 to 45 sec) darkened the ferrite and reduced the contrast. Etch No. 3e colored the ferrite yellow and blue and the carbide brownish violet. Although Beraha recommended[6] nital as the pre-etchant, the orientation sensitivity of this etch degrades tint-etch results, and picral is preferred for these structures.

Figure 1-11 shows the microstructure of gray cast iron (Fe–4%C–0.1%P–1.2%Si) containing the ternary ferrite-cementite-phosphide eutectic. Etching with picral and nital (sequentially) revealed the pearlite matrix and outlined the ferrite in the eutectic. There was no distinction, however, between cementite and phosphide. Boiling alkaline sodium picrate (etch No. 3g) colored the cementite only. Murakami's reagent (etch No. 3j, not illustrated) darkened the phosphide after about 10 sec and lightly

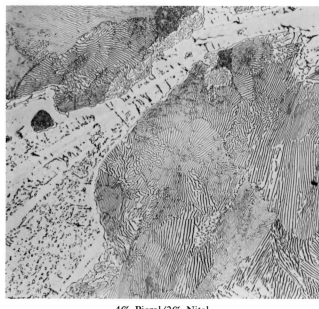
4% Picral/2% Nital
Outlines ferrite but does
not attack Fe₃C/Fe₃P interface

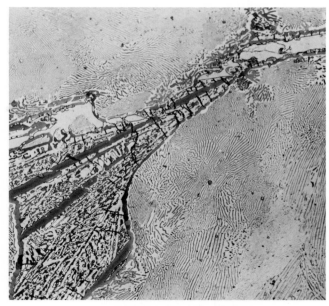
Alkaline Sodium Picrate (etch 3g)
40 sec boiling, Fe₃C brown

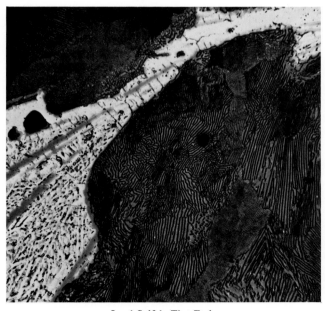
Lead Sulfide Tint Etch
(etch 3h) Fe₃C pale blue

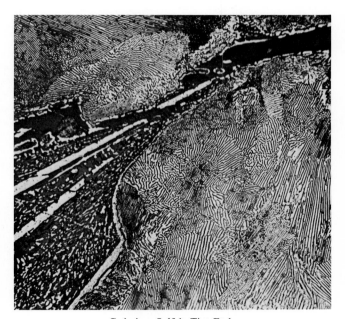
Cadmium Sulfide Tint Etch
(etch 3e) Fe₃C yellow-tan

Fig. 1-11. Examples of selective etching techniques for phase delineation of ternary ferrite–cementite–phosphide eutectic in gray cast iron. (At 500×.) See Table 1-1 for etch compositions.

colored the cementite after about 2 min in the boiling solution. Beraha's lead-sulfide reagent (etch No. 3h) colored both ferrite and cementite (darker of the two). Beraha's cadmium-sulfide reagent (etch No. 3e) colored all three phases: Phosphide shows up quite dark; cementite and ferrite are light with little contrast difference in black and white photography.

Beraha's lead-sulfide reagent (etch No. 3h) is exceptionally useful for preferentially coloring manganese-sulfide inclusions. Figure 1-12 shows an as-polished sample of an alloy steel containing MnS inclusions and one darker oxide inclusion. The sample was etched with nital and then with the lead-sulfide tint etch, which deposited a white film on the sulfides.

Two alkaline chromate etchants[8,9] have been developed to detect and reveal oxygen enrichment associated with

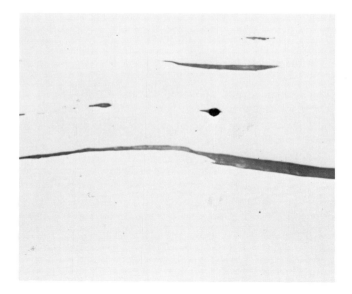

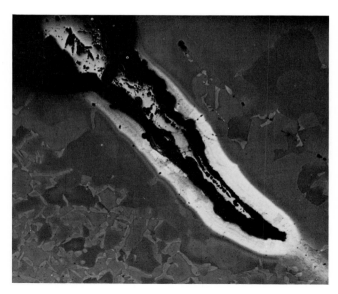

Fig. 1-13. Oxygen enrichment at a forging lap in an alloy steel revealed by Fine's alkaline chromate etch (etch No. 3i). (At 100×.)

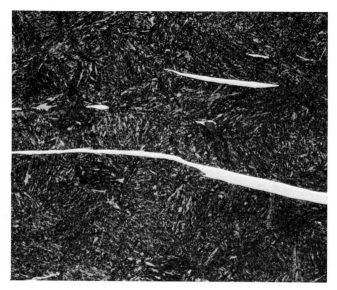

Fig. 1-12. Manganese sulfide can be positively identified and discriminated by the use of Beraha's lead-sulfide tint etch (etch No. 3h). Top: as-polished. Bottom: 2-percent nital pre-etch, lead-sulfide tint etch for 90 sec. (Both at 500×.)

defects in steels present at hot-working temperatures. One such etch[8] (etch No. 3i) was used to reveal oxygen enrichment at a forging lap in an alloy steel, as shown in Fig. 1-13. The oxygen-enriched area is white, whereas the matrix was colored light green and pink.

ELECTROLYTIC ETCHING

Although many electropolishing solutions can be used as etchants by reducing the applied voltage at the end of the polishing cycle, none of these are noted for selectivity. A number of electrolytic etchants have been developed that do provide excellent selective etching action, but none of these are useful for polishing. Electrolytic etching solutions are often rather simple in composition as the applied potential replaces the oxidizer component in immersion etchants.

In nearly all work, the sample is the anode, and direct current is employed. Greater control of the etching process is achievable by selection of the electrolyte composition, etching time, voltage, temperature, and current density. Generally speaking, an underetched sample can be placed back into the electrolyte to strengthen the attack without repolishing. After some familiarity is gained, etching conditions can be easily reproduced.

Stainless Steels

Electrolytic etchants have been widely used in the study of stainless steels. The classic study of Gilman[10] provided a clear understanding of selective etching of such steels. Simple hydroxide solutions will selectively color certain constituents. The stain can be removed by light repolishing, leaving the constituent outlined. The colors produced pass through a well-defined sequence: yellow, green, red, blue, and reddish brown. These colors are produced by interference effects. Strong hydroxide solutions—e.g., KOH—attack sigma phase more readily than carbides, whereas weak hydroxides—e.g., NH_4OH—attack carbides but not sigma. For hydroxides of intermediate strength, changing the voltage controls the attack, carbides being stained at low voltages and sigma and carbide being stained equally at high voltages. To identify phases in austenitic stainless steels, Gilman[10] recommended that samples be pre-etched with Vilella's reagent (100 ml ethanol, 1 g picric acid, 5 ml HCl) to outline the phases. Next, the sample should be electrolytically etched with 10N KOH at 3 V dc for 0.4 sec; doing so colors sigma but not carbide. Then, the sample is electrolytically etched in concentrated NH₄OH at 6 V dc for 30 sec to color the carbides.

Delta ferrite in martensitic or austenitic stainless steels can be preferentially revealed by electrolytic etching in 20-percent aqueous NaOH at 20 V dc for 20 sec (etch No. 3o).[11] Coloration is more uniform than with 10N KOH (etch No. 3n), which also colors delta ferrite (2.5 V dc for 10 sec). Figure 1-14 shows delta ferrite in solution annealed and aged 17–4PH stainless steel (Fe–0.03%C–16.5%Cr–4%Ni–3%Cu–0.3%Nb). Since electrolytic 20-percent NaOH colors only the delta ferrite, the latter is easily observed at any magnification. Immersion etching with Fry's reagent (40 ml HCl, 5 g CuCl₂, 30 ml water, 25 ml ethanol) revealed the martensitic microstructure very well and outlined the delta ferrite, but the results were not as good as the electrolytic etch. Most simple immersion etches attack only the martensite in this sample.

Carbides in austenitic stainless steels are often selectively revealed by electrolytic etching with either 10-percent aqueous oxalic acid (6 V dc for 15 to 30 sec) or 10-percent aqueous sodium cyanide (6 V dc for up to 5 min). Since oxalic acid will also outline matrix phases in austenitic stainless steels and etch out the sigma phase, its action must be carefully controlled.

The electrolytic etchants are simpler and more reliable for phase identification in stainless steels than the Murakami-type reagents (etchants Nos. 3j to 3l). Etching results with Murakami-type reagents vary with etch composition, temperature, time, and phase orientation. When

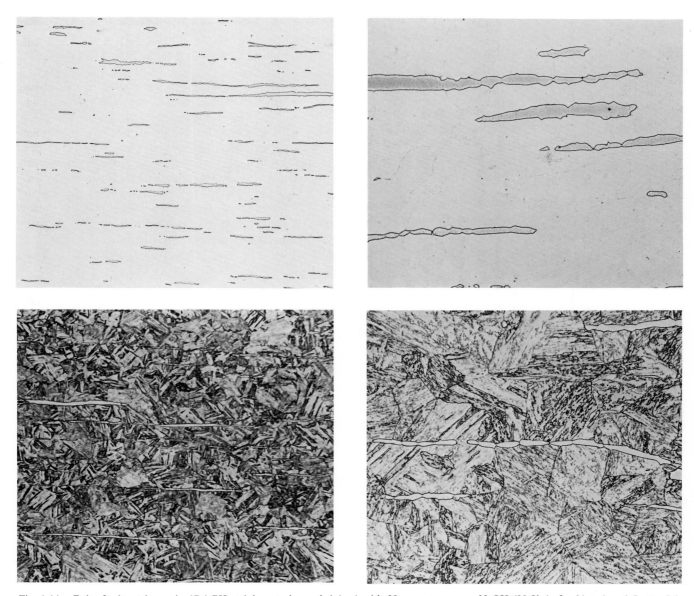

Fig. 1-14. Delta ferrite stringers in 17-4 PH stainless steel revealed (top) with 20-percent aqueous NaOH (20 V dc for 21 sec) and (bottom) by matrix etching with Fry's reagent. [100× (left) and 500× (right).]

using such reagents, conditions must be rigorously controlled, and it is wise to compare etch results with those from control samples of known phase composition. These reagents are quite popular, but the sharpness and uniformity of etching is not always as good as with electrolytic etchants. The latter are preferred for image analysis work.[1]

ANODIZING

Anodizing is an electrolytic etching process used to produce an oxide film on the sample surface to develop interference colors selective to the mirostructure. Although the technique is most commonly associated with grain structure development in aluminum, it has also been used for phase identification in titanium,[12-16] niobium,[17] and zirconium.[18,19] The solutions used for these alloys are rather complex compared to the electrolytes discussed previously.

HEAT TINTING

Heating polished samples in air at relatively low temperatures will oxidize the surface. The oxidation rates of constituents vary, depending on their composition, and produce variations in oxide film thickness that generate colors characteristic of the constituents. Films greater than 30 nm thick will produce visible interference colors.[20] Oxidation rates also vary with crystallographic orientation as demonstrated by heat tinting of single-crystal spheres.[21]

Heat tinting is a very simple process. Prior to heating, the sample should be lightly pre-etched with a general-purpose etchant that does not roughen any of the phases present. Pre-etching yields sharper definition of the structure after heat tinting. Next, the sample is placed face up on a hot plate or in a laboratory furnace. The latter produces more consistent results because of the difficulty in controlling heating on a hot plate. Contrary to some pronouncements, heat tinting results can be quite reproducible.

Examples: Ferrous Alloys

For carbon steels, heat tinting provides no advantage over standard etchants. However, for cast irons, alloy steels, and stainless steels, heat tinting is very useful.[4] For cast irons, temperatures from 300 to 400°C are employed with times up to about 10 min (sufficient to produce red-to-purple macroscopic coloring).

Heat tinting is used with alloy steels, such as tool steels, to darken the matrix but not the carbides (alloy carbides) for image analysis measurements. Figure 1-15 shows the microstructure of AISI D2 tool steel (Fe–1.5%C–12%Cr–0.9%Mo–0.8%V) in the quenched and tempered condition. Etching with 10-percent nital revealed the prior-austenite grain boundaries, carbides, and the martensitic matrix. Heat tinting at 538°C for 5 min darkened the matrix so that the carbide phase was vividly revealed.

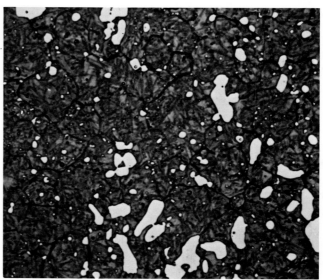

Fig. 1-15. Use of heat tinting to reveal carbides in AISI D2 tool steel. Top: 10-percent nital. Bottom: heat tinted at 538°C for 5 min after nital pre-etch. (Both at 1000×.)

The retained austenite is more clearly observed after heat tinting.

Heat tinting is also useful for phase discrimination in stainless steels.[22,23] If a polished sample is heated at 649°C, austenite is colored first, then delta ferrite, sigma, and carbides. For example, a 25%Cr–20%Ni alloy heated for 20 min at 649°C was colored as follows: austenite, mottled green; sigma, orange; carbides, white (i.e., unaffected). Optimum heating conditions increase with increasing chromium content.[23] Heat tinting has also been used for phase identification in wrought Co-based alloys.[24]

As an example of the use of heat tinting to reveal the structure of austenite stainless steel, Fig. 1-16 shows the microstructure of a stainless steel weldment containing austenite and delta ferrite. The structure has been revealed

PHASE IDENTIFICATION BY SELECTIVE ETCHING

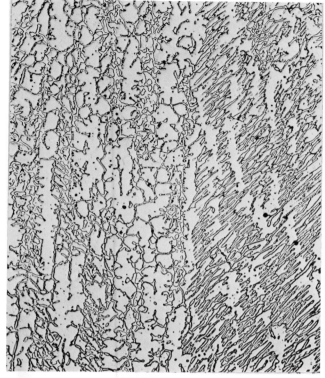
Vilella's Reagent

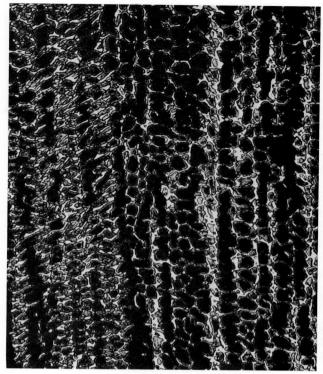
Heat Tinted

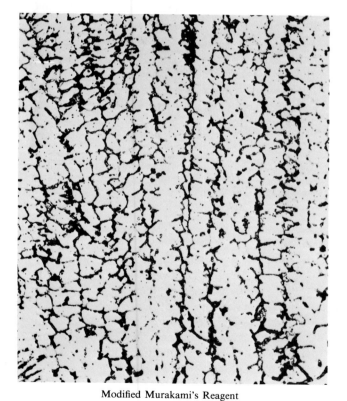
Modified Murakami's Reagent

Fig. 1-16. Example of the use of heat tinting and etching to reveal the microstructure of an austenitic stainless steel weldment containing delta ferrite. (All at 500×.)

by etching and by heat tinting. Vilella's reagent (1 g picric acid, 5 ml HCl, 100 ml ethanol) outlined the delta ferrite (uniformly attacks the austenite), whereas modified Murakami's reagent (etch No. 31) attacked and colored the delta ferrite reddish-brown. This is only one of several modifications of Murakami's reagent that will preferentially color delta ferrite.[4] Heat tinting was conducted after a light pre-etch with Vilella's reagent. The sample was placed in a laboratory furnace at 593°C until the surface was colored red, after about 20 min. This colored the austenite red, but the delta ferrite stayed unaffected.

Examples: Titanium Alloys

Heat tinting produces excellent color development in titanium alloys and is useful for phase identification. The grain structure of all alpha alloys is difficult to reveal by etching, and polarized light is used to improve the image. Heat tinting is helpful with all alpha alloys and has been used to distinguish between primary and secondary alpha.

Osadchuk et al.[25] heat tinted polished titanium samples in air at 260°C until the surface was golden pink. Carbides were bright blue and darkened with additional heating. Alpha remained white for some time but gradually darkened, although not so quickly as carbide. Transformed beta darkens quickly during heat tinting.

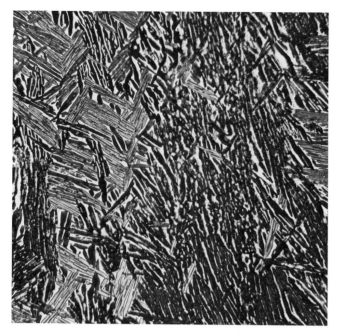

Fig. 1-17. Microstructure of Ti–6%Al–4%V revealed by heat tinting at 704°C for 2 min after a light pre-etch with Kroll's reagent. Alpha platlets colored; retained beta unaffected. (At 500×.)

Figure 1-17 shows the microstructure of Ti–6%Al–4%V given a light pre-etch with Kroll's reagent. (100 ml water, 2 ml HF, 5 ml HNO_3) and heat tinted at 704°C for 2 min; this treatment colored the alpha platelets, but the retained beta was unaffected. Coloration of the alpha phase varied with orientation.

Examples: Sintered Carbides

Heat tinting has been widely used to reveal constituents in sintered carbides.[26-29] Franssen[26] heat tinted samples at 800°C for 90 to 120 sec which colored WC blue, TiC brown to red, and the cobalt binder reddish black. Chaporova[27] used a lower temperature, 399°C, for 15 min to outline and color the cobalt binder, which left the carbides unaffected or lightly colored. Bleecker[28] heat tinted WC–TiC–Co samples at 260 to 316°C for 10 min to darken the Co binder, color the TiC yellowish brown, and leave the WC unaffected. Separation of WC from TiC by etching is difficult.

Figure 1-18 shows preferential darkening of the cobalt binder in a WC–Co carbide sample by heat tinting at 316°C for 5 min. The cobalt binder can also be preferentially darkened by etching in HCl saturated with $FeCl_3$, as shown.[27]

SUMMARY

This chapter has shown how different etching techniques can be used to attack a desired phase in multiphased alloys preferentially. These methods are rather simple to employ and extremely useful for phase identification or quantitative microscopy. Generally speaking, some experimentation will be required, aided by published etchant compilations[2-4,6] and Table 1-1.

Acknowledgment. The author wishes to extend thanks to Professors John N. Hoke, Pennsylvania State University, and Robert R. Jones, Lafayette College, for providing many of the nonferrous samples and for their advice concerning nonferrous metallography.

REFERENCES

1. Vander Voort, G. F. Etching techniques for image analysis. *Microstructural Science*, vol. 9, pp. 135–154. New York: Elsevier North-Holland, 1981.

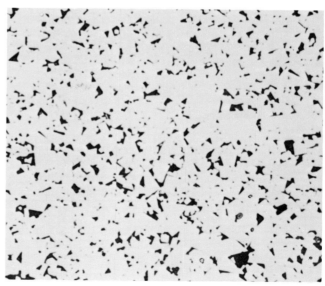

Fig. 1-18. Cobalt binder phase in WC–Co sintered-carbide cutting tool (same sample as in Fig. 1-3) revealed (left) by etching with HCl saturated with $FeCl_3$ (etch No. 7a) and (right) by heat tinting at 316°C for 5 min. (Both at 1000×.)

2. Anderson, R. L. Revealing microstructure in metals. Westinghouse Res. Lab. Sci. Paper 425–C000–P2, Dec. 22, 1961.
3. Petzow, G. *Metallographic Etching.* Metals Park, OH: American Society for Metals, 1978.
4. Vander Voort, G. F. *Metallography: Principles and Practice.* New York: McGraw-Hill Book Co., 1984.
5. Kehl, G. L., and Metlay, M. The mechanism of metallographic etching. I. The reaction potentials of a two-phase brass in various etching reagents. *J. Electrochem. Soc.* 101:124–127 (March 1954).
6. Beraha, E., and Shpigler, B. *Color Metallography.* Metals Park, OH: American Society for Metals, 1977.
7. Lienard, P., and Pacque, C. Analysis of the selective coloring mechanism for identification of different phases in Al–Si–Cu foundry alloys. *Hommes Fonderie* 126:27–35 (June–July 1982).
8. Fine, L. A new etching reagent for the detection of oxygen segregation in steel. *Metal Progress* 49:108–112 (Jan. 1946).
9. Hall, A. M. Metallographic etchant to distinguish oxidation in steel. *Metal Progress* 50:92–96 (July 1946).
10. Gilman, J. J. Electrolytic etching—the sigma phase steels. *Trans. ASM* 44:566–600 (1952).
11. Skidmore, A., and Dillinger, L. Etching techniques for Quantimet evaluation. *Microstructures* 2:23–24 (Aug/Sept. 1971).
12. Ence, E., and Margolin, H. Phases in titanium alloys identified by cumulative etching. *J. Metals* 6:346–348 (Mar. 1954).
13. Hiltz, R. H. Metallographic methods. Color staining of titanium and its alloys. Watertown Arsenal Lab. Report WAL 132/24, Apr. 20, 1956.
14. Hiltz, R. H., and Douglass, R. W. Orientation sensitivity of alpha titanium to electrostaining. *Trans. AIME* 215:286–289 (Apr. 1959).
15. Grosso, J., and Nagel, D. J. Anodizing as a technique for studying diffusion in the TiCb system. *Trans. AIME* 236:1377–1379 (Sept. 1966).
16. Olsen, R. H., and Smith, W. D. Color metallography for texture analysis of titanium. *Metallography* 4:515–520 (1971).
17. Crouse, R. S. Identification of carbides, nitrides, and oxides of niobium and niobium alloys by anodic staining. Oak Ridge National Laboratory Report ORNL–3821, July 1965.
18. Picklesimer, M. L. Anodizing as a metallographic technique for zirconium base alloys. Oak Ridge National Laboratory Report ORNL–2296, 1957.
19. Picklesimer, M. L. Anodizing for controlled microstructural contrast by color. *Microscope* 15:472–479 (1967).
20. McAdam, D. J., and Geil, G. W. Rate of oxidation of steels as determined from interference colors of oxide films. *J. Res. NBS* 23:63–124 (July 1939).
21. Young, F. W., et al. The rates of oxidation of several faces of a single crystal of copper as determined with elliptically polarized light. *Acta Met.* 4:145–152 (Mar. 1956).
22. Emmanuel, G. N. Metallographic identification of sigma phase in 25–20 austenitic alloy. *Metal Progress* 52:78–79 (July 1947); see also *ASTM STP* 110:82–99 (1951).
23. Pinasco, M. R., and Stagno, E. Metallographic structure formation of Fe–Cr–Ni–C alloys. *Mem. Sci. Rev. Met.* 65:627–642 (Sept. 1968).
24. Weeton, J. W., and Signorelli, R. A. Effect of heat treatment upon microstructures, microconstituents, and hardness of a wrought cobalt base alloy. *Trans. ASM* 47:815–852 (1955).
25. Osadchuk, R. et al. Metallographic structures in commercial titanium. *Metal Progress* 64:93–96, 96B (Nov. 1953).
26. Franssen, H. Structure of cemented carbide composites. *Arch. Eisenhutten* 19:79–84 (1948); HB translation No. 2175.
27. Chaporova, I. N. Preparation of metallographic sections and development of microstructures of cemented carbides. *Zav. Lab.* 15:799–805 (July 1949); HB translation No. 3061.
28. Bleecker, W. H. A metallographic technique for cemented carbides. *The Iron Age* 165:71–74 (May 25, 1950).
29. Powers, J. H., and Loach, W. J. Color shows up the unknown in metallography. *Steel* 133:93–96 (Oct. 15, 1953).

2
POTENTIOSTATIC ETCHING

E. E. Stansbury

Department of Materials Science and Engineering
University of Tennessee–Knoxville

Potentiostatic etching is the selective corrosion of one or more morphological features of a microstructure resulting from the metal to be etched being held in a suitable etching electrolyte at a controlled potential. Independent control of potential is accomplished by potentiostats. These are instruments capable of holding the potential of a metal to prescribed values relative to a reference electrode placed in the environment solution. The selected potential determines if dissolution can occur. Measuring the current density, moreover, can establish the dependence of the etching or corrosion rate on the potential. In alloys, the etching rate dependence on the potential may differ significantly for different phases and other microstructural details such as grain boundaries and etch pits. Thus, once a correlation has been established between potential, environment, and selectivity of attack on microconstituents, potentiostatic etching becomes an additional tool for revealing alloy morphology. Representative early applications of the potentiostat to etching are described by Edeleanu[1] and Cihal and Prazak[2]; the use of the method for color metallography was recognized by Lichtenberger[3] and Jeglitsch[4].

ELECTROCHEMICAL BACKGROUND

Understanding of the potentiostatic etching process requires a knowledge of the corrosion processes involved in chemical etching with an extension to another variable—control of potential. To use this variable effectively, however, requires an understanding of how the chemical etching process may be altered by changing the potential of a metallographic specimen relative to its environment. It will be shown that conventionally used etchants may not be suitable under potential control and that a single etchant may lead to a range of selective attacks on microconstituents, depending on the potential. As a background to the selection of potentiostatic etching conditions, the electrochemical processes involved in chemical etching are reviewed. These concepts provide the basis for understanding the factors involved in selection of environments, potentials, instrumentation, and procedures for potentiostatic etching.

Consider the simple model of a metal (M) in contact with a solution containing its ions (M^{z+}) or an inert metal such as platinum in contact with a solution containing dissolved hydrogen gas (H_2) and hydrogen ions (H^+). Reactions such as $M = M^{z+} + ze$ or $H_2 = 2H^+ + 2e$ are referred to as *half-cell reactions* and designated M/M^{z+} and H_2/H^+. Electrical charge separation—electrons in the metal and ions in the solution—occurs at the metal/solution interface, resulting in a difference in electrical potential (in volts) across the interface. This difference cannot be measured experimentally since the probe from the measuring instrument contacting the solution introduces another metal/solution interface and, hence, the instrument indicates only the difference in potential between the metal under investigation and the metal of the probe. A practical solution is provided by a number of metal/aqueous environment combinations that give highly reproducible potentials and thereby function as standard-reference electrodes. More specifically, these are referred to as *standard-reference half-cells* since they must be used in conjunction with the metal/etchant combination to produce an electrochemical cell offering metal leads between which a difference in potential can be determined.

The accepted primary-reference electrode consists of platinum in contact with a solution saturated with hydrogen gas at a pressure of 1 atmosphere and containing hydrogen ions at pH = 0 (unit effective concentration). In practice, the major use of the standard hydrogen electrode (SHE) is for calibration of secondary-reference electrodes, which are more convenient. Two common reference electrodes are the calomel or mercury/saturated mercurous-chloride half-cell with potential +241 mV relative to the SHE and the silver/saturated silver-chloride half-cell with relative potential of +196 mV. Both of these are also saturated with potassium chloride to provide constant chloride ion concentration.

The potential difference between the reference electrode and a simple M/M^{z+} electrode or freely etching sample should be measured with an electrometer; the high impedance of this instrument limits current flow through the cell during measurement to values having negligible influence on the electrochemical reactions. If, on attaching the electrometer leads to the electrode of the cell, a positive reading is shown when the positive lead is attached to the SHE and the negative lead to the M/M^{z+} electrode, then the half-cell potential of the M/M^{z+} electrode is assigned a negative value, or, $E_{M,M^{z+}} < 0$. Otherwise, the half-cell connected to the SHE will be given a positive value. In principle, this scheme is followed for all half-cell reactions, and the values so obtained (for unit effective concentration of dissolved species and for gases at 1-atm pressure) constitute the standard electrochemical series. For reactions such as $Fe^{3+} + e = Fe^{2+}$ and $O_2 + 4H^+ + 4e = 2H_2O$, these species, which are dissolved in the

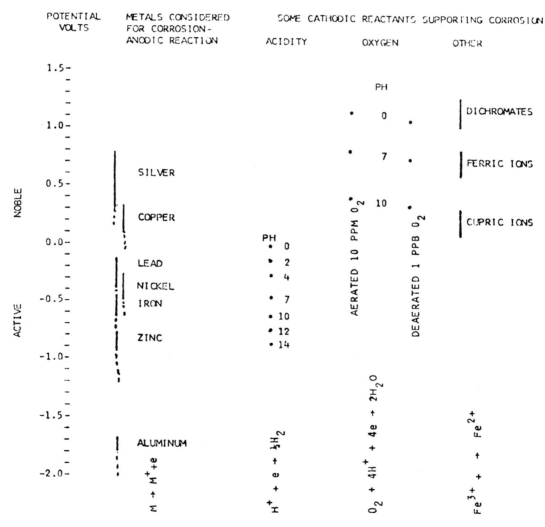

NOTES: VERTICAL BARS REPRESENT RANGE OF POTENTIALS FOR METAL IN CONTACT WITH SOLUTION CONTAINING METAL IONS FROM APPROXIMATELY 10% DOWN TO ONE PPM. PRECIPITATED CORROSION PRODUCTS LOWER RANGE.

HYDROGEN AND OXYGEN REACTION DEPEND ON BOTH PH AND PRESSURES OF HYDROGEN AND OXYGEN GAS. DATA FOR HYDROGEN ARE FOR PRESSURE OF HYDROGEN OF ONE ATMOSPHERE. DATA FOR OXYGEN ARE FOR CONTACT WITH AIR (AERATED) GIVING 10 PPM OXYGEN IN WATER AND FOR WATER DEAERATED TO ONE PPB.

Fig. 2-1. Ranges of half-cell potentials of some electrochemical reactions of importance in corrosion and etching.

aqueous solution, are placed in contact with platinum to allow electron exchange between species in solution and an electron conductor. The half-cell potential is then given as the potential difference between the platinum of the SHE and the platinum in contact with the species whose half-cell potential is being measured; thus $E°_{Fe^{2+},Fe^{3+}} = +0.77$ V.

A convenient overview of relative positions of half-cell potentials of several common metals and half-cell potentials of several possible etchant species is given in Fig. 2-1. To the left, is the scale of potentials in volts relative to the standard hydrogen electrode. The solid vertical lines identified by a metal give the range of half-cell potentials for the metal, extending from unit concentration in moles per 1000 grams of water at the top to a concentration of approximately 1 ppm at the bottom. The dotted extension to lower potentials apply when precipitating or complexing reagents are added that reduce the metal ion concentration below 1 ppm. Reactions that might support corrosion are hydrogen ions, dissolved oxygen and cupric, ferric, and dichromate ions. The potential of the hydrogen ion reaction depends on the pH and is given for the pH range 0 to 14. The potential of the oxygen reaction depends on pH and oxygen concentration; ranges are given for pH of 0 to 10, for 10 ppm oxygen, the approximate concentration in contact with air, and for deaeration to 1 ppb. The other ions will have a range of potentials depending on concentration, as shown by the solid vertical lines on the right.

An electrochemical cell consists of two half-cells, which, for present purposes, would be made up of one of the metals considered as undergoing corrosion and one of the reactions capable of supporting corrosion. If the half-cell potential of the metal is designated as E_M and for a possible etchant reaction as E_C, then defining E_{cell} as

$$E_{cell} = E_C - E_M \quad (1\text{-}1)$$

provides a means of predicting if etching will actually tend to occur. It can be shown that if E_{cell} is positive, then etching or spontaneous corrosion tends to occur; otherwise, attack is not possible. To illustrate, consider iron with Fe^{2+} ion concentration of 1 ppm in contact with a completely deaerated environment (all dissolved oxygen excluded) of pH = 2. The cell potential is given by

$$E_{cell} = -0.1 - (-0.6) = +0.5 \text{ V} \quad (1\text{-}2)$$

These are approximate values within limits of reading values from the figure. Since the cell potential is positive, however, chemical attack should occur. In contrast, applying the calculation to silver in contact with the same solution leads to a negative cell potential, and silver is not attacked at pH = 2. It is emphasized at this point that the cell potential determined in this manner is a limiting maximum value. It does indicate whether or not attack will occur. As will be shown, however actual attack must be associated with current passing between sites on the metal surface where the two reactions occur and, depending on the current density, actual potential differences will be less than indicated here. Where these reactions are occuring across the surface, moreover, movement of a reference electrode across the surface may or may not be able to distinguish where the reactions are occurring; in this case, a single composite potential known as the *corrosion potential* is measured with a value relative to a reference electrode that lies between the values for the individual reactions identified above.

ETCHING AS A CORROSION PROCESS

A schematic representation of the corrosion of a metal, e.g., iron, in an environment containing the symbolic cathodic reactant, C^+, is shown in Fig. 2-2. Electrons from the metal at an anodic site (oxidation) are consumed at cathodic sites by the reduction of species supporting the cathodic reaction, for example, the reduction of H^+, O_2, or Fe^{3+}. Any structural dissimilarities on the surface can give rise to local anodic and cathodic sites. Inhomogeneities in the solution contacting the surface can also give rise to these sites, although this is usually not a controlling factor over small surfaces unless the formation and spread of corrosion products change the local composition of the solution. For pure metals, any relatively localized positions of higher energy are potential anodic sites; these include grain and twin boundaries, dislocations intersecting the free surface, scratches, smeared surface layers resulting from mechanical grinding and polishing, and differences of crystal lattice orientation between grains intersected by the free surfaces. In segregated solid-solution alloys, differences in composition may result in relative anodic sites. In multiphase alloys, local anodes and cathodes result from differences in composition, crystal structure, and lattice orientation of the phases. In all of

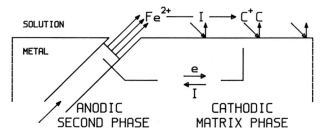

Fig. 2-2. Representation of a local corrosion cell resulting in the etching of a second-phase particle (anodic reaction) with a supporting cathodic reaction on the matrix phase. (1) Anodic reaction—symbolic: $M = M^{z+} + ze$; specific: $Fe = Fe^{2+} + 2e$, and (2) cathodic reaction—symbolic: $C^+ + e = C$; specific: $H^+ + e = 1/2\ H_2$, $O_2 + 4H^+ + 4e = 2H_2O$, and $Fe^{3+} + e = Fe^{2+}$.

these cases, selected areas, such as specific phases, may exhibit high kinetic activity for the cathodic reaction and thereby become cathodes that induce other areas to be anodic and hence corrode or etch relatively. In the limiting case, the electrochemical reactions may be insensitive to these variable interface conditions, in which case the anodic and cathodic reactions would occur homogeneously over the surface with intersite distances approaching atomic dimensions, and the attack becomes uniform to these dimensions.

The electron transport in the metal and the conventional current, I, in the metal and in the solution are also shown in Fig. 2-2. Current in the solution is carried by ions, positive ions moving from anodic to cathodic sites and negative ions in the opposite direction. If the metal were iron and the solution contained hydrogen ions at pH = 2 and ferrous ions at 1 ppm, then E_{cell} would be +0.5 V, as concluded previously, and the iron would spontaneously corrode.

If, as shown in Fig. 2-3, a reference electrode such as the calomel half-cell is placed in the solution, then the potential of the corroding metal relative to this electrode can be measured using an electrometer. If the size of the local anodic and cathodic sites is large relative to the size of the reference electrode, then movement of the reference electrode permits detection of these sites. Limiting examples are the use of very small reference electrodes to measure variations in potential over grains, phases, and interfaces. If the sites are small relative to the size of the reference electrode probe, however, or if the reference-electrode to metal-interface distance is large relative to the dimensions of the local anodic and cathodic sites, a single potential is measured regardless of the position of the reference electrode. In either case, the measured potential will lie between the equilibrium or open-circuit potentials for the anodic and cathodic reactions; interpretation is more complex if two or more anodic and/or cathodic reactions occur simultaneously. For example, in an aerated acid environment in contact with iron, both the reduction of hydrogen ions and of dissolved oxygen are cathodic reactants supporting corrosion.

RELATIONSHIP OF HALF-CELL POLARIZATION TO CORROSION AND ETCHING

The observation that actual half-cell potentials do not agree with those estimated from Fig. 2-1 is a result of current passing through the interface, which, for small values of current density, upsets the charge distribution across the interface. For large current density, an additional influence is the change in composition occuring in the solution near the interface as a result of reactions at the interface that produce or consume ions faster than the bulk solution concentration can be maintained by diffusion. This deviation of the measured potential from the equilibrium potential is defined as the polarization, $\eta(i) = E_{eq} - E_{pol}(i)$, where E_{eq} is the half-cell potential calculated from the standard half-cell potential and $E_{pol}(i)$ is the actual or polarized potential that is a function of the current density, i. On a corroding surface, the polarization of the cathodic reaction is negative and, for the anodic reaction, it is positive. As a consequence, the corrosion current is associated with a decrease in the potential at cathodic sites and an increase at anodic sites. These shifts continue until at steady-state corrosion, with the total anodic current (electrons released) equal to the total cathodic current (electrons consumed), the potential difference is equal to that required to maintain the current through the resistance of the metal-solution-interface path. For relatively high conductivity solutions, ΔE may be less than a millivolt, and the anodic and cathodic reactions polarize to essentially the same potential value. This is then the single corrosion potential that is measured by a reference electrode. If the conductivity of the solution is low and the anode-cathode areas are separated, a measurable potential drop exists between the areas.

The only information obtained by using a reference electrode and electrometer for measurement on freely etching (corroding) metals is the corrosion potential. A single measurement, however, does not provide any quantitative information on either the rate or mode (film forming, faceting, etc.) of etching. Rather, it is necessary to alter the electrochemical reactions on the metal surface by passing current either to or from the specimen under external current control or external potential control. The former is referred to as *galvanic polarization* and the latter as *potentiostatic polarization*. The former can be accomplished with relatively simple equipment but is not applicable to certain types of polarization. For investigation of corrosion phenomena, and, in particular, controlled corrosion for metallographic etching, the potentiostatic polarization technique has advantages; for many alloys, it is the required method.

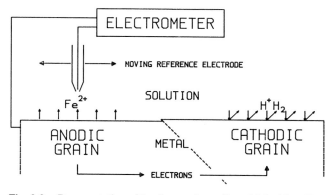

Fig. 2-3. Representation of local corrosion cell established by differently oriented grains. Relative anodic and cathodic areas are detected using an electrometer and moving a reference electrode to measure corrosion potential as a function of position.

POTENTIOSTATIC POLARIZATION

Experimental Method

A schematic representation of the arrangement for potentiostatic polarization or etching is shown in Fig. 2-4. The four components are (1) the cell containing the sample to be studied or etched, a reference electrode, and an auxiliary; (2) an electrometer; (3) the potentiostat, and (4) an ammeter (A) to measure the current to or from the sample. The metallograhically polished sample to be etched, frequently referred to as the *working electrode*, must be immersed in such a way that the conducting leads are insulated from the solution. When a cathodic reaction is being studied—such as the reduction of hydrogen ions, dissolved oxygen, or Fe^{3+} to Fe^{2+}—the working electrode is platinum, on which the reaction permits transport of electrons to or from the potentiostat. The auxiliary must be a material essentially inert to the environment, usually either platinum or high-density pyrolytic graphite. Its function is to complete the circuit through the cell, and although electrochemical reactions occur at its interface, these frequently are of little consequence. The reference electrode is usually a calomel half-cell or a silver/silver-chloride half-cell, although others may be used, particularly when chloride ion contamination of the environment may be a problem. When the switches from the cell and electrometer to the potentiostat are open, the electrometer is used to measure the "open circuit" or corrosion potential of the working electrode or specimen to be etched.

A schematic representation of the major components of the potentiostat circuit are shown in Fig. 2-5, in which the specimen is shown in an alternative position with the polished surface up. The potentiostat is basically an electronic device, one component of which is an internal reference (set-point) potential that is adjusted to the potential to be maintained at the working electrode, the metallographically polished sample that is to be etched. Any potential difference between the reference and working electrodes, as detected by the electrometer, is sensed

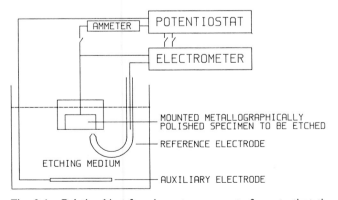

Fig. 2-4. Relationship of equipment components for potentiostatic etching.

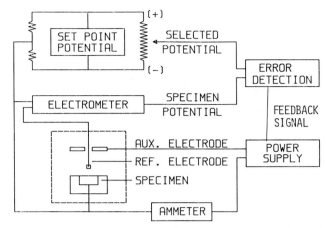

Fig. 2-5. Representation of potentiostat circuit in relationship to reference electrode and specimen undergoing etching.

by an operational amplifier that functions as an error detector to output a feedback signal to a power supply. The response of the power supply is to pass electrons to the working electrode if the potential is to be lowered and removing electrons if the potential is to be raised. Under steady state, the specimen potential is detected by the electrometer, and the current through the working electrode interface is determined by an ammeter. This external current, I_{ex}, is always the difference between the sum of all anodic currents and the sum of all cathodic currents across the interface—a net anodic current passing if anodic currents exceed cathodic currents or conversely. Specifically,

$$I_{ex,anodic} = \Sigma I_{anodic} - \Sigma I_{cathodic}$$

or (1-3)

$$I_{ex,cathodic} = \Sigma I_{cathodic} - \Sigma I_{anodic}$$

The demand or set-point potential at the potentiostat can be adjusted manually or changed either incrementally or continuously to provide a potential-current scan of the working-electrode behavior. The results are usually recorded by direct plotting of potential as a function of log current or log current density.

INTERPRETATION OF POLARIZATION CURVES FOR APPLICATION TO ETCHING

The dependence of dissolution rate of a metal or alloy on the electrochemical potential is represented by the polarization or potential vs current density curve. A representative, experimentally determined curve for a metal that forms corrosion-product films in a deaerated acid environment is shown in Fig. 2-6. Sections of the curve are identified as potential ranges of net cathodic, net active anodic, passive anodic, and transpassive anodic behavior.

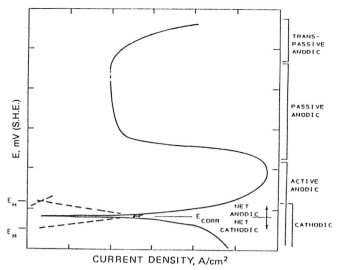

Fig. 2-6. Representative relationship between potential and current for a metal exhibiting active, passive, and transpassive potential ranges of electrochemical reaction.

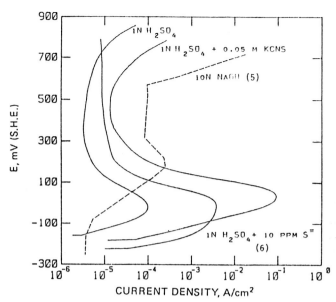

Fig. 2-7. Representative examples of effects of several variables on polarization of AISI 304 stainless steel. Base curve is for 1N H_2SO_4. Effects of CNS^- and $S^=$ (6) in 1N H_2SO_4 and of 10N NaOH (5).

Dashed extensions of the curves indicate the potential and current-density ranges over which cathodic and anodic reactions occur when the net current density is anodic and cathodic, respectively. In the net cathodic potential range, the rate of metal dissolution may be small with little etching. On increasing the potential, the current reverses at $E_{corrosion}$—the natural corrosion potential of the specimen in the absence of a potentiostat—and further increase is accompanied by a net removal of electrons. The entire anodic curve is the potential range of anodic dissolution (oxidation) of the metal to either soluble or insoluble corrosion products. In the cathodic potential range, there is a net flow of electrons to the specimen, with the predominant effect of reducing hydrogen ions to hydrogen gas. In the active anodic region, the dissolution rate increases as the potential increases; etching may occur but corrosion product films may not form. A maximum in the current density results from the initiation and growth of films that cause a reduction in that density until such time as an adherent oxide film characteristic of the passive state forms. Increasing the potential in the passive range results in progressive thickening of the film in such a way that the current density remains relatively constant. In the transpassive range, the passive film becomes unstable relative to soluble species in solution such as $CrO_4^=$; the film disappears and the current density increases.

The polarization curve is sensitive to the composition of the environment, for present purposes, to the etchant composition. Representative examples for AISI type 304 stainless steel are shown in Fig. 2-7. The reference curve is for 1N H_2SO_4, the environment most commonly used for comparison of the corrosion behavior of various materials. The pH is a major variable, and since much of the reported work on potentiostatic etching relates to 1–10 N NaOH, a curve is shown in the figure for a strongly alkaline environment. The curves for 1N H_2SO_4 with additions of $S^=$ and CNS^- ions are examples of additives to the 1N H_2SO_4 to increase the dissolution rate in the active range—an important consideration in increasing the current density to accomplish etching in a reasonable time. Chloride ions have a major influence on the polarization of most active/passive alloys; these and other halide ions increase the current density and may cause breakdown of the passive film at potentials below the transpassive range. This breakdown occurs as localized attack on the passive film in the form of pitting, and, in the limiting case for some alloys, high halide-ion concentrations can prevent formation of the passive film. Therefore, enhancement of potentiostatic etching by chloride ions may be uncertain and difficult to control.

The polarization curve is usually determined by a continuous scan of potentials from the cathodic range or from $E_{corrosion}$ at 6V/hr. The experimental curve is sensitive to the scan rate and surface topology, and films at any potential may be very sensitive to the potential-time history, i.e., whether a specimen is scanned to, or is initially set at, the given potential. Because of this sensitivity, reference to polarization curves in the literature as guides for conditions for potentiostatic etching may be limited to their qualitative value since etching will usually be carried out by directly setting the potential and holding it for a specified time to produce the desired etching response. They do, however, indicate potential ranges of dissolution with and without film formation and readily reflect changes in etchant composition.

PRINCIPLES OF POTENTIOSTATIC ETCHING

An initial assumption in potentiostatic etching (identified by Kurosawa et al.[7] as SPEED, Selective Potentiostatic Etching by Electroclytic Dissolution) is that any micro-region on the surface has a dissolution rate (current density) that is the same as that observed in a bulk sample identical in composition to the micro-region when exposed under the same conditions—e.g., environment and potential. Thus, microconstituents of different composition and structure, regions of different composition within a solid solution (e.g., dendritic segregation), and regions of local composition differences such as those encountered in sensitized stainless steels, all pass into solution at different current densities at a given potential. In principle, these characteristic current densities, as a function of potential, are given by the polarization curves derived from macrosamples representative of the micro-region of concern.

During etching, the external current, which is measured under potentiostatic control, is the difference between the anodic and cathodic currents given by the polarization curves of the respective reactions as described previously. As a consequence, measurement of the external current can be related to the rate of etching of a single microconstituent only if one anodic process is occurring and current densities associated with cathodic reactions are negligibly small. In the usual case of several simultaneous anodic processes, differentiation of microstructural detail will depend on the mode and amount of reaction at each site. The reaction rate is given by the current density, and whenever two microstructural details, A and B, are to be differentiated, Prazak and co-workers[2,8] have defined a differentiation ratio as follows:

$$D_A = (i_A/i_B)_E \qquad (2\text{-}4)$$

where D_A is the differentiation ratio with respect to structural detail A and i_A and i_B are the current densities characterizing the reaction rate over the respective sites at potential E. The ratio has also been used by Greene and Teterin.[9] In using a current density criterion for total reaction, as would be required for etching, time must also be specified. Since the current density may vary, usually decreasing, with time, Gruetzner and Schueller[5] have found specification of difference of charge density—$q = \int I dt$ in units of C cm^{-2}—to be a better criterion for satisfactory etching in that the difference in charge density relates directly to relative total dissolution of each microconstituent. Application of either of these criteria relates to the selection of environments that result in polarization curves from which potentials at which preferential etching can be expected may be specified.

Insight into the factors determining this mode of attack and to the time required for acceptable etching is obtained by examining four metal/solution interface interactions that may be singularly responsible for the etching attack or involved in combination to reveal the microstructure. These are (1) relative uniform etching, (2) localized etching of grain boundaries and dislocations, (3) film and/or irregular corrosion product formation, and (4) film deposition from solution.

General Etching

Uniform attack will be defined as a non-film-forming dissolution that results in the differentiation of grains for purposes of revealing grain size and shape. Development of morphology, therefore, depends upon differences in dissolution rate as a function of lattice orientation of the individual grains relative to the viewed surface and/or relative faceting or other surface roughening leading to grain contrast. The current density required to produce resolvable grain contrast based on differences in elevation can be estimated under several simplifying assumptions. Consider the current density-time interrelationship for removal of a monolayer of atoms. For each atom passing into solution in time t, the charge transferred, q', is

$$q' = z(1.6 \times 10^{-19}) = It \qquad (2\text{-}5)$$

where z is the number of electrons transferred per ion formed (valence change of z); I, the current, in amperes ($C\ s^{-1}$); and t, the time, in seconds; the constant is the charge on the electron in coulombs. The charge transfer per unit area is

$$q = N_s z(1.6 \times 10^{-19}) = it \qquad (2\text{-}6)$$

where N_s is the number of surface atoms per unit area and i is the current density. The number of surface atoms per unit area depends on the atom radii and planar density of atoms for the lattice plane exposed to the environment. For example, with FCC metals, $N_s(100, \text{FCC}) = 1/(4r^2)$ and $N_s(111, \text{FCC}) = 1/(2\sqrt{3}\ r^2)$ for (100) and (111) planes of FCC metals with atom radius r. For close-packed planes of atoms [(111) in FCC], the relationship becomes

$$q = 4.62 \times 10^{-4} z/r^2\ C\ cm^{-2} \qquad (2\text{-}7)$$

per atom plane with r in angstroms.

To use the above relationship qualitatively for purposes of referring to potential-current density curves that characterize the dissolution rate, a value for the difference in elevation of adjacent grains required for useful contrast and the related ratio of dissolution charge densities of adjacent grains providing the differential height must be established. Using a measuring microscope, Greene et al.[10] concluded that a difference in elevation of about 2.5 μm is required to provide metallographic contrast at 100X when areas to be differentiated are not stained.

Resolution of the microstructure depends upon the optical properties of the surface/microscope combination that are critical to the differentiation in elevation—specifically, whether it is a contrast relating to the depth of focus or the ability to resolve the geometry of the grain boundary as the transition from one elevation to the other. Reference may be made to Vander Voort,[11] who discusses the significance of these two factors in limiting the ability to resolve microstructural detail. For an objective with a numerical aperture of 0.4, consideration of both depth of field and resolving power indicates that a difference of elevation of 1 to 2 μm is required for adequate contrast, a figure in general agreement with the experimental observation of Greene et al.[10] Smaller differences in elevation may be sufficient if phase contrast optical methods are used.[12]

The effect of grain orientation on relative rates of dissolution can be estimated from polarization measurements on single crystals, with exposed surfaces having different lattice orientations. Mauvais et al.[13] determined the polarization response in 1N H_2SO_4 of nickel single crystals sectioned to expose (100), (110), and (111) planes. The curves shown in Fig. 2-8 were determined from scans in the positive potential direction starting at the corrosion potential. Large differences in dissolution rate are indicated in the active range where etching would be carried out; for example, near +0.25 V (SCE), the current density for the (111) exposed face is almost two orders of magnitude greater than for the other two faces. Large differences also occur in the passive range of potentials, but passive film formation may prevent useful development of grain contrast by differences in elevation; the current densities are also much smaller, and, in the passive range, the (111) face exhibits the smaller value. Although these differences in dissolution rate have not been correlated with differential attack on polycrystalline nickel, the curves can be used as a guide for potentiostatic etching for grain contrast. Applying Eq. 1-7 to nickel with an atom radius of 1.24 angstroms and etching to produce divalent ions, removal of one atom layer from a square centimeter of surface requires 6×10^{-4} C cm^{-2}. The 2.5-μm difference in elevation required for the resolution of adjacent grains corresponds to the removal of approximately 10^{14} layers of nickel atoms or to a charge transfer density of 6 C cm^{-2}.

According to Fig. 2-8, the etching rate of (111) oriented grains dominates over other orientations and is responsible for essentially the total current. Assuming that 25 percent of a polycrystalline sample contains grains sufficiently close to the (111) orientation to etch at the (111) orientation rate of 60 mA cm^{-2} [corresponding to a potential of +0.2 V(SCE)], then the etching time would be 25 sec. This value is in general agreement with a criterion adopted by Greene et al.[10] that a dissolution current of 10 mA cm^{-2} for 180 sec is adequate for a reasonable etching of nonstaining microconstituents, the exact current density–time relationship being governed by the polarization curve for the material. The selectively etched grains will be sharply outlined against the surrounding faintly etched matrix of grains having orientations of low dissolution rate. Further reference to Fig. 2-8 reveals that the selective attack should reverse at potentials below 0 V (SCE), although the etching rate would be very slow. Systematic studies of the effects of additives to the etchant to alter the magnitude and relative rates of dissolution as a function of lattice orientation have not been reported.

Grain contrast can also be developed by conditions of potential and environment that cause uniform faceting of the surface. Under these conditions, regardless of the orientation of the crystal lattice at the specimen interface, dissolution is preferable to removal of atoms parallel to a particular lattice plane, usually the higher atom density planes such as (100), (110), or (111). Attack initiates uniformly over the surface but propagates along the selected plane, thereby producing a serrated surface. The geometry of the serrations depends upon the angle of intersection of the selected planes with the surface. As a consequence, light is scattered differently from each grain as a function of the orientation of the facets. Representative faceted microstructures developed in the active region for an AISI type 304 stainless steel have been published by Sato[14]; etch pitting on nickel single crystals, by Mauvais.[13]

Color differentiation of microconstituents by etching in the active and high transpassive potential ranges depends on the development of a surface topology that contains irregularities such as facets, etch pits, and differences in elevation, as discussed previously. If the dissolution is uniform, then, within a factor of about 2, a current density of 1 mA cm^{-2} will remove surface at the rate of 50 nm per minute, and since films do not form, this dissolution rate is relatively constant with time. Optical features of the microscope such as sensitive tint and quarter-wave plates and phase-contrast devices can develop color for surface irregularities with widths and depths

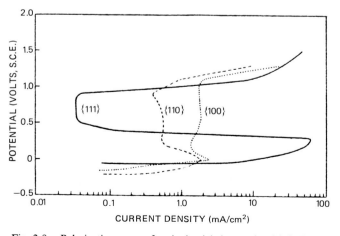

Fig. 2-8. Polarization curves for single nickel crystals with indicated planes in contact with 1N H_2SO_4 (13).

of the order of one-quarter of a wavelength or less. For wavelengths at the lower end of the visible range (violet, 40 nm), the dimension of the irregularities can be in the range of 10 nm. Considering that the exposed surface area per unit of sample increases rapidly as the surface topology becomes progressively irregular, these approximations lead to current densities on the order of 1 mA cm^{-2} for 1 min to produce surfaces with irregularities capable of producing color. Whether the desired surface topology develops depends on the etchant, and, unfortunately, systematic investigations of the interrelationship of these factors in producing useful microstructural contrast, particularly in color, have not been reported.

The potential-current combination to give satisfactory potentiostatic etching of solid-solution alloys is more complex than for pure metals. As a first approximation, the dissolution rate of each component is taken as the one observed for the pure component of the solid solution at the same potential. This approach ignores the effects of interaction between component atoms and the fact that the atoms of one component may exist in a crystal structure differing from the pure metal. More importantly, if the dissolution rates for the component atoms differ significantly, either as a result of a large difference in half-cell potential or response to polarization, then surface enrichment of the component having the smaller current density will occur. Effectively, the latter is functioning as the more noble component, and the effect is greater, the greater the difference in the current densities exhibited by the component atoms.

This enrichment effect has been treated by Mueller[15] and by Steigerwald and Greene.[16] As enrichment progresses, the polarization behavior of the alloy moves toward that of the more noble component, and this change must be taken into account in relating polarization behavior with development of the microstructure. As an example of the magnitude of the effect, consider that if the dissolution rate of the more noble A atoms is negligible compared to that of the B atoms, then the number of atom layers that must be removed to accumulate a layer of the more noble metal is f_B/f_A, or the atom fraction ratio of the component atoms. Thus, for a solid solution of 10 atom percent of the noble metal, total enrichment will occur after dissolution of the more active atoms from nine atom layers. This leads to the conclusion that whenever the development of the microstructure of solid-solution alloys depends on the development of differences of elevation of 10^4 atom layer as concluded earlier, then significant changes in the surface composition of adjacent grains will have occurred.

Localized Etching

If the attack is preferential to the grain boundaries, the overall sample current density required to reveal the boundaries as a function of grain size is approximately as follows. For a simplified model of square grains, 0.01 cm on a side, the exposed grain boundary length is 2×10^2 cm/cm^2 of exposed specimen surface. Taking the limit of resolution of the optical microscope as approximately 3×10^{-5} cm and assuming that dissolution of the grain boundary occurs to this depth, the volume of material to be removed is 1.8×10^{-7} cm^3/cm^2 of surface. Since the volume per atom is M/dN_o—where M is the atomic weight, d the density, and N_o Avogadro's number—then the number of atoms to be removed from grain boundaries per unit area of surface is $1.8 \times 10^{-7}(dN_o/M)$. The charge density to accomplish this removal is

$$q = it = (1.6 \times 10^{-19})z(1.8 \times 10^{-7})dN_o/M$$
$$= 1.7 \times 10^{-2}zd/M \text{ C cm}^2 \quad (2\text{-}8)$$

For divalent nickel, this relationship gives q a value of 5×10^{-3} C cm^{-2}. Since etching of grain boundaries usually occurs along the active dissolutions section of the polarization curve, reference to the polarization curve will give the current density in this section of the curve, and an estimate can be made by use of the above relationship of the time required to reveal the grain boundaries. With the assumptions made for a potential associated with a current density of 1 mA cm^{-2}, etching should be accomplished in 5 sec. Actual times would be longer, since uncertain amounts of general dissolution will occur simultaneously, and the amount of grain boundary attack desired may depend on the purpose for which etching is being conducted.

In both of the previous cases—general attack to provide difference in grain elevation and in preferential grain-boundary etching—formation of passive films may be undesirable in that the rate of attack may be both very slow and nonpreferential. To avoid this possibility, the potential may be restricted to the active region as previously discussed or attempts made to etch in the transpassive region of the polarization curve. The latter conditions are realized in the oxalic etch test specified by ASTM A262[17] for susceptibility of stainless steels to intergranular corrosion, although the test is specified in terms of a current density that places the steel in the transpassive region. In particular, if the current densities in the active range are too small, changes in the etching medium should be investigated to prevent passive film formation and move the polarization curve to higher current densities, as has been described in relation to Fig. 2-7.

Reference has been made to the influence of several anions on the polarization curve. Prazak and co-workers[2,8] used additions of CNS$^=$ in early work, and additions of various kinds have been used more recently for both the electrochemical and metallographic examination of stainless steels for intergranular precipitation of carbides.[18] The use of organic complexing agents to increase the current density for etching of copper and brass in alkaline solutions has been reported by Greene and

Teterin.[9] The influence of chloride ions in destroying passive films is well established in the literature of pitting and stress-corrosion cracking. For example, nickel cannot be passivated in 1N H_2SO_4 above 0.33 N NaCl,[19] and iron with less than 10-percent Cr, even in neutral solution, cannot be passivated above 0.1 N NaCl.[20] With higher chloride-ion concentrations, dissolution rates may be too high to control for effective etching; in fact, polishing rather than etching may tend to occur. At lower concentrations of chloride, pitting may occur rather than desired etching although critical adjustment of conditions can lead to dislocation etch pitting as a desired morphology. Because of these problems, few chloride-containing aqueous potentiostatic etching solutions have been reported. Schwabe and Schmidt[21] have provided evidence that passive film formation depends on the presence of water in the etchant. They show, by means of polarization curves, that as water is replaced by alcohol, for example, the current in the passive range increases, and eventually the polarization curve is continuous from the active range. Similar behavior tends to occur in high concentrations of H_2SO_4 and H_3PO_4. In limiting cases, dissolution rates become governed by diffusion of dissolving cations through a surface layer, and polishing rather than etching occurs. Further modifications to bring about etching under these conditions have been limited but should be investigated. A combination of the effects of chloride ions and non-aqueous media is the prevention of film formation through the use of alcoholic hydrochloric acid solutions as potentiostatic etching media.[21,22,23]

Polarization of Multiphase Alloys

When the potential of a multiphase alloy is established by potentiostat control, the high electrical conductivity of the phases and their close proximity cause each phase to have the potential established on the alloy. The dissolution rate of each phase is then given by the current density characteristic of the pure phase at the given potential. As a consequence, each phase makes a contribution to the current density, which, for the particular environment, depends on the mode and associated current density of dissolution (e.g., active or passive) at the potential. Since phases are usually multicomponent, surface composition changes of the nature discussed previously for single-phase solid solutions can occur.

The polarization curve for a multiphase alloy may be approximated by the volume-fraction weighted sum of the current densities of the phases at each potential. For example, a stainless steel consisting of f_α volume fraction of the α-phase, f_σ volume fraction of the σ-phase, and f_γ volume fraction of austenite, the current density of the alloy is as follows:

$$i_{\text{alloy}} = f_\alpha i_\alpha + f_\sigma i_\sigma + f_\gamma i_\gamma \qquad (2\text{-}6)$$

where i_α, i_σ, and i_γ are the current densities of the individual phases derived from polarization curves for the pure phases. The versatility of the potentiostat to differentiate phases by their dissolution rate was recognized by Edeleanu[1] and by Cihal and Prazak,[2] whose curves for the individual phases—austenite, ferrite, and sigma—are shown in Fig. 2-9. Three things from an experimental standpoint should be noted in these curves. First, as preferred by some investigators, cathodic current densities are assigned negative values and anodic current densities positive values. Second, the current densities are plotted on a linear coordinate scale in units of A cm^{-2}. These plotting methods obscure detail in the curves associated with dissolution current densities below 1 mA cm^{-2} and, in particular, the nature of the passive potential range, here shown as a vertical line. Since potentiostats capable of measurement of currents in the μA range have been available for a number of years, plotting polarization

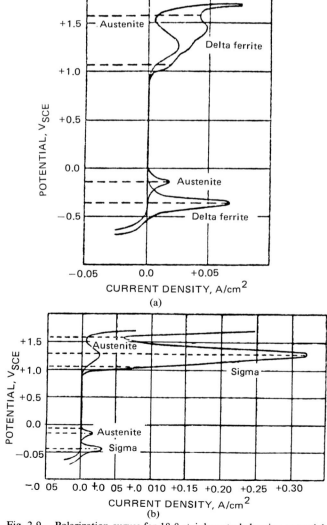

Fig. 2-9. Polarization curves for 18-8 stainless steel showing potential ranges for selective etching of austenite, delta ferrite, and sigma phases.[2]

curves as a function of log of current density more clearly shows the broad dissolution behavior. Third, it should be noted that the curves are for the individual pure phases. Thus, in an alloy containing 10-vol-percent alpha and 10-vol-percent sigma phase, the polarization curve for the alloy would reflect contributions from the dissolution of these two phases of only 10 percent of the values shown by the polarization curves of the individual phases. As a consequence, polarization curves of multiphase alloys may show only small irregularities associated with preferential dissolution of a phase even though the current density of the pure phase may be sufficient to cause significant attack. For this reason, small irregularities that might be ignored in experimentally determined polarization curves may indicate potentials at which useful preferential etching may occur.

Steigerwald and Greene[16] and Greene et al.[10] have examined the validity of constructing polarization curves of alloys from curves for the component phases. Polarization curves for pure tin and zinc in 1N NaOH are shown in Fig. 2-12(a); in Fig. 2-12(b) the calculated curve is shown in comparison to experimental curves for a 51-wt-percent zinc in tin alloy. The alloy is two-phase, the phases being essentially pure metals. It is evident that reasonable agreement is obtained when the polarization curve is calculated based on the relative amounts of the two phases. Some of the factors determining the circumstances under which such agreement may be expected are discussed by Steigerwald and Greene.[16]

Greene et al.[10] reported that zinc etched at -1.45 V (SCE) and -0.60 V and tin at -0.95 V: "Etched tin appears dark due to light refraction from a roughened surface. Tin, which does not form colored corrosion products, may be selectively etched at differentiation ratios of 5 or more. . . . At ratios of less than 5, metallographic contrast decreases to such an extent as to be considered unsatisfactory. Zinc, because of its dark colored corrosion products, may be etched at ratios less than 5, possibly as low as 2."

Film Formation

Development of morphology by etchants that produce corrosion-product films depends on differences of optical properties of the films formed on different phases, the films formed as a consequence of gradients of composition within phases, and the films formed on grains of different orientation. Reference to polarization curves provides indications of ranges of potential and of etchant composition that are conducive to the formation of oxide or other type of films. Optically active films useful in metallography include (1) corrosion-product films that provide interference effects, (2) corrosion-product films with an internal structure that responds to polarized light, and (3) corrosion-product films with a surface topology that responds to polarized light.

Interference effects associated with differences in film thickness relate to the structure of a film, particularly whether it is single-crystal, polycrystalline, or amorphous, and to other optical characteristics such as sensitivity to polarized light. When electromagnetic radiation (for present purposes, with wavelengths in the visible range) impinges on a thin, transparent, adherent film, reflection occurs at the film/air interface and at the film/metal interface. Phase shifts also occur at either or both of these interfaces. As a consequence, selected wavelengths are cancelled between the incident and reflected light, resulting in the reflected light having colors characteristic of those wavelengths that have not been cancelled. Within a good approximation, cancellation of a specific wavelength, λ, occurs for the thickness, $t \simeq N(\lambda/4n)$, where n is the refractive index of the film and N is the integer order of the interference (see, for example, References 24 and 25). If the incident light is monochromatic, then areas of the microstructure having films of thickness given by this relationship will appear dark because of cancellation; other areas will appear light.

As a film thickness is increased, and with incident unfiltered light of all wavelengths, interference occurs first for the shorter wavelengths of the blue limit of the visible wavelength range. The longer wavelengths are reflected, giving a first color of red-yellow. With progressive thickening of the film, the color passes through the spectral range to blue and then repeats for successive values of the order, N. For N greater than 3 or 4, excessive film thickness leads to absorption and poor color development. Considering that light passes into and from the film at an angle, interference for violet light with a wavelength of 400 nm starts to occur for films that are approximately 40 nm thick; these films produce a yellow color. The first blue will occur for films somewhat thinner than 70 nm. Successive color sequences occur for progressively thicker films, but clarity of color based on interference decreases for films thicker than 500 nm. Color enhancement of interference films is frequently accomplished by using polarized light, sensitive tint plates, and phase-contrast devices. These either rely on the ability of some films to alter the plane of polarization or provide a phase shift that is sensitive to wavelength (see References 11, 12, and 25–28).

For color to develop as a result of interference, films 40 to 500 nm thick must be produced. Film thickness is directly proportional to charge density, Q, the integration of the time-current density product to a given time expressed in coulombs cm^{-2}, but only if all metal ions oxidized by the anodic current density remain in the film and do not pass into solution. Otherwise, a correction must be made for this loss. A relationship for film thickness D based on Muller[15] is as follows:

$$D = \frac{M'}{m'z'd'F} Q \alpha \qquad (2\text{-}9)$$

where the primes refer to average values of the quantities:

M' = molecular weight of the oxide
m' = metal atoms per molecule of oxygen
z' = metal ion valence
d' = density of the oxide
F = the Faraday constant

and α is the fraction of the metal that is retained in the film and allows for the preferential loss of selected metal atoms to the environment, such as iron and nickel relative to chromium in an austenitic stainless steel that results in an oxide approaching Cr_2O_3. For this steel, the above relationship reduces to approximately the following:

$$D = 0.5 \, Q \alpha \qquad (2\text{-}10)$$

with D in nm and Q in milliamp sec cm^{-2} or, the equivalent, millicoulombs cm^{-2}.

Theoretical and empirical investigations indicate that the time dependence of the current density during film formation is frequently of the form,

$$\log i = A + \log(1/t^n) \qquad (2\text{-}11)$$

where values of n have been evaluated in the range of 0.6 to 1.[5,15] Further analysis leads to thickening of the films as cubic, parabolic, or logarithmic functions of time, the parameters of the functions depending on the alloy, environment, and potential range in which dissolution occurs. The rate of thickening therefore decreases with time and may lead to excessively long etching times to form films capable of giving interference effects. A limiting thickness may also be reached if the growth rate becomes sufficiently slow for additional growth to be balanced by dissolution of the film into the etchant. These growth-rate characteristics make it difficult to estimate from a conventional polarization curve the time required to form a film of the thickness, in the range of 40 to 500 nm, that is required for interference contrast.

In the passive potential range of most stainless steels and nickel-base alloys, the passive film in acid environments usually attains a steady thickness less than 10 nm and therefore too thin to produce interference colors. In general, as will be shown, good color contrast has been developed by etching in strong NaOH (5- to 40-percent) in potential ranges just above the current density peak or in the early stages of the transpassive potential range. Since the rate of dissolution of the film increases with the increase in potential in the transpassive range, careful control of potential and time is required to obtain the desired film properties. A significant factor that correlates with the formation of thicker films on stainless steel in strongly alkaline solutions is the preferential loss of chromium and the formation of iron- and nickel-rich films that are in contrast to the chromium-rich films that are observed in acid solution.

As an example of these variables, Grutzner and Schuller[5,29] potentiostatically etched a 27.7-wt-percent Cr in iron alloy at 540 mV (SHE) and obtained in 20 sec a yellow color with an estimated thickness of 35 nm; in 60 sec a brown, at 38 nm; in 2 min an orange, at 40 nm; in 6 min a purple, at 44 nm; and in 20 to 60 min a blue, at 48 nm. They relate in detail the interrelationship between several stainless steels in terms of composition, potential, charge density, and current density as a function of time and developed color for a 10N NaOH etching solution. Observations were correlated with potentiostatic polarization curves obtained by holding the alloys at successive potential intervals for 5 min. Based on the polarization curve so derived, a 27.7-percent Cr ferritic steel developed a golden yellow at 440 mV (SHE) in 5 min, corresponding to a charge density of 106 mA sec cm^{-2}. As an example of the decay of the current density with time (during the time required to produce this charge density), the decrease was from 10 mA cm^{-2} at 10 sec to 0.1 mA cm^{-2} at 5 min. Grutzner and Schuller[5] also discuss the difference in charge density required to give color contrast between the ferrite and austenite phases in a two-phase alloy. They observe that after 5 min at +240 mV (SHE), the charge density of a 44.77-percent Cr σ-phase alloy is 208 mA sec cm^{-2} greater than that for the 27.7-percent Cr α-phase alloy. In a two-phase alloy of 60-percent ferrite and 40-percent sigma, the ferrite was blue and the σ-phase was brown (a carbide phase was light yellow). These observations are consistent with the polarization curves shown in Fig. 2-10, in which the current density for the σ-phase is greater than for the α-phase and therefore would produce a thicker film. The curve for the two-phase (α/σ) alloy (Fig. 2-10) usually lies between the curves for the individual phases. The effect of the higher chromium content of the σ-phase

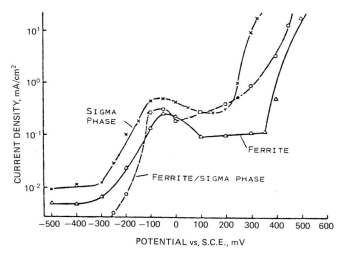

Fig. 2-10. Polarization curves for Fe–Cr alloys in 10N NaOH based on Grutzner and Schuller:[5] Ferrite—27.74-percent Cr; sigma—45-percent Cr; Ferrite/sigma—31-percent Cr + 3-percent Si.

in lowering the potential for the onset of transpassivity is evident when curves for the high- and low-content alloys are compared in the potential range of 200 to 400 mV. Thus, at 250 mV the difference in current density is large and corresponds to excessive attack of the σ-phase in the 5-min holding time used in generating these data. The curves do suggest that useful etching might result for shorter times but that the selection and control of the potential becomes critical.

Deposit Films

Although films can be produced by the deposition of corrosion products, color differentiation can be obtained from films that produce interference contrast by means of the controlled potential oxidation or reduction of species in the etch solution. The method depends on depositing films with thicknesses and/or properties that are sensitive to the substrate phase and to its crystal-lattice orientation. Again, for interference color development, these films must attain thicknesses of 40 to 500 nm, although films that are optically active may be thinner. Examples are the anodic (oxidation) deposition of PbO_2 and MnO_2 according to the following reactions:

$$Pb^{2+} + 2H_2O \rightarrow PbO_2 + 4H^+ + 2e \quad (2\text{-}12)$$

and

$$Mn^{2+} + 2H_2O \rightarrow MnO_2 + 4H^+ + 2e \quad (2\text{-}13)$$

For example, Grutzner and Schuller[29] obtained a yellow film in 1 min at 660 mV (SHE) in a lead-acetate solution; blue was developed in 3 min, and the next order of yellow at 4 min. Potentiostatic deposition of MnO_2 from a 10-percent $MnSO_4$ solution has been reported by Helbach and Bullock.[30] Soluble species of higher valence can be reduced (cathodic deposition) to insoluble film-forming species such as MoO_2 in accordance with the reaction,

$$MoO_4^= + 4H^+ + 4e \rightarrow MoO_2 + 2H_2O \quad (2\text{-}14)$$

Although the formation of deposit films by immersion using similar reagents, including the formation of sulfide films, has been described (see Reference 31), any investigation of deposition by control of potential appears to be limited. Since the film-forming species are in solution, an advantage of the technique is that the growth occurs at the film/solution interface without the necessity of diffusing cations or anions through the film. As a consequence, the current density, and hence film growth rate, is constant and does not decrease with time as it does during thickening of corrosion-product films.

Problems may be encountered if the potential required for the formation of deposit films lies in the range of rapid dissolution of the substrate. The problem is alleviated by the fact that solutions used for deposit films may be relatively neutral and hence not as aggressive in the required potential range as they would be if film formation required extreme values of pH.

COMMENTS ON EXPERIMENTAL PROCEDURE

The major advantage of potentiostatic etching is that the electrochemical potential of the metallographic specimen is an additional independent variable. With this added variable, many etchants that are not effective as, and frequently less aggressive than, chemical etchants can accomplish comparable results and are therefore easier to use. In fact, the use of the potentiostatic technique to enhance conventional chemical etchants is frequently unsuccessful, and if the latter act rapidly, the time to establish the potential for etching potentiostatically and remove the specimen may permit unwanted chemical attack.

A typical arrangement for potentiostatic etching was shown in Fig. 2-4. The specimen should be mounted in epoxy polymer, with particular care given for good adherence between the metal specimen and mounting material. Otherwise, under some etching conditions, crevice attack along the poor bond will lead to uneven etching; in the extreme case, the attack will be confined to the crevice. Various arrangements for electrical contact have been used, the two most common being a screw soldered or projection welded to the sample and the sample/screw combination embedded in the mounting material with the screw projecting from the back. Alternatively, the mount can be drilled and threaded to make contact with the back of the specimen. In both of these arrangements, a thick-walled glass rod with gasket seal and enclosed threaded rod or cylinder is used to compress the gasket. This type of specimen holder can be configured for a sample face down in the etchant or face up, the latter being convenient for observation of the progress of the etching. In general, allowance must be made for quickly inserting and removing the specimen from the etchant, and, under some conditions, this must be accomplished while providing reasonable purging of oxygen from the solution by an inert gas. Whatever the arrangement, it is critical to avoid contact between the electrical connection to the specimen and the etchant solution. If the metal of the electrical contact exhibits a large current density in the etchant at the selected potential, then its dissolution rate from a small exposed area can dominate the current demand on the potentiostat and lead to a large underestimate of the time required to etch the specimen.

Several precautionary factors should be recognized in selecting the potentiostat. The choice depends on the range of applications and performance required. The characteristics to be specified include the range of set-point potentials; the range of current that will be demanded to attain the potentials; the accuracy, stability and reproducibility of the set-point potentials; and the response time to changes in set-point potential and to variations in potential caused by electrochemical changes occurring in the specimen. The response characteristics may be critical for those applications requiring that a specimen attain quickly and then hold a closely specified potential. Slow response in attaining and maintaining the required potential may allow the specimen to pass through potential ranges that can cause adverse attack. A large range of laboratory applications can be satisfied by an instrument with a set-point range of ±5 V and a current range from 1 µA to 5 A. Many applications are satisfied with smaller potential and upper current ranges, but, in particular, a larger capacity instrument will be required when large specimens and electrolytic polishing applications may be involved. Instruments with wider set-point potentials and larger maximum current are available.

Although a potentiostat set manually to the required potential for a particular metal/etchant combination is sufficient for routine etching, the ability to scan prescribed potential ranges for any combination is usually a necessity to establish the polarization curve from which the potential or potentials may be determined that provide the desired selective etching. This is accomplished by either mechanical or electronic components that are integral with or external to the potentiostat proper and that allow selection of scan starting and stopping potentials and scan rates. The results of the measurement are recorded graphically in terms of the potential and usually in terms of the log of the current or current density. Reference is again made to problems that relate the polarization curves from single potential scans to etching conditions. Scanning is frequently practiced over large potential ranges, including both net cathodic and anodic regions of potential. This may lead to observed current densities at a given potential that are significantly influenced by the potential/time history of the specimen before reaching this potential, and therefore a current density/etching effect is observed that differs from that resulting from direct exposure at the same potential. Polarization curves established by observing the current density as a function of time at a series of preset potentials are needed, but few measurements of this type have been reported.

SURVEY OF ETCHANTS USED FOR POTENTIOSTATIC ETCHING

Table 2-1 lists etchants reported in the References as being applicable to potentiostatic etching of the indicated materials. The wide range of potentials, times, and temperatures reported precludes reasonable inclusion of these variables in tabular form. Frequently, multiple potentials or multiple etchants have been used to differentiate microconstituents. The information in the table provides an overview to the etchants that have been used, and the foregoing discussion provides a guide to the variables that should be investigated to establish proper techniques.

EXAMPLES OF POTENTIOSTATICALLY ETCHED MICROSTRUCTURES

Examples of potentiostatic etching to enhance microstructural detail are shown in Figs. 2-11 through 2-15. Figure 2-11 shows the influence of potential on the etching of a 50 wt-percent Zn–Sn alloy in 1 N NaOH.[9] This is a two-phase alloy of essentially pure zinc and pure tin. The potential dependence of the etch attack is consistent with Fig. 2-12 and the previous discussion relating to this figure. The zinc is the dispersed phase in a tin matrix. At −1.45 and −0.60 V (SCE), the curve for the zinc phase

Table 2-1. Potentiostatic Etching: Solutions, Materials, and Morphology Developed.

SOLUTION	MATERIAL	MORPHOLOGY DEVELOPED	REF.
1 N H_2SO_4	Ni:5–32Al	γ and β phases Dislocation etch pits	32
	Fe:18Cr,8–10Ni	Martensite formed by subcooling	33
	Fe	Dislocation etch pits	34
10% H_2SO_4	Fe:18Cr,8Ni	Austenite grain bdry. σ-phase, $Cr_{23}C_6$	35
20% H_2SO_4	Fe:18Cr,9Ni,3Mo	Austenite grain bdry. α-ferrite, Cr depletion	1
50% H_2SO_4	Fe:18Cr,10Ni	Intergranular ppt. $Cr_{23}C_6$	36
5% H_2SO_4 + 0.1 g/l NH_4SCN	Fe:0–43Cr,0–42Ni,Mo,Ti	Austenite grain bdry. δ-ferrite, σ-phase	2
20% H_2SO_4 + NH_4SCN	Fe:0–25Cr,0–20Ni,N	α, δ-ferrite, σ-phase, P′-phase	38
	SS304,316 welds	Weld and heat affected, zone etched	37
1 N NaOH	Fe:0.45C	Phosphide inclusions	23
	Sn:0,15,50,75,100Zn	Primary Sn, Zn, and eutectic	16,10
2 N NaOH + complex. agents	Cu:5–30Zn	Grain boundary, dendritic segregation	9

Table 2-1 (continued)

SOLUTION	MATERIAL	MORPHOLOGY DEVELOPED	REF.
5 N NaOH	Fe:18Cr,8Ni	Austenite grain bdry., σ-phase, $Cr_{23}C_6$	35
	Fe:var C and N	Fe_3C, ϵ-nitrides	39
10 N NaOH	Fe:0–62Cr,4–8C	Fe_3C, $M_{23}C_6$ and M_7C_3 distinguished by color as pure phases and as dispersed in austenite	42
	Fe:18Cr,8Ni	Austenite grain bdry., σ-phase, $Cr_{23}C_6$	35
	Fe:38.7–44.7Cr	Same	35
	Fe:25Cr,2Mo,9Ni	Same	35
	Fe:18–41Cr,2.5–39Ni	Martensite, austenite α-ferrite, σ-phase, color	41
	Fe:0.32C	Fe_3C, nitrides in carbonitrided steel	39
	Fe:0.1–4.8C,0–7.85N	Fe_3C, nitrides	39
	Fe:25–45Cr,2Mo,6.4Ni	$Cr_{23}C_6$, α-ferrite, σ-phase, color	40
	Fe:13Cr,1.5Ti,4V	TiC, M_7C_3	3
	Fe:17–45Cr,0–10Ni, 0–2Mo	Austenite, α-ferrite, σ-phase	5
	Al:Cu	Grain contrast, segregation	4
10% NaOH	Co:20Cr,20Ni,4 (Nb,W,Nb)	Color differentiation of M_6C, NbC, color	43
	Co:31Cr,13W,2.2C (cast)	Color differentiation of M_6C, $M_{23}C_6$, color	43
20% NaOH	Fe:2.6Si	Grain orientation	23
	Fe:4.42C,2.6Si(cast)	Si segregation; nodular and flake graphite, interference colors	44
40% NaOH	Fe:low alloy	Differentiation of bainite and martensite, color	3,43
	Fe:27Cr	Ferrite, σ-phase, color	43
	Co:20Cr,20Ni (Mo,W,Nb,Fe)	M_6C, color	43
10 N KOH	Fe:17–34Cr,24–30Ni 0–3Al,0–2Ti	Austenite, ferrite, σ-phase, color	45
10% NH_4OH	Fe:6–17Cr,1.9–7C (cast)	M_3C, M_7C_3, $M_{23}C_6$	40
$Pb(Ac)_2$,sat	Fe:13Cr,4V	VC and TiC versus M_7C_3	3
1% CrO_3	Aluminum alloys	Si, Al_8Mg_5, Mg_2Si Al-Fe-Si, Al_3Ni, Al_6Cu_3Ni	46
10% Na_2CO_3	Fe:Cr,Ni,Mo	$M_{23}C_6$, color	43
10% H_3PO_4	Fe:18Cr,8Ni	Austenite grain bdry. σ-phase, $Cr_{23}C_6$	35
85% H_3PO_4	Cu:Be,Zr,Ni	Grain bdrys., dendritic	47
		Segregation, dispersed phases, color	48
5M citric acid	U:3Nb	α-martensite, dislocation etch pits	49
1N $Na_2S_2O_3$	Fe:Si	Segregation, color	23
1N $(NH_4)_2SO_4$	Zinc		23
	Copper		23
	Fe:low carbon	Grain boundaries	23,50
	Fe:0.45% C	Pearlite	23,50
10% $MnSO_4$	Fe:25Cr,20Ni	M_7C_3, $M_{23}C_6$ MnO_2, color, stain	30
10% Oxalic acid	Fe:18Cr,8Ni	$Cr_{23}C_6$, σ-phase	35
MeOH–3HNO_3–2$HClO_4$	Fe:25Ni,19Cr,5Al	NiAl,δ-ferrite, austenite	51
MeOH–1% Me_4NCl, 10% acetylacetone	Stainless steel	$Cr_{23}C_6$(TiP-TiC)	52
	Ni-base	Cr_7C_3	52
	Fe:19Cr,2Mo,O,O9Nb	AlN, Ti(C,N), z-phase	53

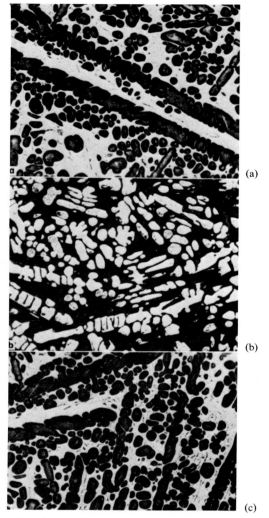

Fig. 2-11. 50 wt-percent Zn–Sn alloy potentiostatically etched in 1N NaOH at: (a) −1.45 V (SCE) (zinc in dark dispersed phase); (b) −0.90 V (SCE) (tin is dark continuous phase); (c) −0.60 V (SCE) (zinc is dark dispersed phase). (Taken from Ref. 9; all at 100×.) Effect of potential relates to polarization curves of Fig. 2-12.

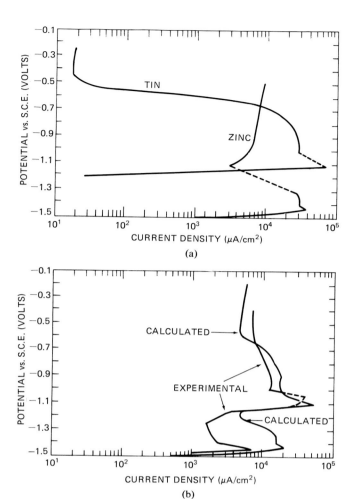

Fig. 2-12. (a) Polarization curves for pure zinc and tin in 1N NaOH, and (b) comparison of calculated and experimental polarization curves for an alloy of 51 wt-percent zinc containing two phases with essentially the composition of pure metals. (Redrawn from Ref. 16.)

lies at a higher current density than for the tin phase; as a consequence, it is preferentially attacked and appears dark in a light matrix of tin. At −0.90 V, the polarization curves predict that the tin phase should be preferentially attacked and thus appear as a dark matrix in which the light, unetched zinc phase is dispersed.

The effects of carbon content on the tendency toward intergranular attack in two cast stainless steels are shown in Figs. 2-13(a) and 2-13(b). The steel (Fe:19Cr,9Ni), containing about 10-percent ferrite in an austenite matrix, was quenched after 1 hr at 1120°C and reheated to 650°C for 1 hr. Chromium depletion near the ferrite/austenite interface tends to occur because of $Cr_{23}C_6$ precipitation. Because of the lower carbon content of the CF3 steel (0.03% C max), the interface in Fig. 2-13(a) has undergone slight chromium depletion, and etching has lightly delineated the ferrite regions. The depletion is greater in the higher carbon CF8 steel (0.08% C max) in Fig. 2-13(b), and the interface is significantly attacked. Austenite/austenite grain boundaries are not etched under these circumstances since the chromium depletion is not large enough relative to the austenite/ferrite interface to make them subject to attack. It is noted that the charge density producing the etching in Fig. 2-13(a) was 8 mC cm^{-2}, which is to be compared to the much larger charge density of 900 mC cm^{-2} that etched the higher carbon steel. KCNS was added to the etchant in this case to move the polarization curve to larger current densities, as was shown in Fig. 2-7.

Figures 2-14 and 2-15 are taken from the publication of Gahm and Jeglitsch[26] on general color methods in metallography. Figure 2-14 shows the results of potentiostatic etching in NaOH to produce gradations in color correlating with silicon segregation in a nodular cast iron. The same technique was used to produce the microstructure of the aluminum alloy in Fig. 2-15. Again, gradations

POTENTIOSTATIC ETCHING

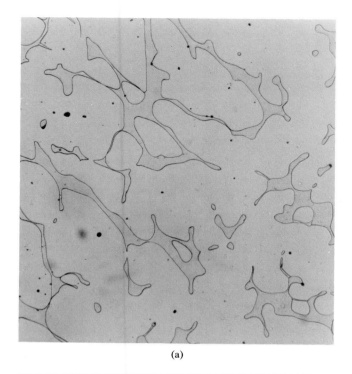

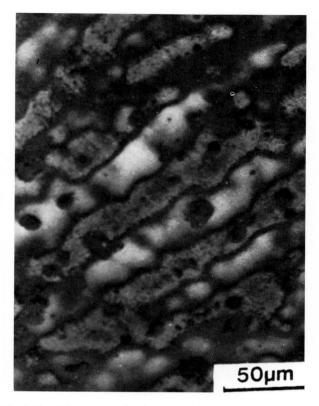

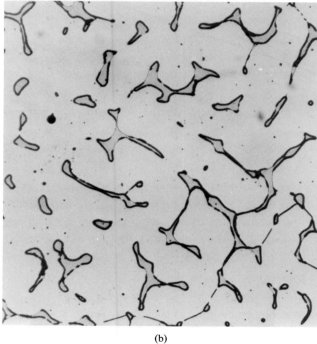

Fig. 2-13. CF3 and CF8 cast stainless steels (Fe:19Cr,9Ni) quenched and reheated 1 hr at 650°C to induce precipitation of $Cr_{23}C_6$ in austenite grain boundaries and potentiostatically etched near +0.05 V (SHE) in 1N H_2SO_4 + 0.05 M KSCN. (a) CF3 with 0.03-percent max., 8 mC cm^{-2}; and (b) CF8 with 0.08-percent C max., 900 mC cm^{-2}. (See text for details; from S. J. Pawel, Univ. of Tenn.)

Fig. 2-14. Silicon segregation in nodular graphitic cast iron, potentiostatically etched in NaOH. (Reproduced from Ref. 26; original source, Ref. 54.)

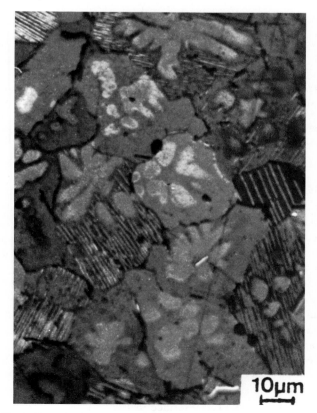

Fig. 2-15. Aluminum-copper alloy showing primary segregation and grains with dispersed phase. (Reproduced from Ref. 26; original source, Ref. 55.)

in color reveal segregation as well as grain contrast and structural detail within grains.

REFERENCES

1. Edeleanu, M. A. The potentiostat as a metallographic tool. *J. Iron Steel Inst., London* 185:482–87 (1957).
2. Cihal, V., and Prazak, M. Corrosion and metallographic study of stainless steels using potentiostat techniques. *J. Iron Steel Inst., London* 193:360–367 (1959).
3. Lichtenegger, P.; Kulmburg, A.; and Bloch, R. Betrag zum potentiostatischen Atzen von Stahl (A contribution to the potentiostatic etching of steel). *Prakt. Metallogr.* 6:535–539 (1969).
4. Jeglitsch, F. Ober die Anwendung elektrolytisch-potentiostatischer Atzverfaheren bei Aluminiumlegierungen. *Aluminium* 45:45–49 (1969).
5. Grutzner, G., and Schuller, H. J. Untersuchungen über das elektrolytisch-potentiostatische Atzen von nichtrostenden Stahlen in 10 N NaOH unter Anwendung eines Coulometers. *Werkst. Korros.* 20:183–194 (1969).
6. Greene, N. D., and Wilde, B. E. Variable corrosion resistance of 18Cr–8Ni stainless steels: Influence of environmental and metallurgical factors. *Corrosion* 26:533–538 (1970).
7. Kurosawa, F.; Taguchi, I.; and Matsumoto, R. Observation of precipitates and metallographic grain orientation in steel by a nonaqueous electrolyte-potentiostatic etching method. *Nippon Kinzoku Gakkaishi* 43:1068–77 (1979).
8. Prazak, M.; Cihal, V.; and Holinka, M. Uber die Differenzierung der Strukturphasen beim metallographischen Atzen. I. Elektrolytisches Atzen mit festgelegtem Potential. *Collection Czechoslov. Chem. Commun.* 24:9–15 (1959).
9. Greene, N. D., and Teterin, G. A. Development of brass etchants by electrochemical techniques. *Corros. Sci.* 12:57–63 (1972).
10. Greene, N. D.; Rudaw, P. S.; and Lee, L. Principles of metallographic etching. *Corros. Sci.* 6:371–379 (1966).
11. Vander Voort, G. F. *Metallography: Principles and Practice.* New York: McGraw-Hill Book Co., 1984.
12. Richardson, J. H. *Optical Microscopy for the Materials Sciences.* New York: Marcel Deker, Inc., 1971.
13. Mauvais, C. J.; Latanision, R. M.; and Ruff, A. W. On the anisotropy observed during the passivation of nickel monocrystals. *J. Electrochem. Soc.* 117:902–903 (1970).
14. Sato, N. The passivity of metals and passivating films in passivity of metals. In *Passivity of Metals*, pp. 29–58. Princeton, NJ: The Electrochemical Society, 1980.
15. Mueller, W. A. Derivation of anodic dissolution curve of alloys from those of metallic components. *Corrosion* 18:73t–79t (1962).
16. Steigerwald, R. F., and Greene, N. D. The anodic dissolution of binary alloys. *J. Electrochem. Soc.* 109:1026–1034 (1962).
17. *Annual Book of ASTM Standards. Standard A262-1981.* Philadelphia: Am. Soc. for Testing and Mat., 1984.
18. Majidi, A. P., and Streicher, M. A. Potentiodynamic reactivation method for detecting sensitization in AISI 304 and 304L stainless steels. *Corrosion* 40:393–408 (1984).
19. Szklarska-Smialowska, Z. Effect of the ratio of chloride/sulphate in solution on the pitting corrosion of nickel. *Corros. Sci.* 11:209–221 (1971).
20. Hodge, F. G., and Wilde, B. E. Effect of chloride ion on the anodic dissolution kinetics of chromium-nickel binary alloys. *Corrosion* 26:146–150 (1970).
21. Schwabe, K., and Schmidt, W. Der Einfluss des Wasser auf die Passivierbarkeit von Nickel in schwefelsaurer Losung. *Corros. Sci.* 10:143–155 (1970).
22. Ludering, H. Beitrag zum elektrolytisch-potentiostatischen Atzen mit besonderer Berücksichtigung der coulometrischen Bestimmung der Atziefe. *Arch. Eisenhuttenwes.* 30:605–611 (1959).
23. Ludering, H. Das elektrolytisch-potentiostatische Atzen. *Radex-Rundschau* 3/4:650–656 (1967).
24. Herbsleb, G., and Schwaab, P. Fundamentals of the potentiostatic development of structures using high-alloy steels as an example. *Prakt. Metallogr.* 15:213–223 (1978).
25. Beraha, E., and Shpigler, B. *Color Metallography.* Metals Park, Ohio: American Society for Metals, 1977.
26. Gahm, H., and Jeglitsch, F. Color methods and their applications in metallography. *Microstruct. Sci.* 9:65–80 (1981).
27. Modin, H., and Modin, S. *Metallurgical Microscopy.* London: Butterworths, 1973.
28. Phillips, V. A. Modern Metallographic Techniques and Their Applications. New York: Wiley-Interscience, 1971.
29. Grutzner, G., and Schuller, H. J. Potentiostatic color etching of stainless steels. *Prakt. Metallogr.* 6:246–258 (1969).
30. Helbach, P., and Bullock, E. Potentiostatic Etching of Carburized Steels. Petten, Neth.: Comm. Eur. Communities, (Rep) EUR, 1982.
31. Vander Voort, G. F. Tint etching. *Metals Progress* 127:31–41 (March 1985).
32. Sullivan, C. P.; Jensen, J.: Duvall, D. S.; and Field, T. T. Potentiostatically controlled etching of nickel-aluminum alloys. *Trans. Am. Soc. Met.* 61:582–591 (1968).
33. Belo, M. C., Berge, P., and Montuelle, J. Metallographie—Attaque selective, a l'aide du potentiostat, des phases en presence dans an acier inoxydable biphase. *C. R. Acad. Sc., Paris* 7:570–573 (1964).
34. Khaldeyev, G. V.; Knyazeva, V. F.; and Kuznetsov, V. V. Selective potentiostatic etching on dislocations in iron. *Zashchita Metallov Korroz* 11:729–731 (1974).
35. Roschenbleck, B., and Buss, E. K. Potentiostatisches, differentielles Atzen der Gefugebestandteile in 18/8 Cr-Ni-Stahlen. *Werkst. Korros.* 4:261–269 (1963).
36. Tsinman, A. I., Degtyareva, V. K., Neiman, N. S., Kassinskaya, L. L., Kuzub, V. S., and Murashkina, A. A. Exchange of experience determination of tendency in chromium–nickel steel Kh18N10T towards intercrystalline corrosion by potentiostatic etching method. *Zashchita Metallov Korroz* 6:475–478 (1970).
37. Voeltzel, J.; Henry, G.; Manenc, J.; and Plateau, J. Utilisation du potentiostat electronique pour l'attaque micrographique. *Memoires Scientifiques Rev. Metallurg.* 62:129–134 (1965).
38. Gooch, T. G., Honeycombe, J., and Walker, P. Potentiostatic study of the corrosion behaviour of austenitic stainless steel weld metal. *Br. Corros. J.* 6:148–154 (1971).
39. Langenscheid, G., and Naumann, K. Die Unterscheidung von Eisenkarbid und Eisennitriden durch Atzen. *Arch. Eisenhuttenwes.* 36:505–508 (1965).
40. Naumann, F. K., and Langenscheid, G. Die Unterscheidung von Eisenund Chromkarbiden druch Atzen. *Arch. Eisenhuttenwes.* 38:463–468 (1967).
41. Schaarwachter, W., Ludering, H., and Naumann, F. K. Die elektrolytische Atzung mehrphasiger Eisen-Chrom-Nickel-Legierungen in Natronlauge. *Arch. Eisenhuttenwes.* 31:385–391 (1960).
42. Naumann, F. K. Beitrag zum Nachweis der Alpha-Phase und zur Kinetik ihrer Bildung und Auflosung in Eisen-Chrom- und Eisen-Chrom-Nickel-Legierungen. *Arch. Eisenhuttenwes.* 34:187–194 (1963).
43. Bloch, R., and Lichtenegger, P. Die selektive Darstellung von gefugebestandteilen mittels potentiostatischer Atzung (The use of potentiostatic etching to reveal microstructural constituents selectively). *Prakt. Metallog.* 12:186–193 (1975).
44. Ludering, H. Versuche zum Atzen von Siliziumseigerungen in Eisen-Kohlenstoff-Legierungen. *Arch. Eisenhuttenwes.* 2:153–159 (1964).
45. Jones, J. D., and Hume-Rothery, W. Constitution of certain austenitic stainless steels, with particular reference to the effect of aluminum. *J. Iron Steel Inst., London* 203:1–7 (1966).
46. Roschenbleck, B.; Fecht, D.; and Koslowski, W. Potentiostatisch differentielles Atzen von Aluminiumlegierungen (Potentiostatic dif-

ferential etching of aluminium alloys). *Prakt. Metallogr.* 18:376–384 (1981).
47. Mance, A. Potentiostatic etching and polishing of copper and its alloys. *Metallog.* 4:287–296 (1971).
48. Mance, A.; Perovic, V.; and Mihajlovic, A. Potentiostatic controlled etching and polishing of copper and its alloys. *Metallog.* 6:123–130 (1973).
49. Mihajlovic, A., and Mance, A. The potentiostatic method for electrolytic etching of U-Nb Alloys. *J. Nucr. Mater.* 26:267–272 (1968).
50. Ludering, H. Uber das elektrochemische Anatzen verschiedener Gefugebestandteile und von Mischkristallen unterschiedlicher Zusammensetzung. *Werkst. Korros.* 8:665–668 (1966).
51. Kurosawa, F.; Taguchi, I.; and Matsumoto, R. Observation and analysis of beta phase in steel using nonaqueous electrolyte-potentiostatic method. *Nippon Kinzoku Gakkaishi* 45:165–173 (1981).
52. Kurosawa, F.; Taguchi, I.; and Matsumoto, R. Studies on observations and analysis by the SPEED method. 5. *Nippon Kinzoku Gakkaishi* 44:1288–1295 (1980).
53. Kurosawa, F.; Taguchi, I.; Tanino, M.; and Matsumoto, R. Observation and analysis of nitrides in steels using the nonaqueous electrolyte-potentiostatic etching method. *Nippon Kinzoku Gakkaishi* 45:63–71 (1981).
54. Jeglitsch, F. Siliziumseigerungen in kugelgraphitschem Gusseisen und ihr Nachweis. *Microchem. Acta* 3:479–493 (1965).
55. Jeglitsch, F. Ober die Anwendung elektrolytisch-potentio-statischer Atzverfahren bei Aluminiumlegierungen. *Aluminium* 45:45–49 (1969).

3
APPLICATIONS OF THE INTERFERENCE LAYER METHOD

Hans-Eugen Bühler

Senior Vice President, DIDIER-Werke AG
Wiesbaden, FR Germany

and

Irfan Aydin

Deloro Stellite GmbH
Koblenz, FR Germany

OBJECTIVES OF METALLOGRAPHY

The objective of metallographic examination is to classify all phases of the microstructure of a material according to their shape, distribution, and size (1) in a good light/dark or color contrast, and (2) in a true geometric representation. In addition, contrast development should be reproducible. Of the chemical and physical methods utilized in microstructure development, only the latter fulfill the requirements for contrast, contrast reproducibility, and true geometric representation. The interference layer method is especially suited to meet these requirements.

Figure 3-1 presents a survey of the standard preparation methods in metallography (chemical etching, potentiostatic etching, ion etching, interference layer coating).

OPTICAL CHARACTERISTICS OF METALLIC AND NONMETALLIC PHASES

The various materials and phases that are to be identified in metallographic polished sections exhibit conspicuously deviating optical behavior in the wavelength zone of visible light. As examples, Fig. 3-2 schematically illustrates the optical characteristics of (1) α-iron and γ-iron, (2) aluminum and gold, and (3) oxides and sulfides.

One will notice that the metallic materials (aluminum, α-iron, and γ-iron) have a very high reflectivity (60 to above 90 percent). With the exception of copper and gold, the reflectivity of metals is almost constant in the range of the visible spectrum (0.4 to 0.7 μm), a fact that produces the metallic sheen characteristics of most of them. Nonmetallic phases, for example, sulfides and especially oxides, have less reflectivity than metals and consequently can be clearly distinguished from the other phases on a polished section without any additional preparation to intensify the light/dark contrast. On the other hand, the human eye cannot clearly distinguish reflection differences of approximately 4 percent between α-iron and γ-iron (58- and 62-percent reflectivity, respectively).

Because the α- and γ-phases cannot be distinguished by observing the polished metal surface under a light microscope, their contrast must be enhanced by etching methods. As can be seen in Fig. 3-1, the etching methods used do not always present sufficient light/dark contrast between the various phases present. In addition, the true appearance of the shape and size of individual phases, especially those in fine-grained microstructures, may be distorted. This fact is clearly illustrated once again in Fig. 3-3, using an austenitic-martensitic microstructure as an example. The microstructure area of Fig. 3-3(a) is also shown after etching and coating in Fig. 3-3(b). In the latter case, the residual austenite stands out light

PREPARATION METHOD	CONTRAST	CONTRAST REPRODUCIBILITY	TRUE GEOMETRIC APPEARENCE
ETCHING	+	–	–
POTETIOSTATIC ETCHING	+	(+)	–
ION ETCHING	+	+	(+)
INTERFERENCE LAYER METHOD	+	+	+

Fig. 3-1. Preparation methods in metallography.

42 APPLIED METALLOGRAPHY

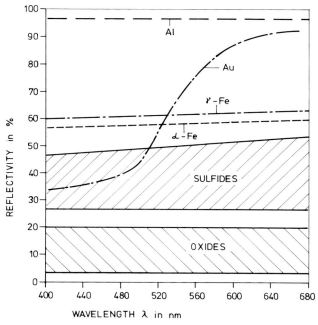

Fig. 3-2. Optical characteristics of different materials (steel, aluminum, gold, sulfides, oxides).

against the transformation structure, and differentiation between the austenite and the martensite is easily made. The true shape and size of the individual microstructure components are not changed by the preparation.

BASICS OF THE INTERFERENCE LAYER METHOD (CONTRAST ENHANCEMENT)

Interference layer metallography is based on the presumption that a polished and smooth material surface is usually available for microscopic observation. On this polished and undisturbed surface, a thin layer of a refractive coating material that permits light penetration (nonabsorbent) is placed by a suitable method, either vaporization or sputtering.

The effect of this interference layer can be described by reference to Fig. 3-4. Light ray E falls on a coated material surface and is refracted at the surface air/layer interface. The ray penetrates the layer and is once again refracted at the second interface, i.e., at the layer/material interface. The light ray is partially reflected at this point. Depending on the absorption capability of the phase, the ray will penetrate the material surface to a certain degree. The wave P coming out of the surface at the air/layer interface is composed of the wave trains reflected at both border surfaces. These interfere in such a way that the amplitude of wave P becomes significantly smaller than the amplitude of incident wave E. This, however, indicates that the reflectivities of the individual phases are drastically lowered by the coating. Figure 3-5 shows the measured reflectivity of two phases in hot-dip aluminum-coated sheet steel (α-iron and the carbide Fe_3AlC_x) after

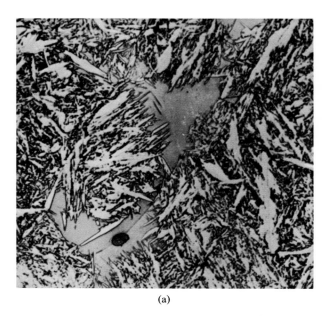

(a)

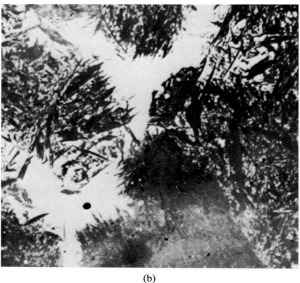

(b)

Fig. 3-3. Light/dark contrast between austenite and ferrite (steel with 1-percent C and 2-percent Mn): (a) etched in 3-percent alcoholic nitric acid, and (b) coated with ZnSe. (Both at 500×.)

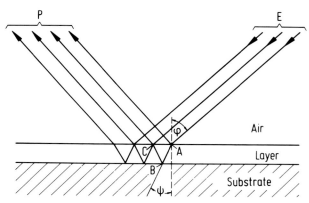

Fig. 3-4. Reflection of a light ray, E, at a coated metal surface.

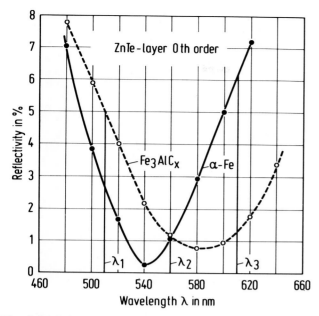

Fig. 3-5. Reflectivity of α-iron and of iron-aluminum carbide (Fe₃AlCₓ), with a vacuum-deposited ZnTe layer, as a function of wavelength.

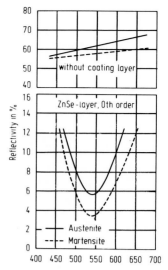

Fig. 3-8. Reflectivity as a function of observed wavelength both in a specimen's uncoated condition and after being coated with a vacuum-deposited ZnSe layer (light/dark contrast between austenite and martensite).

interference layer coating with ZnTe. One notices that the high reflectivity at the onset—between 60 and 70 percent for the α-iron—drops to values below 1 percent. The same behavior is exhibited by the carbide. It is also remarkable that the reflectivity shows a distinct minimum in contrast to the polished metal surface (see Fig. 3-2).

The location of the reflectance minimum differs from phase to phase. As a result, because of the effects of the interference layer, a different part of the visible wavelength spectrum is eliminated for each phase. As Fig. 3-6 shows, this explains why various colors become visible under the light microscope after an interference layer is applied (color contrast). In Fig. 3-7, the color development from an uncoated material with high reflectivity (a) to a coated material with little reflectivity (b) is clearly illustrated. (See color insert for Figs. 3-6 and 3-7.)

The occurrence of a good light/dark contrast is illustrated in Fig. 3-8. The contrast of the α-iron and γ-iron phases in Fig. 3-3 is defined (where $R_1 > R_2$) as follows:

$$K = \frac{R_1 - R_2}{R_1}$$

In the polished condition, this becomes

$$K_{\text{polished}} = \frac{62 - 58}{62} = \frac{4}{62} \approx 0.06$$

After coating with ZnSe, the following, however, results:

$$K_{\text{coated}} = \frac{5.8 - 3.5}{5.8} = \frac{2.3}{5.8} \approx 0.4$$

In this case, the light/dark contrast is improved as a result of coating by a factor of about 10, and the phases can be clearly seen by the human eye. If both requirements for contrast improvement prevail (color contrast and light/dark contrast), then the possibilities of a quantitative and reproducible metallography exist because the physical characteristics of the metal surface, applied coating material, and polychromatic light used do not change.

PREPARATION OF SPECIMENS (GRINDING AND POLISHING)

To obtain optimum results with the interference layer method, an intensified effort is necessary to guarantee that the specimen surface is as physically clean as possible. Very careful work when preparing the surface is an absolute must. Figure 3-9 shows the spectral reflectivity of α-iron and Fe₃C after polishing with different methods. Electrolytically polished and diamond-polished surfaces have a significantly higher reflectivity.

In agreement with Fig. 3-9, Fig. 3-10 shows the results of the influence of different polishing methods on the reflectivity of Cu, Au, Ag, Bi, Ni, and Sb₂S₃. Here, too, it is obvious that an Al₂O₃ polish will often provide unfavorable results. The best results are obtained with an electrolytic polish. Planing the specimen with a microtome will produce excellent reflectivity, but this method cannot be utilized with many materials.

The high reflectivity of electrolytically polished surfaces shows that all types of smeared layers, deformation layers, passive coats, and residual humidity must be avoided. The following examples give a clear illustration of characteristic mistakes made during specimen preparation.

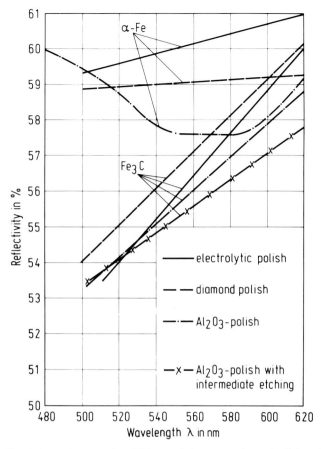

Fig. 3-9. Influence of polishing techniques on the reflectivity of α-iron and Fe₃C.

Fig. 3-11. Chill-cast chromium steel with 2-percent C and 17-percent Cr, coated with ZnSe, and photographed at 540 nm (at 500×).

Fig. 3-12. Same microstructure area of same alloy with same preparation as in Fig. 3-11, but with a smeared surface layer as a result of improper polishing (at 500×).

Polishing and grinding scratches on a specimen surface will become noticeable in a particularly unpleasant manner after an interference layer is applied. The shade effect of scratches will be intensified decisively by the reduction of the reflectivity of a polished surface.

Smeared layers, deformation layers, and long hygroscopic exposure of layer materials to air are also typical preparation mistakes. Smeared layers can consist of oxidation products that form on a specimen surface during its preparation if it has a high affinity for oxygen, or they can consist of a very thin layer of deformed surface particles. In both cases, the presence of these layers changes the optical characteristics of the specimen surface. The same microstructure area in a chill-cast chromium steel is shown after careful specimen preparation in Fig. 3-11 and with the presence of a smeared layer in Fig. 3-12.

Too forceful pressure on a specimen during preparation can cause a strong degree of cold deformation of the surface. In some cases, phase transformation will occur, together with a change of optical characteristics. A typical example is the transformation of residual austenite to martensite in chromium-nickel stainless steels. Figures 3-13 and 3-14 show the microstructure of an austenitic steel with 18-percent Cr and 9-percent Ni. Because of excessive deformation, the surface regions have been transformed to martensite in Fig. 3-13 (see color insert). In contrast, a flawlessly polished section after etching reveals only austenite in Fig. 3-14. In Fig. 3-13, the surface is deformed to such an extent that the microstructure bears no resemblance to the actual microstructure.

Some of the materials used for the interference layer method react hygroscopically after short exposure to humid air, causing drops to appear on the surface. These drops have a very fine distribution. An example occurs in the evaporation treatment with zinc selenide illustrated in Fig. 3-15 (see color insert). In addition to zinc selenide, zinc telluride is also a hygroscopic layer material and

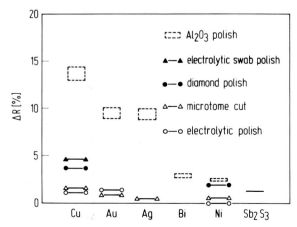

Fig. 3-10. Influence of polishing techniques on the relative change in reflectivity ($\Delta R = 0$ for ideal surface).

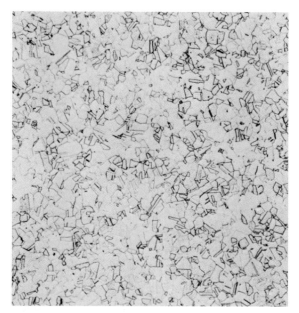

Fig. 3-14. Steel X5 CrNi 18 9, austenite, etched with a solution of 80 g HCl, 5 g HNO$_3$, 20 g glycerin, and 1 g CuCl$_2$ (at 200×).

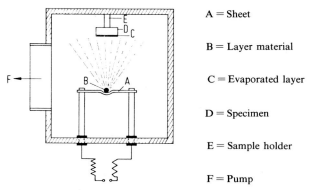

A = Sheet
B = Layer material
C = Evaporated layer
D = Specimen
E = Sample holder
F = Pump

Fig. 3-18. Schematic illustration of apparatus for coating a specimen by vaporization treatment.

should consequently not be stored over a long period, especially if exposed to moist air.

Drying stains can result if a specimen has not been carefully cleaned and dried. The microscopic picture can be substantially distorted. Figure 3-16 shows the microstructure of an Armco iron specimen that was not properly cleaned and then received evaporation treatment with zinc selenide. In most cases, after the completion of the grinding and polishing work, the cracks and pores on the specimen surface contain residual moisture that will spread over the surface after drying. The presence of this moisture will produce false interference colors after coating. An example is shown in Fig. 3-17. Here a hot-aluminized thin sheet is shown embedded in synthetic resin. Because of the shrinkage of the resin, a gap was formed that caused the moisture ring. Rainbow colors are characteristic of drying stains and moisture. (See color insert for Figs. 3-16 and 3-17.)

APPLICATION OF INTERFERENCE LAYERS

Interference layers can be produced by evaporation treatment or by sputtering.

The evaporation treatment to produce interference layers takes place in a vacuum at approximately 10^{-5} bar. Figure 3-18 shows a schematic illustration of the apparatus required. The vapor source consists of one or more holders for boats or simple sheets made from metals with a high melting point, e.g., tungsten or molybdenum (these metals are heated by resistance). Such apparatus is available on the market in a wide variety of models. To implement the evaporation treatment, the clean and dry polished specimen is mounted in the specimen chamber vertical to the vapor source in order to obtain uniform layer thickness. It is possible to interrupt the treatment to observe the layer thickness on the specimen surface. Visual inspection during the treatment is likewise possible. There are also several layer thickness measurement systems that permit automatic and continuous measurement of the layer thickness.

The interference color will depend on the layer thickness. Favorable conditions for contrast are usually obtained when the specimen surface appears to have a purple-violet interference color.

The evaporation equipment must always be kept clean. When it is opened, the layers deposited during evaporation treatment can easily absorb water. This moisture is very hard to remove by evacuation. The result can be a deterioration of the vacuum or even reactions with the layer materials. The apparatus should be kept under vacuum when not in use. The equipment should not be cleaned with acids (for example, etching solutions). Reactions with zinc selenide or zinc telluride can lead to toxic reactions. After one or two evaporation treatments, the glass cylinder should be cleaned with a dry leather cloth.

On first looking at a specimen, it can be useful to produce a variable layer thickness in order to make a preliminary estimation of good light/dark contrasts or color contrasts. In such cases, a wedge-shaped evaporation treatment is recommended. The specimen should not be positioned vertically to the vapor source as in Fig. 3-18 but in an inclined position. The various interference colors that will result from an inclined position are shown in Fig. 3-19 (see color insert).

In addition to the evaporation process, interference layers for creating contrasts can be produced by sputtering. The sputtering treatment consists of atomizing material by bombarding its surface with highly energetic particles. Figure 3-20 is a schematic of the sputtering apparatus. The specimen that is to receive the interference layer functions as the anode and the target as the cathode. Between cathode and anode there is a dc voltage of 1 to 5 kV. During atomization, it is usually not possible for the cathode material to react with the atmosphere

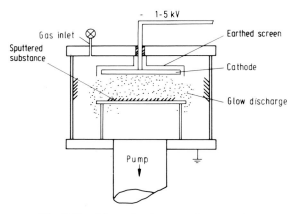

Fig. 3-20. Schematic of a sputtering apparatus.

because of the presence of inert gases. Such reactions are consciously utilized in the metallographic contrast procedure, however. The layers produced very often have good optical characteristics. Oxygen is used (in most cases air is sufficient) so that the atomized target material (cathode material) will be oxidized. Oxidized layers will be produced whose optical characteristics are known.

The sputtering equipment must be of such design that it is possible to maintain a specific pressure of reaction gas during the reaction, for example, via a needle valve with fine regulation. Constant test conditions must be maintained at all times because changing the voltage or gas volumes will alter the optical characteristics of the oxide layers and consequently the desired contrast. Iron, copper, platinum, palladium, lead, and other metals have been used as cathode materials. The interference layers then consist of the oxides of these metals.

The materials that have been used to date in interference layer metallography for the evaporation treatment and the cathode materials used for sputtering are listed in Table 3-1. Tables 3-2 and 3-3 provide a survey of the optical characteristics of the layer materials.

CHARACTERISTICS OF LAYER MATERIALS AND LAYERS

Table 3-1 shows that the layer materials used for evaporation have a refractive index range between 1.35 and 3.25 and an absorption coefficient between 0 and 0.5.

For light/dark contrast and color contrast, the choice among the various layer materials will depend on the corresponding specimen material. For highly reflecting metallic materials, one will need layer materials that either have a high refractive index and are nonabsorbing or have a suitable ratio of refractive index to absorption coefficient as a result of their optical characteristics. For low-reflecting materials with weak absorption (oxides or sulfides), layer materials with a refractive index less than 2.5 are used. In Fig. 3-21, two optical constants—the absorption coefficient, k, and the refractive index, n—of various pure metals are listed. The circular arcs in this figure mark those areas in which the reflectivity of a material is decreased to zero. A coating with zinc tellu-

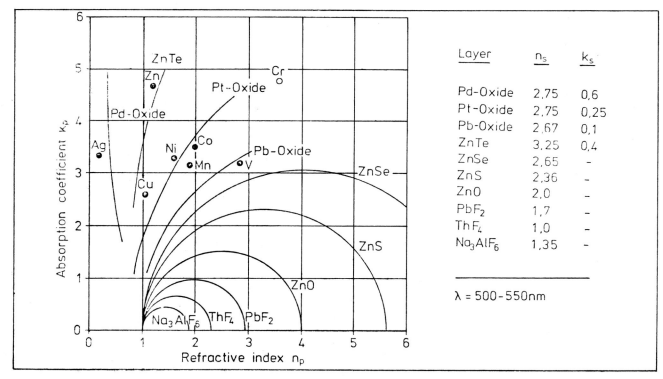

Fig. 3-21. Optical constants of pure metals ($\lambda = 550$ nm).

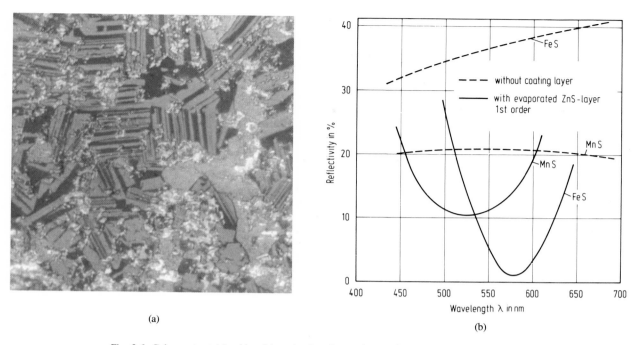

Fig. 3-6. Color contrast (a) achieved by using interference layers (b), to discriminate FeS from MnS, both of which formed on the surface of a steel after extensive service in a sulfur-rich atmosphere (at 1000X).

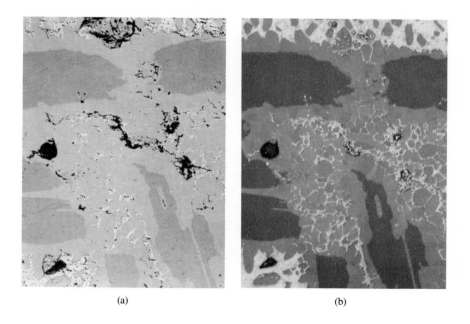

Fig. 3-7. Reflectivity of polished (a) and coated (b) phases in an aluminum–30-percent cobalt alloy (at 100X).

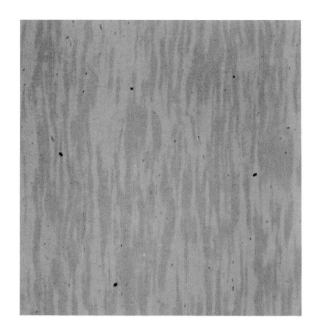

Fig. 3-13. Excessive deformation during polishing of an X5 CrNi 18 9 austenitic steel produced stress-induced martensite. This sample was coated with ZnSe and examined with bright-field illumination (at 200X)

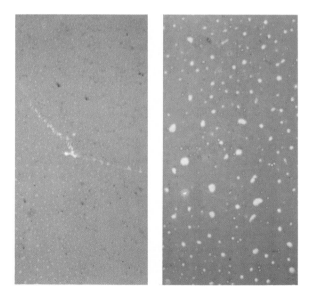

Fig. 3-15. Armco iron with drying stains; coated with ZnSe and examined with bright field illumination (at 100X).

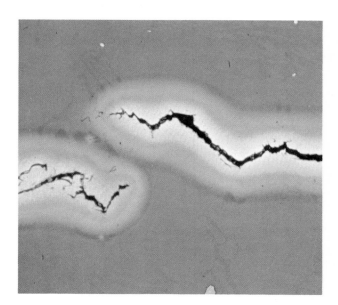

Fig. 3-16. Armco iron with drying stains as a result of moisture from cracks; coated with ZnSe and examined with bright-field illumination (at 200X).

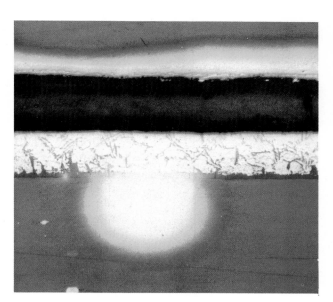

Fig. 3-17. Hot-aluminized sheet embedded in a synthetic resin and exhibiting drying stains from moisture seeping out of the gap between the specimen and the embedment; coated with ZnSe and examined with bright-field illumination (at 200X).

Fig. 3-19. Specimen coated with wedge-shaped, vaporization-treatment layer (5X).

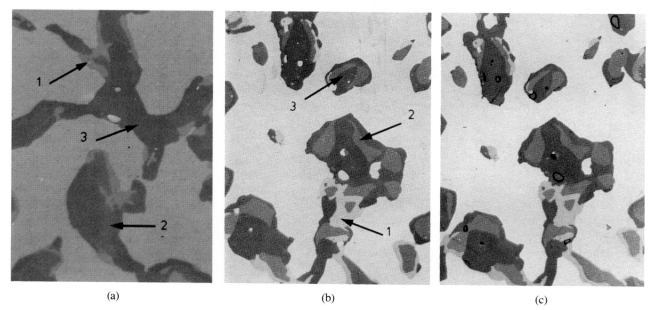

Fig. 3-26. Carbides in austenitic cast–alloy steel X45 CrNiSi 25 30 after 45,000 hours in a nitriding atmosphere: (a) sputtered with iron cathode in oxygen (under bright field; austenite matrix, light brown), where (1) is $M_{23}C_6$ (yellow-red), (2) is $M_2(C,N)$ (red), and (3) is $M_6(C,N)$ (violet-red); (b) same material and same area of microstructure but coated with ZnSe, where phase 1 is $M_{23}C_6$, phase 2 is $M_2(C,N)$, and phase 3 is $M_6(C,N)$; and (c) same material and same area of microstructure but sputtered with copper cathode in argon, photographed at 570 nm. (All at 500X.)

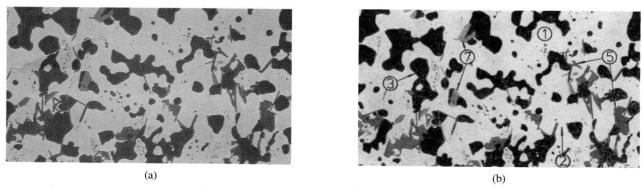

Fig. 3-28. Color contrast (a) and light/dark contrast (b) of sulfides in specimen of IN 713LC coated with ZnS and corroded in 0.9 Na, 0.1 K_2SO_4. (Both at 500X.)

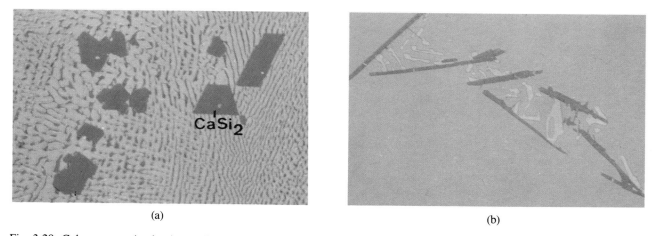

Fig. 3-29. Color contrast in aluminum alloys: (a) aluminum alloy with 7.5-percent Ca, 0.8-percent Si, and 0.2-percent Ti, chill cast and coated with ZnTe, where $CaSi_2$ is red-violet and Al_4Ca is blue; and (b) aluminum alloy with 2.3-percent Ti and 1.4-percent Si, sand cast and coated with ZnTe, where Al_3Ti is violet and silicon is pale green. (Both at 540X.)

Table 3-1. Characteristics of Materials for the Production of Evaporated Layers.

NO.	MATERIAL	REFRACTIVE INDEX N_s ($\lambda = 550$ NM)	ABSORPTION COEFFICIENT K_s ($\lambda = 550$ NM)	EVAPORATION TEMPERATURE RANGE (°C)	SUPPORT SHEET MATERIAL	REMARKS
1	Na_3AlF_6	1.35	0	900–1200	Mo	Very suitable
2	MgF_2	1.38	0	1200–1600	Mo, Ta	Evaporates from the melt; very suitable
3	ThF_4	1.52	0	1300–1600	Mo, Ta	Suitable, but radioactive, care needed with handling
4	LaF_3	1.6	0	1300–1600	Mo, Ta	Evaporates from the melt; suitable
5	CeF_3	1.6	0	1300–1600	Mo, Ta	Evaporates from the melt; very suitable
6	PbF_2	1.75	0	800–1100	W, Al_2O_3	Evaporates from the melt; very suitable
7	ZnS	2.4	0	800–1200	Mo	Reproducible refractive index; very suitable
8	ZnSe	2.65	>0	500–800	Mo, W	Very suitable; slightly hygroscopic
9	TiO_2	2.7	0.04	>1500	W	Suitable, but evaporation not possible directly; very high evaporation temp.; white heat
10	ZnTe	3.25	0.4	800–1200	Mo, Ta	Very suitable; slightly hygroscopic
11	Sb_2S_3	2.3–2.4	0.07	300–400	Mo, Ta	Strong absorp. at <600 nm; otherwise suitable
12	CdS	2.4	0.14	800–1200	W, Mo, Ta	Does not condense well on polished surface; not so suitable
13	CdSe	~2.6	>0.05	800–1000	Mo, Ta	Decomposition can be avoided by careful evaporation; in this case suitable
14	CdTe	3.3	0.2	800–1000	Mo, Ta	Partial decomposition possible; unsuitable
15	InP	3.4–3.5	>0.2	—	W	Evaporates from the melt; partial decomposition; unsuitable
16	GaAs	3.4	1.6	—	—	Decomposes on evaporation; unsuitable
17	GaP	3.45	0	—	—	Decomposes on evaporation; unfortunately unsuitable

Note: Care must be taken when working with compounds of cadmium because of their toxic nature.

Table 3-2. Optical Constants of Metallic Oxide Layers Produced by Sputtering in Oxygen.

OXIDE	450 nm		550 nm		650 nm	
	N_s	K_s	N_s	K_s	N_s	K_s
Hg	2.77	0.21	2.66	0.04	2.66	0.01
Pb	2.68	0.25	2.55	0.06	2.43	0.01
Fe	2.55	0.30	2.42	0.10	2.32	0.02
Pt	2.50	0.24	2.45	0.12	2.40	0.06
Au	2.28	0.34	2.25	0.23	2.22	0.12
Cu	2.03	0.44	2.09	0.37	2.11	0.31
Ni	1.88	0.22	1.87	0.19	1.87	0.18
Error	±3%	±20%	±2%	±15%	±1%	±10%

Table 3-3. Optical Constants of Evaporated Layers.

λ (nm)	ZnO		CdS		ZnS		Sb_2S_3		ZnSe		TiO_2		ZnTe		CdTe		GaAs		GaP		InP	
	N_s	K_s	N_s	K_s	N_s	K_s	N_s	K_s	N_s	K_s	N_s	K_s	N_s	K_s	N_s	K_s	N_s	K_s	N_s	K_s	N_s	K_s
400			2.01	0.54								0.05	3.48	1.1	3.9	1	2.45	2.36		0		
450	2.11	0	2.04	0.48	2.48						2.99	0.06	3.37	0.75	3.63	0.6	2.92	2.26		0	3.7	
500	2.06	0	2.28	0.33	2.42				2.74		2.7	0.05	3.31	0.54	3.43	0.32	3.29	1.94		0	3.5	
550	2.02	0	2.13	0.14	2.4		2.5	0.07	2.65		2.73	0.04	3.24	0.4	3.33	0.2	3.4	1.63	3.44	0		
600	2.0	0	2.01	0.12	2.36				2.61		2.75	0.03	3.2	0.28	3.3	0.12	3.45	1.42	3.35	0		
650	1.99	0	2.0	0.1			2.5	0.01			2.72	0.01	3.06	0.21	3.23	0.08	3.48	1.2	3.29	0		
700	1.90	0	1.96	0.1	2.33						2.69	0.01	2.98	0.19	3.18	0.05	3.5	1.05	3.24	0	3.4	

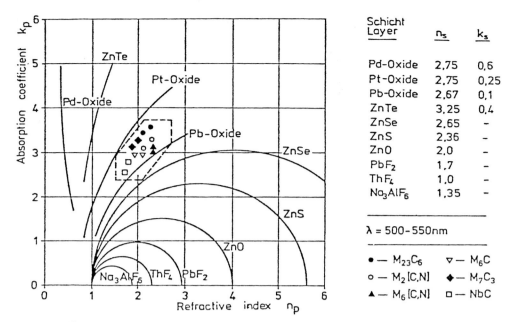

Fig. 3-22. Optical constants of carbides and carbonitrides in steels (λ = 550 nm).

ride, for example, decreases the reflectivity of zinc and copper. A coating with platinum oxide effects a reduction of the reflectivity of nickel and cobalt to zero. The coating of lead oxide has the same effect for vanadium.

Figure 3-22 lists the measured optical constants of carbides and carbonitrides. As a supplement to Fig. 3-21, it is noticed that a coating with platinum oxide or lead oxide (sputtering) or the evaporation treatment with zinc telluride will provide the best light/dark contrast. All those materials and phases for which optical constants have been measured are listed in the bibliography. These include the steels, phase groups of borides, carbides, nitrides, sulfides, and oxides. In addition, several aluminum compounds have been measured. If the optical characteristics of these compounds are known, it will be possible to select those layer materials listed in Tables 3-1 to 3-3 that will guarantee the best contrast. In addition to Figs. 3-21 and 3-22, Figs. 3-23 and 3-24 list the results for sulfides in steels and for some aluminum compounds in aluminum.

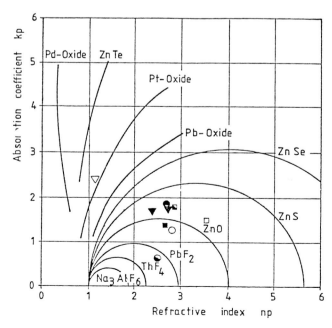

Fig. 3-23. Optical constants of sulfides in steels (λ = 550 nm).

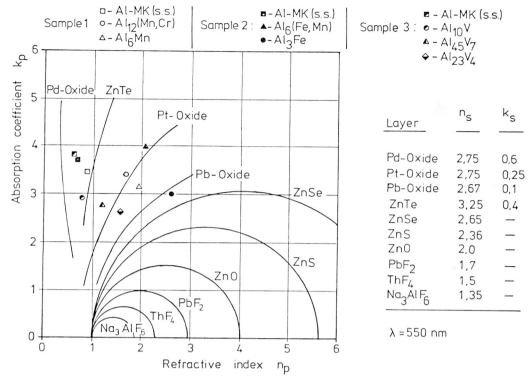

Fig. 3-24. Optical constants of aluminum phases ($\lambda = 550$ nm).

APPLICATIONS OF THE INTERFERENCE LAYER METHOD

In the following discussion, some examples of the application of the interference layer method are given, as well as advice for obtaining the largest possible light/dark contrast and ways to attain color contrast. In order to deal with the various alloy systems in a representative manner, we will present examples in the following fields:

1. Steels
2. Nonmetallic inclusions (sulfides)
3. Aluminum alloys

APPLICATIONS INVOLVING STEELS

The separation of retained austenite from the transformation microstructure of a sample after etching and coating with a layer capable of interference was shown in Fig. 3-3. The separation of fine grained retained austenite, transformation microstructures, and carbides is shown once more in the case of steel 14 NiCr 14 carburized to a carbon content of 1.5 percent. The specimen shown in Fig. 3-25 was subjected to heat treatment at 830°C for 15 minutes and then quenched in oil. It was coated by evaporation with zinc selenide [Fig. 3-25(a)] and finally etched in 3-percent alcoholic nitric acid [Fig. 3-25(b)]. It is clearly noticeable that after evaporation the retained austenite appears white, and the transformation microstructure plus carbides are dark grey to black. In the etched picture, the martensite is white or black, depending on the carbon content. Within the martensitic areas, the retained austenite cannot be seen after etching. At this point, it must be mentioned that austenite and ferrite—both of which are modifications of iron—appear only in a light/dark contrast and not in a color contrast. The reason for this is revealed by Fig. 3-8. After a specimen has been coated, it can be seen that the minima of the measured reflective curves of both phases are at the same wavelength. This does mean, however, that the phases do not differ in regard to interference color. The same is true of carbide $M_{23}C_6$ (M = metal) and carbide Fe_3C. Both these phases have almost the same interference minima as α-iron and γ-iron and, consequently, can hardly be distinguished from them in a color contrast.

Matters are different for carbides with a different structural and chemical composition. Figure 3-26 (see color insert) shows the differentiation of austenite: $M_{23}(C,N)_6$, $M_2(C,N)$, and $M_6(C,N)$. The steel X45 CrNiSi 25 30 failed after 45,000 service hours in an atmosphere rich in nitrogen. The three mentioned carbonitrides can be clearly distinguished from one another. In the color micrograph, the $M_{23}(C,N)_6$ appears yellow-red (phase 1); the $M_2(C,N)$, dark red (phase 2); and the $M_6(C,N)$, a typical violet color (phase 3). In the color contrast and light/dark contrast, all three carbonitrides can be contrasted very well. The same is true for the matrix of austenite. For both black/white micrographs, different evaporation layers

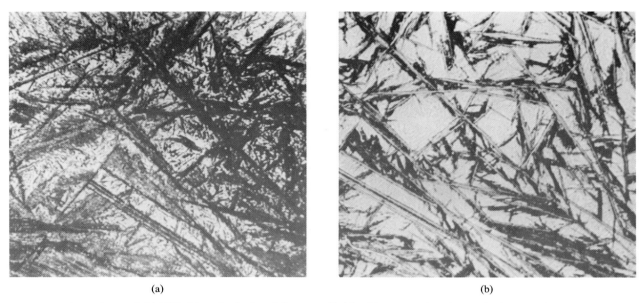

Fig. 3-25. Case-hardening steel 14 NiCr 14, carburized to 1.5-percent C: (a) with vacuum-deposited ZnSe (under bright field), and (b) same specimen and microstructure area etched in 3-percent alcoholic nitric acid (martensite, white and dark; carbides, light grey). (Both at 500×.)

were used. Both an evaporation treatment with zinc selenide and sputtering with a copper cathode in argon with low oxygen content (copper-oxide formation on specimen surface) will give good results for the contrast work. This example shows that both processes—evaporation and sputtering—make sense and can be utilized side by side.

APPLICATIONS INVOLVING NONMETALLIC INCLUSIONS (SULFIDES)

The sulfides will be taken as the example for this phase group. Sulfides belong to those compounds that have a reflectivity between 20 and approximately 45 to 50 percent. Consequently, they have different optical characteristics so that the choice of the best layer material appears to be more difficult than it is for metallic materials.

Figure 3-27 shows the reflectance curves for the sulfides Ni_3S_2, $(Nb,Ta)S_2$, $(Mo,Al,Cr)_3S_4$, $(Cr,Ti)_3S_4$, $(Ni,Cr)_3S_4$, and the reflectance behavior of a nickel-base alloy, IN 713 (γ-Ni solid solutions). Furthermore, an unknown phase X was noticed and its optical behavior also recorded. The results show that the reflectance minima are located at different wavelengths, a fact that again points to a different interference color after coating. Figure 3-28 proves that this is actually the case. In the color contrast and light/dark contrast, a total of seven phases

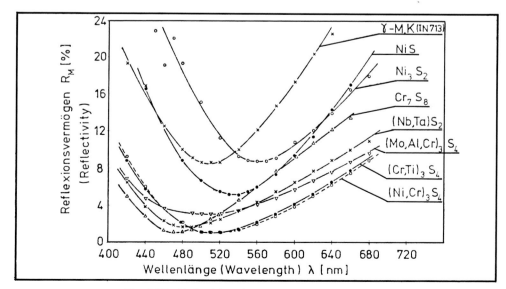

Fig. 3-27. Reflectivity of sulfides after coating with ZnS.

are clearly seen. The grey-scale values of the black/white micrographs agree with the reflectance measurements. (See color insert for Figs. 3-28 and 3-29.) For all measured sulfides, the optical constants, refractive index, and absorption coefficient are recorded in Fig. 3-23. Here, the different character of the various sulfides can be observed. Maximum light/dark contrasts can be obtained for Ni_3S_2, for example, by coating with platinum oxide.

The sulfides M_3S_4 (M = metal) require layer materials with a significantly lower refractive index. For example, zinc sulfide (ZnS) or zinc oxide (ZnO) will be sufficient. The manganese sulfides or iron sulfides MnS and FeS, usually found in steels, are contrasted best by selecting thorium fluoride (ThF_4), zinc oxide (ZnO), or zinc sulfide (ZnS).

APPLICATIONS INVOLVING ALUMINUM ALLOYS

Aluminum and aluminum compounds often have very high reflectance values, e.g., between 55 and above 90 percent. In Fig. 3-24, the optical constants of several aluminum compounds are recorded in an $n-k$ diagram. Lead oxide, platinum oxide, zinc telluride, and palladium oxide must be used for achieving the best contrast.

Aluminum compounds are ideal objects for good color contrast. Figure 3-29 shows aluminum alloys with 7.5-percent Ca, 0.8-percent Si, and 0.2-percent Ti (chill casting) or 2.3-percent Ti and 1.4-percent Si (sand casting). In the first alloy, the metallic compound $CaSi_2$ has a red-violet color, whereas Al_4Ca is blue after coating with ZnTe. In the alloy with 2.3-percent Ti and 1.4-percent Si, an intermetallic aluminum–titanium compound has formed (Al_3Ti). This appears in the form of rods and has a violet color after being coated with ZnTe. The pale-green phases are pure silicon, whereas the aluminum solid-solution matrix appears as a light violet. The good color contrasts show the possibility of obtaining a very good light/dark contrast because of the variable location of the interference minima. With the assistance of interference-layer metallography, there will be no difficulties in contrasting aluminum compounds.

SUMMARY

This chapter has described the application possibilities of the interference layer method in the metallographic preparation of materials with different optical characteristics. After a brief description of the objectives of metallography and a characterization of the optical behavior of metallic and nonmetallic phases, the basics of the interference layer method to improve contrasts were described. Next, the preparation problems caused by grinding, polishing or application of interference layers were presented, and, finally, a number of applications were examined. Selected contrast possibilities were also shown in the field of steels, nonmetallic inclusions (sulfides), and aluminum compounds.

This chapter has made it plain that with the assistance of interference layer metallography, all phases of a polished section, independent of shape and size, can be revealed by a sufficient light/dark contrast and also by color contrast. This statement is generally valid for all materials. Color contrasts cannot be expected, however, if the reflectivity of a material is very low (as in glasses or oxides).

Because the interference layer method has a purely physical basis, the parameters of specimen surface, layer material, layer thickness, and wavelength of the observing light must all be kept constant. The operator creates reproducible observation conditions that will produce contrast reproducibility so long as the preparation method is unchanged. In contrast to all other preparation methods used in metallography, this method achieves not only good contrast and true geometric appearance, but the question of quantitative metallography with sufficient contrast reproducibility receives a definitely positive answer as well. The attached bibliography will provide the interested reader with additional details.

REFERENCES

1. Bühler, H. E., and Hougardy, H. P. *Atlas of Interference Layer Metallography.* Oberursel: Deutsche Gesellschaft für Metallkunde, 1980.
2. Bühler, H. E., and Jackel, G. Tafeln zur quantitativen Metallographie mit aufgedampften Interferenzschichten. *Archiv Eisenhüttenwesen* 41:859 (1970).
3. Pepperhoff, W., and Ettwig, H. H. *Interference Layer Microscopy.* Darmstadt: Dr. D. Steinkopf-Verlag, 1970.
4. Zogg, H.; Weber, S.; and Warlimont, H. Optische Gefügeentwicklung von Al-Legierungen durch aufgedampfte Interferenzschichten. *Praktische Metallographie* 14:553 (1977).
5. Bühler, H. E. Interference layer metallography–State of knowledge and development trends. *Microstructural Science* 9:55 (1981).
6. Jackel, G.; Bühler, H. E.; and Robusch, G. Interferenzschichten-Mikroskopie an opaken und transparenten Phasen. *Thyssen-Forschung* 4:67 (1972).
7. Bühler, H. E.; Jackel, G.; and Thiemann, E. Möglichkeiten der quantitativen Beschreibung von oxidischen Phasen mit Hilfe der ätzfreien Gefügeentwicklung nach Pepperhoff. *Archiv Eisenhüttenwesen* 41:405 (1970).
8. Ettwig, H. H.; Bühler, H. E.; and Jackel, G. Absorbierende Schichtwerkstoffe in der Interferenzschichten-Metallographie. *Archiv Eisenhüttenwesen* 41:957 (1970).
9. Wei Tao Wu; Aydin, I.; and Bühler, H. E. Die optischen Konstanten von Sulfiden. *Praktische Metallographie* 19:322 (1982).
10. Aydin, I., and Bühler, H. E. Optische Konstanten von Carbiden und Carbonitriden in hochhitzebeständigen Gußlegierungen. *Praktische Metallographie,* special issue 12, p. 35 (1981).

4
MAGNETIC ETCHING

R. J. Gray

Formerly, Oak Ridge National Laboratory

In the application of magnetic etching as a metallographic tool, we should be aware that magnetism is, in its most basic form, an atomic phenomenon.* The atoms in most materials have electrons that are paired and spinning in opposite directions so that the net magnetism is cancelled. In some materials, however, there are atoms with one or more unpaired electrons, and these produce a net magnetic moment.

Materials that exhibit strong magnetism are classified as *ferromagnetic;* materials that display significant magnetism are classified as *ferrimagnetic.* Materials showing weak magnetic response are *paramagnetic;* materials having no net magnetism are *antiferromagnetic (diamagnetic).*

The origin of magnetic etching must be attributed to the efforts of Bitter[1] and independently of von Hamos and Thiessen[2] during the same year—1931. The original "Bitter patterns" were obtained by sprinkling magnetic powder on the surface of a material in a magnetic field and observing the distribution of the particles. The concept remains valid today, but many advances have been made. Although Bitter[3] improved his methods in 1932, real progress in the early development of microscope techniques for the observation of magnetic patterns was made by McKeehan and Elmore.[4] Utilizing magnetic particles in colloidal suspension, they successfully observed well-defined and reproducible magnetic domain patterns on the faces of iron crystals. In the words of the *Handbook of Chemistry and Physics,* "The magnetization of a ferromagnetic or a ferrimagnetic material tends to break up into regions called domains, separated by thin transition regions called domain walls."[5] Magnetic etching offers a technique for the microscopic observation of these domain patterns.

Further progress has been reported during the interim years[6-11] in the use of the colloid technique for observing magnetic domain patterns and ferromagnetic phases and constituents in microstructures, although several problems have continued to hamper progress in the field of magnetic etching. Nearly every report has presented a recipe for making the colloid, but the anticipated stability of those made in the laboratory has generally not been fulfilled. Since the lab procedure for making colloids is quite time-consuming, satisfactory colloids may not be readily available. Magnetic etching has not been as popular as it might have been had a more suitable colloid been easily obtained. In addition, metallographers have reported that specimens have corroded after being treated with certain formulas of laboratory-prepared colloids.

Improvements in magnetic etching have been reported.[12-16] These have included the use of a commercial colloid, Ferrofluid,[17] and the use of a larger coil magnet to separate the coil heat from the specimen and also provide a larger central area where the flux is normal to the specimen surface. (The magnetic coils previously in use for magnetic etching have had a diameter that just fit the 1- or 1¼-in.-diameter right-cylindrical specimen mount. Since the observation of a specimen often extends to 30 minutes or more, the heat generated in the coil could thus evaporate the aqueous carrier and terminate the examination.) This chapter offers some suggestions for simplifying the use of this technique and applying it in the metallographic studies of several materials.

RESULTS

Ferrofluid and Its Application

We continue to find Ferrofluid to be an extremely stable colloid. With its water base and saturation magnetization of 200 gauss, it is highly recommended. This saturation measurement is an indication of the colloid concentration, which we have found to be two to four times the particle density most suitable for magnetic etching. A 50- to 100-gauss-concentration work solution is achieved merely by diluting the 200-gauss stock solution with distilled water.

---
* Research for this chapter was sponsored by the U.S. Department of Energy under Contract Number DE-ACO5-84OR21400 with Martin Marietta Energy Systems, Inc.

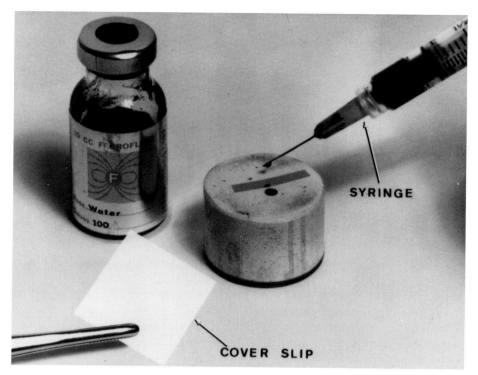

Fig. 4-1. Application of ferrofluid: Less than one-quarter drop (approximately 5 microliters) of the iron colloid is applied to the specimen. The cover glass is positioned over the colloid to form a thin fluid layer between the glass and specimen surface.

The lower colloid density has usually been the most satisfactory since it reduces "clouding" over the microstructural details by the colloid particles, thus allowing the true chemically etched microstructure (substrate) to be observed through the colloid and the analog colloid patterns to be related to the microstructure. Dilution of only about 1 milliliter of the Ferrofluid is strongly recommended for two reasons: (1) There is a possibility that the less dense colloid particles might not remain in suspension and settle, and (2) there is always the possibility of accidentally contaminating or losing the diluted solution. Our experience, however, has been that the particles in the diluted solution remain in suspension for several months.

Application of the Ferrofluid on the specimen follows the technique shown in Fig. 4-1. We have found the most desirable specimen size to be ½ to 2 cm square. The specimen should be mechanically polished so as to be free of detectable scratches, then lightly electropolished to provide a scratch-free, undisturbed surface. The best mechanically polished specimen retains a sufficiently disturbed surface to preclude sharp magnetic patterns. Since a mechanically polished surface is usually in slight relief to the surrounding mount, a light (10–15 sec) electropolish must be limited to removing only the relief effect and any evidence of the mechanically polished surface. A specimen profile that is lower than the specimen mount will cause the colloid layer to be too thick to allow the colloid pattern to be related to the substrate. The recommended specimen size is a distinct advantage when following these very important preparation steps. The specimen mount should be very dense. An epoxy resin is very suitable, but Bakelite is too porous and will absorb the colloid carrier.

The dispensing syringe (see Fig. 4-1) should have a small volume. The plastic tuberculin-type syringe with a 1-cc capacity is easily controllable for dispensing the required small volume—less than one drop, or approximately 5 microliters. We have found that the cover glass, or cover slip, should be the No. 0 type (22 mm² × 0.085 to 0.12 mm thick). An objective (lens) specifically designed for viewing a specimen through a cover glass might be more desirable for this application. Although metallographic objectives are usually not classified or designed for this use, our work has been conducted on a Bausch and Lomb Research metallograph with very satisfactory results.

After the ultrasonically cleaned cover glass has been positioned over the colloid and specimen, the colloid layer should remain within the perimeter of the cover glass. If the colloid fluid boundary extends beyond the perimeter and collects on top of the cover glass, the glass should be removed, the specimen surface cleaned with detergent, and the process repeated. By confining the colloid within the perimeter of the glass, the fluid surface tension secures the cover glass in place on the specimen mount. Generally speaking, microscopic examinations of magnetic patterns are made at high magnifications using immersion oil. If

MAGNETIC ETCHING 55

Fig. 4-2. (A) Magnetic etching coil with iron filings display, and (B) coil and specimen in position on metallograph stage. The aluminum spool—⅛ in. wall, 4½ in. ID, 2⅞ in. high—has 800 turns of #18 AWG, double glass-covered copper wire. The coil, energized by direct current supplied from a rectifier, is operated at 4.0 A, 37 V. Current polarity is controlled by a switch.

the cover glass is not held in place by the surface tension of the colloid, the glass will float between the two fluids and the examination becomes difficult.

Optical Examinations

An iron-filings display of the flux pattern of a magnet coil with an 11.5-cm inside diameter and the position of the coil on a metallograph are shown in Fig. 4-2. A schematic display is shown in Fig. 4-3, with the various components in place for the examination.

In conventional etching, a metallographic microstructure is in a static state. A very noticeable difference is observed by the metallographer in the use of magnetic etching. Although the colloid particle size is less than 300 Å and is not resolvable with the optical microscope, motion can be faintly observed at 1000X. This motion is due to Brownian movement—the random thermal agitation produced by impact of colloid particles with the molecules of the liquid. A gross motion is also observed if the magnet is energized and de-energized with specimens that are fully or partially ferromagnetic. Masses of colloid particles collect over a ferromagnetic phase or constituent in a magnetic field and then disperse in a variety of patterns when the magnet is de-energized. This particle mobility can offer a most interesting and informative display.

An example of magnetic domain patterns as observed on delta ferrite in an austenitic matrix is shown in Fig. 4-4.

Interpretations of the response of the colloid to a ferromagnetic phase or constituent in a paramagnetic matrix

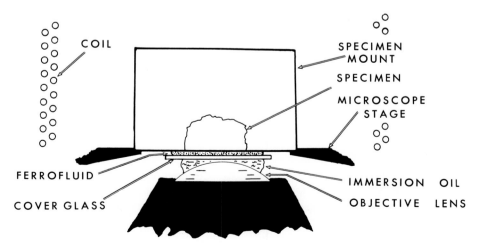

Fig. 4-3. Schematic display of specimen and coil in position on the metallograph stage.

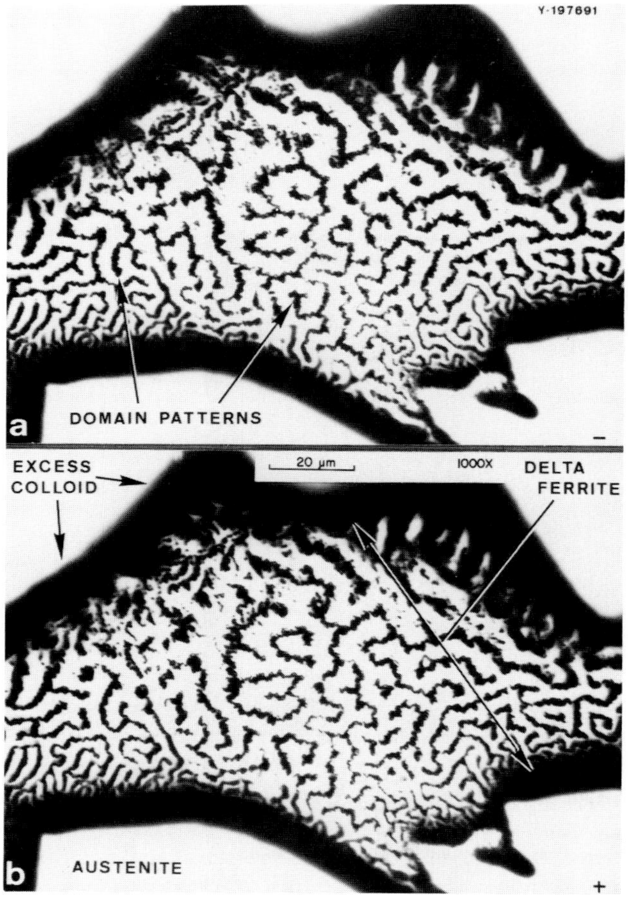

56 APPLIED METALLOGRAPHY

Fig. 4-4. "Coral" type domain patterns on delta ferrite. The magnetic delta ferrite in paramagnetic austenite offers an ideal display of the sensitivity of magnetic etching. Note the pattern is reversed with change in polarity of the direct current to the coil.

must not be based merely on the respective activity or nonactivity of the particles. If the matrix is paramagnetic austenite with some ferromagnetic ferrite or martensite present, the relative amounts of these two constituents are quite important. If the principal microstructure is paramagnetic austenite with a minor amount of a ferromagnetic constituent, the colloid response will follow the pattern seen in Fig. 4-5. Although the austenite is paramagnetic, the colloid particles respond to form "colloid colonies." The inexperienced eye might first misinterpret this activity over the austenite to indicate a ferro- or ferrimagnetic substrate. The four photomicrographs of Fig. 4-5 show the colloid attracted to a ferromagnetic ferrite stringer, with a majority of the colloid collected in colonies over the paramagnetic austenite. These colloid colonies display restricted Brownian movement in the magnetic field. The size of the colonies and the confinement of the colloid over a paramagnetic area are inversely related to the magnetic flux.

An example of the colloid sensitivity is shown in Fig. 4-6. This microstructure represents a sufficient distribution and size of delta ferrite to cause the colloid particles, in the absence of a magnetic field, to line up from one phase to another in ribbon-like patterns. By rotating the metallograph stage 90 degrees, the colloid bands would break up and regroup in the north–south direction as a result of the magnetic attraction of the earth.

Magnetic etching is a very sensitive technique that will detect the formation of strain-induced martensite in some types of austenitic stainless steel. A type 304 stainless-steel tensile specimen tested to failure was magnetically examined from the undeformed shoulder area to the highly deformed area near the fracture, and photomicrographs were made of selected areas, as shown in Fig. 4-7. The heterogeneous transformation of the austenite to martensite in the form of bands is vividly displayed even very near the fracture. Although the deformation is very severe near the fracture, some austenite can be seen.

The presence of 3- to 7-percent delta ferrite is known to be beneficial for preventing hot cracking in welds of some austenitic stainless steels. Various tools can be used to measure gross amounts of delta ferrite or other magnetic phases extremely well, including: the Magne Gage, Elcometer, Ferritescope, and Severn Gage.[18] X-ray diffraction can also be used to detect the presence of different phases, such as delta ferrite or martensite in austenite. All of these methods have serious limitations, however, in that only gross segregations of the ferromagnetic phase(s) can be detected, and amounts of delta ferrite under 1 percent are only marginally detected. The X-ray method is questionable even under 10-percent volume. In addition, none of these methods can relate the ferromagnetic phase to the microstructure.

With the magnetic etching technique, however, a metallographer can observe delta ferrite or martensite in an austenitic matrix that makes up as little as 0.5 percent of the microstructure. This technique is particularly useful in evaluating ferromagnetic conditions in amounts of 5 to 10 percent. An example of the sensitivity of this method has been reported[19] and is shown in Fig. 4-8. The results from use of a conventional chemical etchant on a commercial-grade type 316 stainless-steel sheet are shown in Fig. 4-8(a), which reveals the grains and the cellular structure in a gas tungsten arc-weld pass; a crack can be seen at the grain boundary. Magnetic etching [Fig. 4-8(c)] shows that the amount of delta ferrite in this centerline area of the weld pass is very low (much under 3 to 7 percent) and helps explain the probable cause of the grain-boundary crack formation.

Although magnetic etching is not accurate in quantitative evaluation, because of the halo effect of the colloid [Fig. 4-8(b)], this technique can be a useful complementary tool with the other magnetic measuring devices described above.

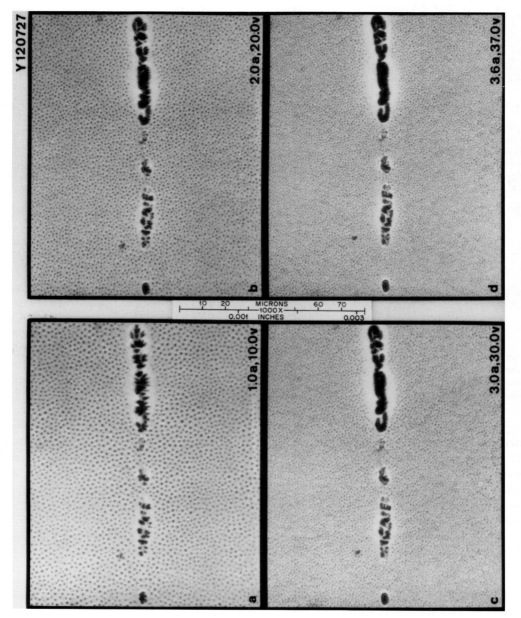

Fig. 4-5. Colloid patterns on paramagnetic austenite and ferromagnetic ferrite with increasing magnetic flux. Note the influence of the magnetic flux on the size and distribution of the colloid colonies and the confinement of the colloid particles over the ferrite; A = amperes; V = volts.

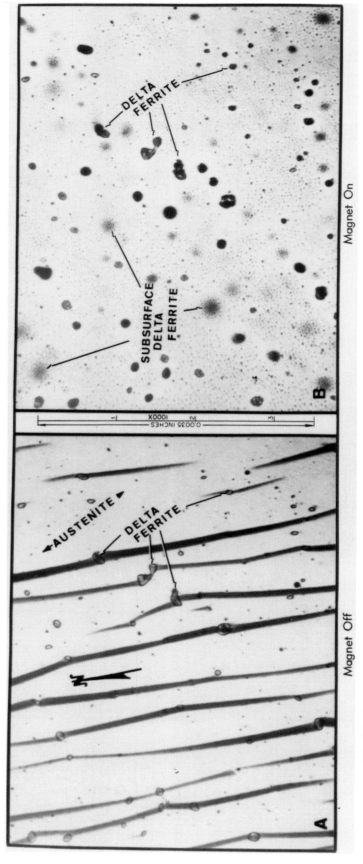

Fig. 4-6. Demonstration of colloid response to weak magnetic field: A type 304 stainless steel specimen was heat-treated at 1950°F (1066°C) for one hour and water-quenched to convert the ferrite to austenite. With the magnet off (A), some ferrite remained and served to align the colloid in bands as per the magnetic field of the earth. With the magnet on (B), the colloid is attracted to the delta ferrite.

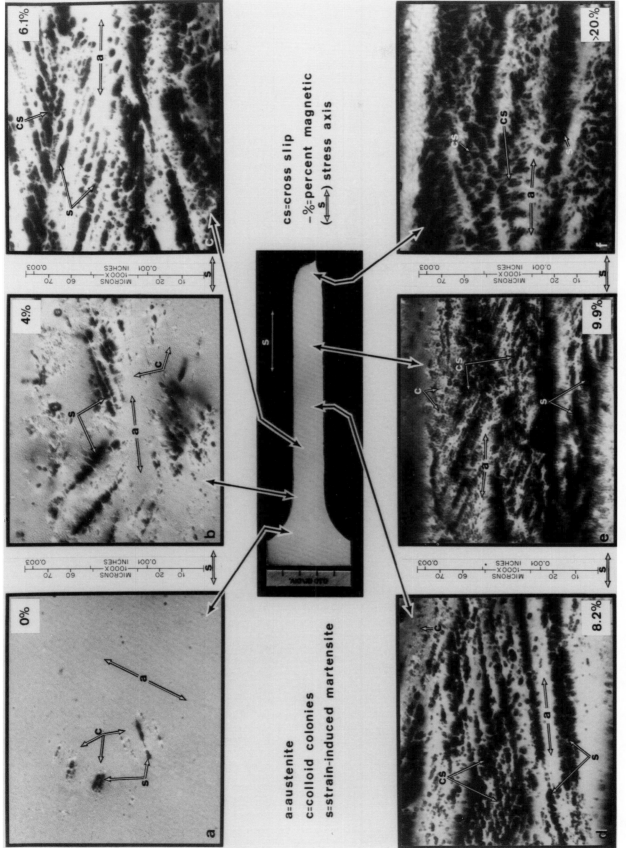

Fig. 4-7. Detection of strain-induced martensite in Type 304 stainless steel tensile specimen: (a) Shoulder area of specimen—location of minor amount of strain and martensite, (b) location of increased strain and martensite, (c) location of an increased amount of strain and martensite, (d) increased strain and martensite, (e) high density strain and martensite, (f) highest density strain and martensite. Note the banding of the slip and cross-slip. Some retained austenite can be seen near the fracture. The percent of magnetism was measured with a magne gage.

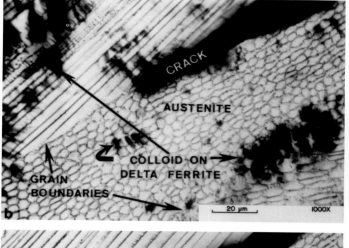
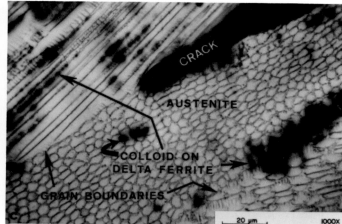
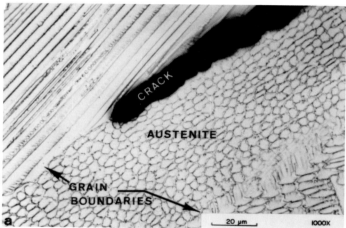

Fig. 4-8. The use of magnetic etching to reveal minute amounts of delta ferrite in a weld centerline, type 316 stainless steel sheet, 0.025 mm thick: (a) Conventional chemical etch, glyceregia; (b) same as (a), colloid applied, magnet off; (c) same as (a), colloid applied, magnet on.

CONCLUSIONS

Magnetic etching is one of the more potent tools in the metallography laboratory. The availability of a commercial colloid and the use of a larger magnetic coil greatly increase the potential for this application over that for previous ones. The observations of microscopic ferromagnetic phases and constituents are now limited to the resolution of the optical microscope. Delta ferrite, strain-induced martensite, and several other ferromagnetic constituents can be seen in a totally different mode, one that is complementary to conventional chemical and electrolytic etching techniques.

Magnetic etching is a very fascinating tool. Much of its attraction is the dynamic display generated as the colloid patterns are formed. As metallographers become more aware of this method, it will assume its rightful place in the laboratory.

Acknowledgements. The author is grateful to R. S. Crouse for his suggestions and to C. A. Valentine for her assistance in preparing the manuscript.

REFERENCES

1. Bitter, F. On inhomogeneities in the magnetization of ferromagnetic materials. *Phys. Rev.* 38:1903 (1931).
2. von Hamos, L., and Thiessen, P. A. Uber die Sichtbarmachung von Bezirken verschiedemen ferromagnetrsihen Zustandes festen Korpen. *Z. Phys.* 71:442 (1931).
3. Bitter, F. Experiments on the nature of ferromagnetism. *Phys. Rev.* 41:507 (1932).
4. McKeehan, L. W., and Elmore, W. C. Surface magnetization in ferromagnetic crystals. *Phys. Rev.* 46:226 (1934).
5. *Handbook of Chemistry and Physics*, 48th ed. Ohio: The Chemical Rubber Co., 1967–68.
6. Avery, H. S.; Homerberg, V. O.; and Cook, E. Metallographic identification of ferromagnetic phases. *Metals and Alloys* 10:353–355 (1935).
7. Elmore, W. C. Ferromagnetic colloid for studying magnetic phases. *Phys. Rev.* 32:309–310 (1938).
8. Harvey, E. A. M. Metallographic identification of ferro-magnetic phases. *Metallurgia* 32:71–72 (June 1945).
9. Weinrich, P. F. Microferrographic technique. *Australasian Engr.*, pp. 42–44 (November 1948).
10. Fisinai, George F. Magnetic oxide etchant. *Metals Progr.* 17:120–122 (October 1956).
11. Carey, R., and Isaac, E. D., eds. *Magnetic Domains and Techniques for Their Observation*. New York: Academic Press, 1966.

12. Gray, R. J. Revealing ferromagnetic microstructures with ferrofluids. *Proceedings of the International Microstructural Analysis Society,* Denver, Colo., Sept. 21–23, 1971. International Microstructural Analysis Society, Northglen, CO 80221.
13. Gray, R. J. Revealing ferromagnetic microstructures with ferrofluid. ORNL–TM–368, Oak Ridge National Laboratory, Oak Ridge, Tenn. (March 1972).
14. Gray, R. J. The detection of ferromagnetic phases in types 304 and 301 stainless steels by epitaxial ferromagnetic etching. In *Microstructural Science,* eds. Gray, R. J. and McCall, J. L. New York: American Elsevier Publishing Co., Inc., 1:159–175, 1973.
15. Gray, R. J. Detection of ferromagnetic phases in types 304 and 301 stainless steels by epitaxial ferromagnetic etching. *Metallographic Review* 2:2 (1973).
16. Gray, R. J. Magnetic etching with ferrofluid. *Metallographic Specimen Preparation.* eds. McCall, J. L. and Mueller, W. M. New York: Plenum Publishing Corp., 155–177, 1974.
17. *Ferrofluid.* Nashua, N. H.: Ferrofluidics Corp.
18. DeLong, W.; Ostrom, G.; and Szumachowski, E. Measurement and calculation of ferrite in stainless steel weld metal. *Welding J. (Research Supplement)* 35:526–33, (Nov. 1956).
19. Gray, R. J.; Holbert, R. K., Jr.; and Thrasher, T. H. Microstructural analysis for series 300 stainless steel sheet welds and tensile specimens. In *Microstructural Science,* eds. Northwood, D. O., White, W. E., and Vander Voort, G. F. Metals Park, OH: American Society for Metals, 12:345–370, 1985.

Fig. 5-3. Gas-contrasting chamber on microscopic cross stage (in action).

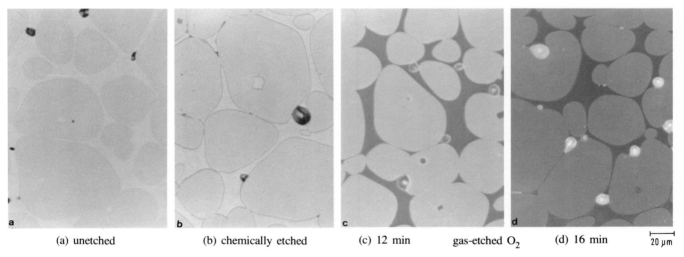

(a) unetched (b) chemically etched (c) 12 min gas-etched O$_2$ (d) 16 min 20 µm

Fig. 5-6. Gas-contrasted Cu–W infiltration alloy (for example: Cu = violet, W = orange; oxide = blue)

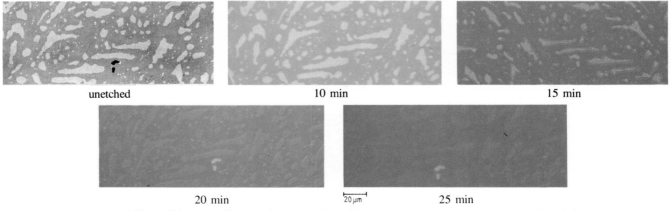

unetched 10 min 15 min

20 min 25 min

Fig. 5-7. Gas-contrasted U$_3$Si$_2$–U$_3$Si intermetallic two-phase material: U$_3$Si is the matrix phase and U$_3$Si$_2$ is the included phase (oxygen as contrasting gas; steel cathode; 1.05kV).

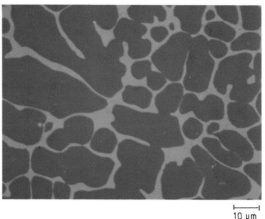

Fig. 5-8. Gas-contrasted HfC–Co hard metal (HfC = brown; Co matrix phase = gold).

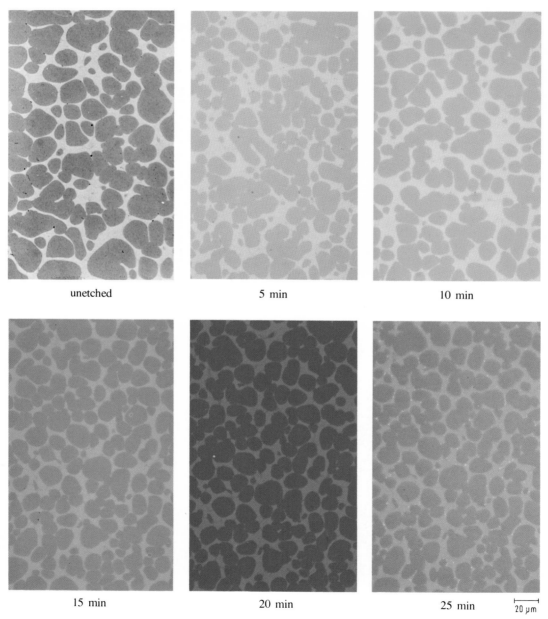

Fig. 5-9. Color sequence in HfC–Co hard metal by gas contrasting (oxygen as contrasting gas; steel cathode; 1.05kV).

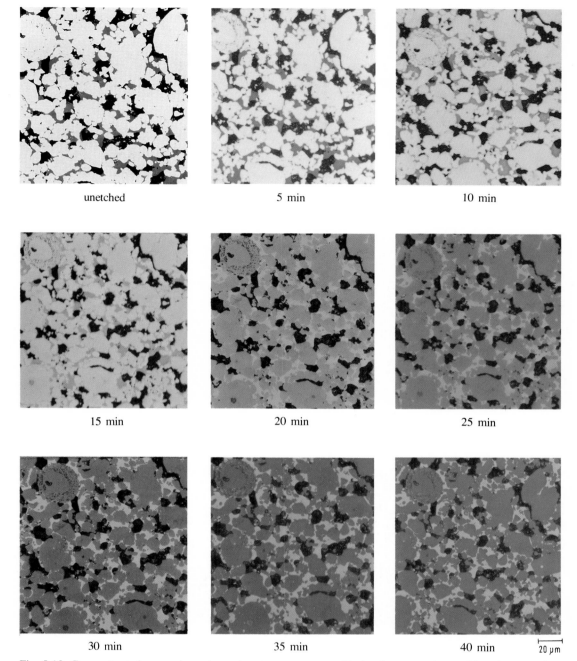

Fig. 5-10. Gas-contrasted porous iron-glass microstructures: pores, black; glass, grey; iron, white when unetched (oxygen as contrasting gas; steel cathode; 0.9 KV).

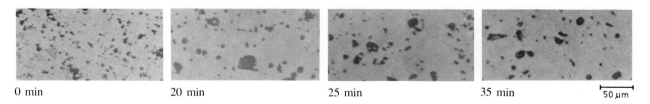

Figure 5-11. Gas-contrasted porous Al_2O_3–ceramic (carbon dioxide as contrasting gas; steel cathode; 1.5kV)

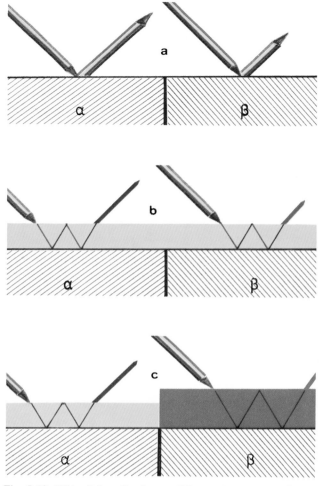

Fig. 5-12. White-light reflection on different phases, α and β: (a) without and (b) with layers of constant, and (c) with different thicknesses.

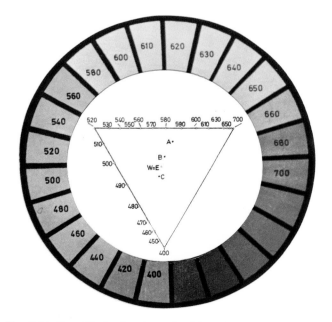

Fig. 5-14. Simplified color chart after Maxwell-Newton-Ostwald (numbers in nm).

5
THE GAS CONTRASTING METHOD

Dr. Gerhard Ondracek

University and Nuclear Center
Karlsruhe, FR Germany

It is essential to the qualitative and quantitative microscopic analysis of a material's microstructure that its constituents be distinguishable. Distinctions between the phases, crystallites, and boundaries of the polished section are based on their differential interaction behavior—reflection and absorption—with optical light rays in optical microscopy or with electron beams in electron microscopy. To obtain an image that can be analyzed microscopically—i.e., which is rich in contrast—these differences in the reflection or absorption behavior of the microstructural constituents must be "intensified." The relevant methods of preparation are classed under the term, *contrasting*. Especially in quantitative microstructural analysis, as schematically summarized in Fig. 5-1 this contrasting step is of particular importance.

The microstructural constituents in the microscopic image can be differentiated either by a bright/dark contrast or by a color contrast. The bright/dark contrast is based on intensity differences of the light waves (amplitude condition); the color contrast, on differences in wave length (phase condition). Methods of contrasting are based either on the formation or removal of surface layers or on the simple use of optical aids. The use of optical aids (polarized, brightfield and darkground illumination, filters, etc.) is restricted, however, to certain properties of the specimens (e.g., polarizability of phases), but it has the advantage of keeping the surface of the specimen unchanged.[1,2]

The removal of surface layers—as, for example, by cathodic ion etching[3,4,5]—changes the polished surface but may bring out an even more "true image" of the microstructure, since deformed surface parts disappear. On the

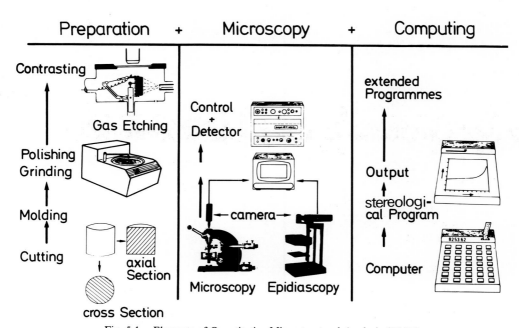

Fig. 5-1. Elements of Quantitative Microstructural Analysis (QMA).

64 APPLIED METALLOGRAPHY

other hand, however, the contrasting effect remains restricted. This is the reason why layer formation is usually the most effective way to contrast. Contrasting layers are produced either by vapor phase deposition,[6,7,8] where practically no interaction occurs between the layer and the sample surface, or by chemical or electrochemical etching. In the latter case, interaction between the etching agent and the polished section surface takes place.[9,10,11] Gas contrasting, as described in the next section, belongs to the layer forming methods.[12,13,14,15]

THE GAS-CONTRASTING APPARATUS

Gas contrasting takes place in a vacuum chamber that is externally similar to a hot stage attached to the mechanical stage of a microscope. The working principle is illustrated in Fig. 5-2[14,15]; the chamber, in Fig. 5-3 (see color insert). The sample is handled in one of three ways:

1. It is connected separately between two electrodes (cathode, anode).
2. It serves itself as anode if it is a conducting material.
3. It is directly fixed on the adjustable anode, which can be swiveled and therefore alternatively moved into the position opposite the cathode or below the microscope objective (see Fig. 5-2). Doing so permits the contrasting process to be controlled directly by tilting the specimen below the microscope objective for *in-situ* observation.

The gas-contrasting chamber prototype (Fig. 5-3) takes normal-sized, mounted metallographic specimens (~ 25 mm in diameter).[16,17] Its airtightness is increased by the pneumatic tilting device because the latter is operated by the same gas used for etching. Figure 5-2 illustrates its mode of operation. In the upper diagram, two needle valves (1 and 2) are shut and one (3) is open. This disposition sets up an excess pressure in the cylinder that moves the piston and lever system. The specimen is pneumatically tilted under the objective and automatically seated by resting against the cover glass. The piston stroke and,

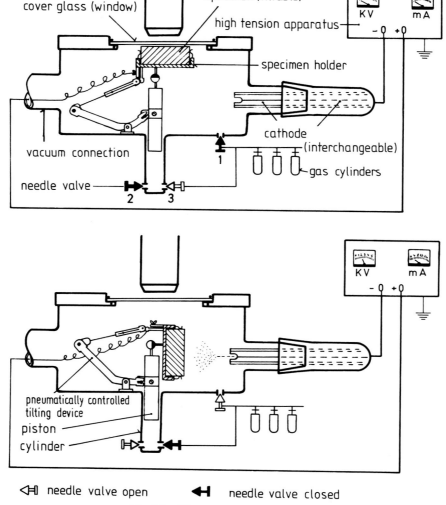

Fig. 5-2. The gas-contrasting chamber.

hence, the specimen height (11 to 15 mm) and diameter (< 27 mm) can be varied by means of adjusting screws. If the position of the valves is altered to correspond with the lower diagram in Fig. 5-2, the piston falls. Since the specimen with its holder is attached to the piston eccentrically through a ball joint, it tilts back under its own weight into a vertical position opposite the cathode.

THE PRINCIPLE OF GAS CONTRASTING

After the pumping unit connected to the pump nozzle is used to evacuate the gas-contrasting chamber, as shown in Fig. 5-2, it is filled via the needle valves with the etching gas or gas mixture. The ignited electron beam between the electrodes may be focussed by direct observation because the chamber (see Fig. 5-3) is transparent (it is made of quartz glass); the focussing additionally prevents discharging of the beam to chamber parts if they are nonconducting and different from the anode (quartz).

While the etching gas is being ionized by the electron beam, elements of the cathode material evaporate and mix into the gaseous etchant. The degree of ionization and evaporation—and with them, indirectly, the duration of contrasting—depends on the variable electrical data of current intensity and potential as well as on the distance between the sample anode and cathode and the gas pressure.[12] The gaseous ions may be produced either by accumulation of electrons on the neutral gas molecules— that is, by surface ionization (negative ions)[18]—or by impact ionization when accelerated electrons transfer their kinetic energy to neutral gas molecules via impact processes. In the latter case, positive ions are formed that are accelerated in the electrical field. They hit the cathode, causing sputtering of the cathode material.[19,20,21] The result is a mixture of atoms and/or ions of the etching gas and the cathode material; the mixture sublimates on the specimen's surface and forms interference layers that color different microstructural constituents differently. This effect is well known.[3]

Basically, no restriction exists with respect to various groups of materials for this type of layer formation. It takes place on the surfaces of the following:

1. Metallic materials, including intermetallic compounds
2. Composites, including hard metals
3. Ceramic materials, including glasses

The interaction tendency between the various materials and the etching gas is different, however, and this is why layer formation occurs in one of the following ways:

1. By interaction (reaction, solid solution) between the gaseous etchant and the material's constituents on the specimen's surface ("interaction layers")
2. By adsorption or sublimation of the gas atoms, ions, or molecules onto the material's constituents on the specimen's surface without interaction ("adsorption layers")

If the layer formation is based on interaction, it is controlled by the free energies of reaction or solution and their temperature dependencies, and these differ for the various microstructural constituents. In this case, differences in chemical composition and X-ray structure and differences in layer growth rate on the different microstructural constituents have to be taken into account. Interactions promote temperature increases in gas contrasting (Fig. 5-4).[14]

If the specimen is not directly connected to the anode, sublimation might be especially favored because heating of the specimen will not then occur, thereby promoting adsorption layer formation. Adsorption layers for all microstructural constituents should have approximately identical chemical composition, structure, and thickness because their growth rate is constant across the specimen surface.

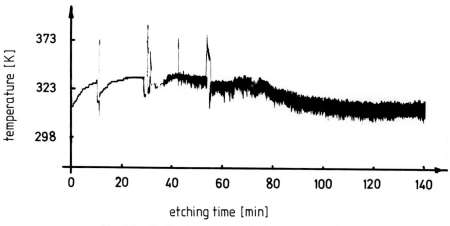

Fig. 5-4. Surface temperatures during gas contrasting.

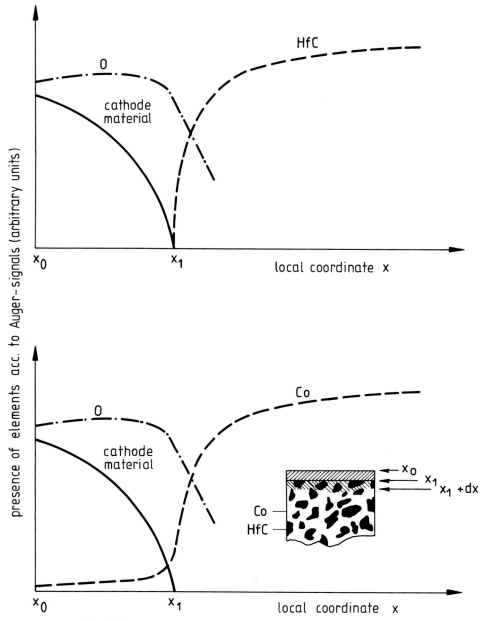

Fig. 5-5. Layer composition of gas-contrasted HfC-Co hard metal.

A study of gas-contrasted HfC–Co hard-metal microstructures using Auger electron spectroscopy[22] recently demonstrated the following:

1. A layer grows on the original, uncoated surface ("grown zone") that is predominantly formed by the elements of the gaseous etchant (etching gas and evaporated atoms from the cathode material).
2. Elements of the original specimen material may migrate into the growing layer, but this depends on the type of constituent.
3. Elements of the etching gas penetrate the interface between the layer and the original, uncoated surface, thereby altering its nature at least slightly ("diffusion zone").

Subsequently, a layer formed by gas contrasting is subdivided into, and is inhomogeneous within, two zones—a grown zone and a diffusion zone. The layer may be composed differently on different phases or constituents.

These results are summarized graphically and schematically in Fig. 5-5. Similar information about an inhomogeneous and nonstoichiometric layer structure was formerly obtained by Auger analysis of layers after chemical etching.[23]

GAS-CONTRASTED MICROSTRUCTURES

In order to get a more general assessment about gas contrasting, experiments have been performed with various classes of materials, as follows:

1. Single and multiphase metals, including intermetallic compounds
2. Composites, including hard metals
3. Ceramic materials

The microstructures of some of these have been selected to serve as examples. Others are shown in the quoted literature.

During the experiment for contrasting the polished structure of an austenitic steel by means of oxygen ions, a certain color sequence occurred with increasing contrasting time.[14] An improvement of the original contrast between crystallites and grain boundaries produced by chemical etching did not occur.

The same phenomenon occurs in the example shown in Fig. 5-6 of a two-phase infiltrated alloy of copper and tungsten. Figure 5-6(a) shows the polished microstructure as it appeared under the microscope. Chemical etching (10-percent aqueous ammonium-persulphate solution) leads to a negligible increase in the contrast differentiation between the two phases [Fig. 5-6(b)], which, however, is not comparable with the color contrast obtained by means of gas contrasting in oxygen [Figs. 5-6(c) and (d)]. In addition, this color contrast can be controlled with the contrasting duration. It is true that the same characteristic color sequence as in steel is produced for both phases, but the same color sequence occurs in copper after a different contrasting duration than that for tungsten. The color sequence is also present at the oxide inclusions [Figs. 5-6(c) and (d), blue]. (See color insert for Figs. 5-6 through 5-11.)

Differentiation of the phases is considerably improved by means of gas contrasting, but grain boundaries or crystallites do not become visible. Figure 5-7 shows intermetallic phases of uranium and silicon (U_3Si_2, U_3Si) together in one microstructure, where U_3Si is the matrix phase and U_3Si_2 are the discrete second-phase particles. As made clear by Fig. 5-7, gas contrasting provides definite contrast distinction between the phases. The bright and dark contrast distinction required for quantitative microstructural analysis may be easily obtained by use of color filters.[24] All phases again traverse (after differential contrasting time) the characteristic color sequence as in the other examples, finally entering an achromatic state that is no longer subject to change.

Hard metals form an important group among composites, for which gas contrasting usually provides satisfying contrast. Figure 5-8 illustrates the microstructure of HfC-Co hard metal, which was the subject for the investigation of the layer structure (see Fig. 5-5). As demonstrated in Fig. 5-9, both phases—the metal binder as well as the carbide phase—traverse the characteristic color sequence but shift mutually in time. This statement also holds true for oxide cermets (Fig. 5-10) as well as single-phase ceramics (Fig. 5-11), where the complete color sequence occurs slower than it does in metals. The results of these experiments can be summarized as follows:

1. Gas contrasting produces a characteristic color sequence with increasing contrasting duration, and this is independent of the material, the structure, the type of gas, and other contrasting conditions.
2. The colors assumed by the various phases or constituents of a microstructure after a certain duration of contrasting under otherwise identical conditions differ; i.e., the relative speed with which the various constituents traverse the characteristic color sequence depends on the material.
3. For "two-dimensional" features, such as grain or phase boundaries, gas contrasting does not provide contrast improvement.
4. The absolute speed of sequence of the colors depends on the etching conditions such as the type and pressure of gas, intensity and potential of current, and distance between electrodes.
5. The color sequence always ends in an achromatic, unchangeable state.
6. The relatively different speed of the color sequence in various constituents permits their contrasting through reproducible and controlled color contrast.
7. For the purpose of quantitative microstructural analysis, the color contrast may be converted into bright-dark contrast by using interference filters at the microscope.

THE INTERPRETATION OF THE GAS-CONTRASTING EFFECT

As mentioned at the beginning of this Chapter, it is a necessary presupposition for the optical differentiation and quantitative microscopical description that sufficient contrast exist between the various constituents of a microstructure. Bright-dark contrast is mathematically defined to be

$$K = \frac{R_\alpha - R_\beta}{R_\alpha} \qquad (5\text{-}1)$$

where reflectance is defined as the ratio between reflected and incident (original) radiation intensity, and the reflection coefficient, r (see Eqs. 5-2, 5-3, and 5-4), is defined by the root of the reflectance, R ($r_\alpha = \sqrt{R_\alpha}$; $r_\beta = \sqrt{R_\beta}$). R_α and R_β are the reflectances of constituents α and β, respectively.

With the limiting condition,

$$R_\alpha > R_\beta$$

Eq. 5-1 establishes the following:

$$0 \leq K \leq 1$$

Maximum bright-dark contrast is obtained with minimum reflectance of microstructural constituent β, or

$$K \to K_{max} \quad \text{for} \quad R_\beta \to R_{min}$$

or

$$\lim_{R_\beta \to 0} K = 1$$

Taking into account that contrast enhancement by gas contrasting is based on layer formation, the theoretical consideration has to start with the reflectance of the "coated" constituent β, that is, $R_{\beta S}$, which, according to the Fresnel equation for normal incidence, has the form[7,25,26,27]

$$R_{\beta S} = \frac{r_0^2 + r_1^2 + 2\, r_0 r_1 \cos \Delta\lambda}{1 + r_0^2 r_1^2 + 2\, r_0 r_1 \cos \Delta\lambda} \quad (5\text{-}2)$$

where $r_0\ (=\sqrt{R_0})$, the reflection coefficient at the interface environment/layer, is defined as follows:

$$r_0 = \sqrt{\frac{(n_S - n_O)^2 + k_S^2}{(n_S + n_O)^2 + k_S^2}} \quad (5\text{-}3)$$

and $r_1\ (=\sqrt{R_1})$ the reflection coefficient at the interface layer/constituent, as:

$$r_1 = e^{-\left(\frac{4\pi k_S}{\lambda}\right)d_S} \sqrt{\frac{(n_\beta - n_S)^2 + (k_\beta - k_S)^2}{(n_\beta + n_S)^2 + (k_\beta + k_S)^2}} \quad (5\text{-}4)$$

and the optical phase shift between incident and reflected beam, $\Delta\lambda$, as

$$\Delta\lambda = \left(\frac{4\pi n_S}{\lambda}\right)d_S + \delta_{OS} - \delta_{S\beta} \quad (5\text{-}5)$$

In Eqs. 5-3, 5-4, and 5-5,

n_O, n_S, n_β = refractive index of environment (O), layer (S), and constituent (β), respectively
k_O, k_S, k_β = absorption coefficient of environment (O), layer (S), and constituent (β), respectively
λ = wavelength of optical beam
d_S = layer thickness on constituent β of microstructure

The first term, $\left(\frac{4\pi n_S}{\lambda}\right)d_S$, in Eq. 5-5 refers to the optical path difference between the incident and reflected light ray transformed into arc units, $\frac{2\pi}{\lambda}$, that results from the optical path through the interference layer, $2n_S d_S$, whereas the second, δ_{OS}, and the third, $\delta_{S\beta}$, terms describe the optical phase shiftings as transient conditions for the reflections at the interfaces of environment/layer and layer/constituent, respectively.[25]

In air or vacuum ($n_O = 1$); transmissive, nonabsorbing layers [$n_O < n_S < n_\beta$; $k_S = 0$; $r \neq f(d_S, \lambda)$]; and almost nonabsorbing microstructural constituents ($k_\beta = 0$),[25] the optical phase shifts become

$$\delta_{OS} = \pi \quad (5\text{-}6)$$
$$\delta_{S\beta} = \pi$$

and Eq. 5-5 becomes

$$\Delta\lambda = \left(\frac{4\pi n_S}{\lambda}\right)d_S + \pi - \pi \quad (5\text{-}7)$$

Additionally, we observe that extinction interference on constituent β occurs if the optical phase shift for the beam becomes

$$\Delta\lambda = (2m + 1)\pi \quad (5\text{-}8)$$

where $m = 0, 1, 2, \ldots$

From Eq. 5-2, one obtains

$$\boxed{R_{min} = \left[\frac{r_0 - r_1}{1 - r_0 r_1}\right]^2} \quad (5\text{-}9)$$

and from Eq. 5-7,

$$\boxed{d_S = \frac{(2m+1)\lambda}{4n_S}} \quad (5\text{-}10)$$

Equation 5-10 is the well-known condition for optical interference extinction. Both Eqs. 5-9 and 5-10 provide the crucial basis for interpreting gas contrasting; that is, maximum bright-dark contrast is either (1) due to the fitting of the optical constants of the layer (n_S and k_S) and of the microstructural constituent β (n_β and k_β) in such a way that for *nonabsorbing* layers and independently of layer thickness, Eq. 5-9 results in the value, $R_{min} = 0$; or (2) takes place independently of the optical constants of the microstructural constituent β and with a definite layer thickness, d_S, on this constituent, in which case, with monochromatic light, λ, and given refraction index n_S, of the *nonabsorbing* layer material, Eq. 5-10 is fulfilled.

Microscopical microstructural analyses are usually made with white light, which is not monochromatic. Although differently reflected from various "uncoated" microstructural constituents, white light does not provide sufficient contrast [Fig. 5-12(a)]. After layer formation, however, a defined spectral color will suffer extinction by reflection according to Eq. 5-9 or 5-10; the color is different for different constituents. The rest of the spectral colors mix, resulting in nonwhite complementary colors, obviously different for the various constituents and therefore providing color contrast [Figs. 5-12(b) and (c)].

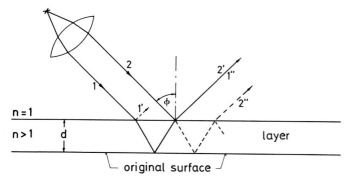

Fig. 5-13. Reflection of monochromatic light on a coated surface (principle of thin-layer interference).

First, assuming that the growing layer on the various microstructural constituents achieves equal *thickness* whether or not identical *structure* [Fig. 5-12(b)], contrast—including color contrast—may occur as a result of the different interaction between the optical constants of the constituents and those of the layer (Eq. 5-9).

Second, assuming that the growing layer on the various microstructural constituents achieves identical or nonidentical structure but *unequal* thickness [Fig. 5-12(c)], contrast and color contrast follow from Eq. 5-10 even when the role of the optical constants of the constituents is neglected. (See color insert for Fig. 5-12.)

The existing results are inadequate to determine which mechanism controls gas contrasting or to what extent different mechanisms take part in it. To explain the gas-contrasting effect using an engineering approach, the following consideration must be restricted to a model with different layer thicknesses. This simplification does not state that the effect of the optical constants of the various microstructural constituents on contrasting may be generally neglected. In Fig. 5-13, a gas-contrasted specimen is imagined to be covered by nonabsorbing, transmissive layers of different thicknesses on different constituents. The light rays (1 and 2) coming from the objective are partially reflected from (1' and 2') and partially transmitted through the layer. The ray penetrating the layer changes its direction corresponding to the refractive index (n_S) of the layer. Ray 1, for example, is reflected again at the interface layer/constituent β and returns in the form of ray 1" to the environment. There it interferes with the primarily reflected part 2' of the original ray 2. Extinction occurs under the following condition:[28]

$$d_S = \frac{(2m + 1)\lambda}{4\sqrt{n_S^2 - \sin^2 \phi}} \quad (5\text{-}11)$$

This equation leads to Eq. 5-10 in the case of normal beam incidence ($\phi = 0$). If the etched specimens are observed in white light, spectral colors satisfying the conditions in Eqs. 5-1 or 5-2 will suffer extinction. The remaining colors are superimposed to produce complementary or mixed colors. Since layer thickness is continually increasing during gas contrasting, it reaches the value (480/2 nm) at which extinction occurs for the blue component of white light first. The corresponding complementary color can be obtained from the color triangle in Fig. 5-14 by drawing a straight line from the wavelength that has suffered extinction through the white point (W). The intersection of this straight line with the opposing side of the triangle gives the wavelength of the mixed color, which can then be compared on the color ring. If, for example, blue suffers extinction, the complementary color, yellow, is obtained. The white point for daylight (E) is displaced for artificial illumination according to the type of source (points A, B, and C in Fig. 5-14). As spectral colors of longer wavelength suffer extinction, the intersection point moves to the uncalibrated side of the triangle, the so-called *purple line*. It represents the purple tones that are not spectral colors themselves but a mixture of red and violet. (See color insert for Fig. 5-14.)

The characteristic color sequence during gas contrasting can be read off from the color chart. If the layer begins to form, the short wave colors interfere. Thus, the color sequence begins with yellow and traverses the colors shown on the color ring as they appear during gas contrasting. It must be taken into account that certain large layer thicknesses satisfy the condition for the destructive interference of certain colors but simultaneously satisfy the condition for the constructive interference of others,[7] as follows:

$$d = \frac{(m + 1)\lambda}{2\sqrt{n_S^2 - \sin^2\phi}} \quad (5\text{-}12)$$

or for normal incidence,

$$d = \frac{(m + 1)\lambda}{2n_S} \quad (5\text{-}13)$$

where $m = 0, 1, 2, 3, \ldots$. These colors make a reinforced contribution to the mixed color. When the layer thickness is considerably greater than the wave-length, the interference condition can be satisfied for several wavelengths simultaneously with the result that, as the color sequence is repeated over and over again, the colors in the sequence become modified. When the layer thickness is finally so great that light of all colors suffers extinction or reinforcement, the coloration fades because the total reflected light combines to colorless. The different phases then appear correspondingly colorless.

SUMMARY

The contrast between various constituents of multiphase materials is the most important factor for quantitative microstructural analysis. How to achieve it by gas contrasting is the subject of the present chapter, which starts with the description of the gas-contrasting chambers and the technique of gas contrasting, continues with the demonstration and explanation of gas-contrasted metal, composite, and ceramic microstructures, and finishes with the interpretation of the gas-contrasting effect on the basis of the theory of thin interference layers.

Acknowledgment. Professor Kahle from Karlsruhe University has given considerable advice concerning theory; Mrs. Karcher, Mrs. Triplett, Mr. Janzer, and Mr. Spieler assisted technically. The author gratefully appreciates the support of all of them.

REFERENCES

1. Petzow, G., and Knosp, H. *Handbuch der Mikroskopie in der Technik*, III-2, p. 27. Frankfurt-Main: Umschau-Verlag, 1969.
2. Rinne, F., and Berek, M. *Anleitung zu optischen Untersuchungen mit dem Polarisationsmikroskop*, p. 71. Stuttgart: Schweizerbart'sche Verlagsbuchhandlung, 1953.
3. Hilbert, F. *Neue Hütte* 6:368 (1962); 7:416 (1962).
4. Rexer, J., and Vogel, M. *Prakt. Metallographie* 5:361-368 (1968).
5. Schwaab, P. *Z. Metallkunde* 55:199 (1965).
6. Bühler, H. E., and Jäckel, G. Tafeln zur quantitativen Metallographie mit aufgedampften Interferenzschichten. *Arch. Eisenhüttenwesen* 41-9:859 (1970).
7. Pepperhoff, W., and Ettwig, H. H. *Interferenzschichten-Mikroskopie*. Darmstadt: Dr. Dietrich Steinkopff-Verlag, 1970.
8. Pepperhoff, W., and Schwab, P. *Handbuch der Mikroskopie in der Technik*, Bd. 3, Teil 2, p. 65. Frankfurt/Main: Umschau Verlag, 1969.
9. Petzow, G. *Metallographisches Ätzen* 5. Auflage, Berlin-Stuttgart: Gebrüder Borntraeger, 1976.
10. Petzow, G., and Exner, E. *Handbuch der Mikroskopie in der Technik*, III-1, p. 37. Frankfurt: Umschau-Verlag, 1968.
11. Weck, E., and Leistner, E. Metallographic instructions for color etching by immersion, p. 77 Deutscher Verlag für Schweiβtechnik, 1970.
12. Bartz, G. *Prakt. Metallographie* X-5: 311 (1973).
13. Bartz, G. German Patent DP 2130605 (1973).
14. Ondracek, G., and Spieler, K. *Prakt. Metallographie* 6:324 (1973).
15. Ondracek, G., and Spieler, K. *Leitz-Mitt. Wiss. u. Techn.* VI-6:224 (1976).
16. Ondracek, G., and Spieler, K. German Patent DP 2433690 (1983).
17. Spieler, K. German Patent DP 2313801 (1973).
18. Westphal, W. *Kleines Lehrbuch der Physik*, p. 129. Berlin-Göttingen-Heidelberg: Springer-Verlag, 1948.
19. Flügge, S., and Trendelenburg, F. Ergebnisse der exakten Naturwissen-schaften, Bd. 35. Berlin: Springer-Verlag, 1964.
20. Hass, G. *Physics of Thin Films*. Vol. 3. New York: Academic Press, 1966.
21. Kohlrausch, M. *Prakt. Physik*, Bd. 2. Stuttgart: Verlag B. G. Teubner, 1962.
22. Nold, E., and Ondracek, G. *Prakt. Metallographie* 23:8 (1986).
23. Gahm, H.; Jeglitsch, F.; Hörl, E. M. *Prakt. Metallographie* 19:369 (1982).
24. Nazare, S., and Ondracek, G. *Prakt. Metallographie* 6:742 (1969).
25. Bergmann, L., and Schaefer, C. *Lehrbuch der Experimentalphysik*, Bd. III-Optik. Berlin-New York: Walter de Gruyter, 1974.
26. Heavens, O. S. *Optical properties of thin solid films*. New York: Dover Publications, 1965.
27. Vasicek, A. *Optics of Thin Films*. Amsterdam: North Holland Publ. Co., 1960.
28. Gerthsen, C. *Physik*. Berlin-Göttingen-Heidelberg: Springer-Verlag, p. 383, 1956.

6
TECHNIQUES FOR STAINLESS STEEL MICROSCOPY

Robert S. Crouse and B. C. Leslie

Oak Ridge National Laboratory

It is not the intent of this chapter to be a treatise on basic preparation techniques for specimens except as certain of them apply, or are peculiar to, stainless steels. There are many books that cover all of the basic principles and general laboratory techniques of metallography; a representative sampling may be found in References 1 through 6.

HISTORICAL BACKGROUND

According to Zapffe, stainless steel "is not an alloy—it is a name inherited by a great group of alloys, a special classification of special steels, and a field of study in itself."[7] Fortunately, the techniques and laboratory practices involved in the metallography of stainless steels are not as extensive and complex as the alloys themselves. As a group of materials, these alloys only occasionally present the metallographer with serious difficulties.

Although the first stainless-steel alloys of significance were not developed until the first decade of the twentieth century, the effect of chromium on the corrosion resistance of iron was known as early as 1821. Unfortunately, all of the early work failed to consider the effects of carbon, and since all of the iron-chromium alloys were high in carbon (1 to 1.5 weight-percent), they resulted in alloys that failed in sulfuric-acid and sea-water tests, these being thought to be representative of corrosion in general.

The most notable workers in the area of stainless steels in the early 1900s were Leon B. Guillet, said to be the "discoverer" of stainless steels,[8] and Borchers and Monnarty in Germany, who patented a stainless alloy in 1910.[9] Commercial usefulness of stainless steels seems to be above all due to the work of Harry Brearly in England in 1913. He discovered, in the course of routine metallographic study, that high-chromium steels did not etch with the customary steel etchants and did not rust. This had been noted by others, but he seemed to be the one to recognize the significance of it. He developed an alloy with 9- to 16-percent chromium and less than 0.7-percent carbon and recommended it for use as table cutlery. This is practically the same alloy in use for that purpose today.

THE NATURE OF STAINLESS STEELS

The term *stainless steel* applies to any of several steels that contain from 12 to 30 weight-percent chromium as a major alloying element. In general, they exhibit a passivity to corrosion in aqueous environments. These are not simple binary iron–chromium alloys but are considerably more complex, with as many as seven, or more, elements deliberately added in varying amounts. The fact that nickel, a major alloying element, may be added in amounts ranging as high as 37 weight-percent allows for a very large number of alloys, all listed under the label "stainless steels." Fortunately for the metallographer, this complexity does not necessarily translate into extreme difficulties in metallographic preparation.

TYPES OF STAINLESS STEELS

Stainless steels are generally classified into four types: martensitic, ferritic, austenitic, and precipitation-hardening steels. They may be either in wrought or cast form, with wrought alloys being the ones most commonly encountered by metallographers. Here again, whether wrought or cast, the problems for the metallographers are not serious. Table 6-1 gives the composition of some of the better known stainless steels. It will be noted that most austenitic stainless steels are numbered in the three hundreds, and all contain substantial nickel as well as chromium. Ferritic and martensitic alloys are numbered in the four hundreds and are "straight chrome" steels. Precipitation-hardening steels have the nominal chromium-nickel composition, plus "PH" as a designation.

Martensitic Stainless Steels

These Fe–C–Cr alloys can be hardened by heat treatment, producing a predominantly martensitic microstructure.

Table 6-1. Compositions of some popular stainless steels.

Bar above element means that the amount listed is the maximum amount allowed.

AISI TYPE	COMPOSITION, WEIGHT-PERCENT							
	C	Mn	Si	Cr	Ni	P	S	OTHER
Austenitic types								
302	$\overline{0.15}$	$\overline{2.00}$	$\overline{1.00}$	17.0–19.0	8.0–10.0	$\overline{0.045}$	$\overline{0.03}$	
304	$\overline{0.08}$	$\overline{2.00}$	$\overline{1.00}$	18.0–20.0	8.0–10.5	$\overline{0.045}$	$\overline{0.03}$	
304L	$\overline{0.03}$	$\overline{2.00}$	$\overline{1.00}$	18.0–20.0	8.0–12.0	$\overline{0.045}$	$\overline{0.03}$	
308	$\overline{0.08}$	$\overline{2.00}$	$\overline{1.00}$	19.0–21.0	10.0–12.0	$\overline{0.045}$	$\overline{0.03}$	
309	$\overline{0.20}$	$\overline{2.00}$	$\overline{1.00}$	22.0–24.0	12.0–15.0	$\overline{0.045}$	$\overline{0.03}$	
310	$\overline{0.25}$	$\overline{2.00}$	$\overline{1.50}$	24.0–26.0	19.0–22.0	$\overline{0.045}$	$\overline{0.03}$	
316	$\overline{0.08}$	$\overline{2.00}$	$\overline{1.00}$	16.0–18.0	10.0–14.0	$\overline{0.045}$	$\overline{0.03}$	2.0–3.0 Mo
316L	$\overline{0.03}$	$\overline{2.00}$	$\overline{1.00}$	16.0–18.0	10.0–14.0	$\overline{0.045}$	$\overline{0.03}$	2.0–3.0 Mo
317	$\overline{0.08}$	$\overline{2.00}$	$\overline{1.00}$	18.0–20.0	11.0–15.0	$\overline{0.045}$	$\overline{0.03}$	2.0–4.0 Mo
321	$\overline{0.08}$	$\overline{2.00}$	$\overline{1.00}$	17.0–19.0	9.0–12.0	$\overline{0.045}$	$\overline{0.03}$	Min Ti = 5X%C
347	$\overline{0.08}$	$\overline{2.00}$	$\overline{1.00}$	17.0–19.0	9.0–13.0	$\overline{0.045}$	$\overline{0.03}$	Min Nb+Ta = 10X%C
Ferritic types								
409	$\overline{0.08}$	$\overline{1.00}$	$\overline{1.00}$	10.5–11.75	—	$\overline{0.045}$	$\overline{0.03}$	Min Ti = 6X%C
430	$\overline{0.12}$	$\overline{1.00}$	$\overline{1.00}$	16.0–18.0	—	$\overline{0.04}$	$\overline{0.03}$	
442	$\overline{0.20}$	$\overline{1.00}$	$\overline{1.00}$	18.0–23.0	—	$\overline{0.04}$	$\overline{0.03}$	
446	$\overline{0.20}$	$\overline{1.50}$	$\overline{1.00}$	23.0–27.0	—	$\overline{0.04}$	$\overline{0.03}$	$\overline{0.25}$N
Martensitic types								
410	$\overline{0.15}$	$\overline{1.00}$	$\overline{1.00}$	11.5–13.0	—	$\overline{0.04}$	$\overline{0.03}$	
440A	0.6–0.75	$\overline{1.00}$	$\overline{1.00}$	16.0–18.0	—	$\overline{0.04}$	$\overline{0.03}$	$\overline{0.75}$ Mo
502	$\overline{0.10}$	$\overline{1.00}$	$\overline{1.00}$	4.0–6.0	—	$\overline{0.04}$	$\overline{0.03}$	0.40–0.65 Mo
504	$\overline{0.15}$	$\overline{1.00}$	$\overline{1.00}$	8.0–10.0	—	$\overline{0.04}$	$\overline{0.04}$	0.9–1.1 Mo
Precipitation-hardening types								
PH13–8 Mo	$\overline{0.05}$	$\overline{0.10}$	$\overline{0.10}$	12.25–13.25	7.5–8.5	$\overline{0.01}$	$\overline{0.008}$	2.2–2.5 Mo, 0.90–1.35Al, $\overline{0.01}$N
15–5 PH	$\overline{0.07}$	$\overline{1.00}$	$\overline{1.00}$	14.0–15.5	3.5–5.5	$\overline{0.04}$	$\overline{0.03}$	2.5–4.5 Cu, 0.15–0.45 Nb+Ta
17–4 PH	$\overline{0.07}$	$\overline{1.00}$	$\overline{1.00}$	15.5–17.5	3.0–5.0	$\overline{0.04}$	$\overline{0.03}$	3.0–5.0 Cu, 0.15–0.45 Nb+Ta
17–7 PH	$\overline{0.09}$	$\overline{1.00}$	$\overline{1.00}$	16.0–18.0	6.5–7.5	$\overline{0.04}$	$\overline{0.03}$	0.75–1.5 Al

The carbon content is deliberately kept sufficiently high (0.15 to 1.2 weight-percent) for it to make its full contribution. Figure 6-1 is a martensitic stainless-steel microstructure, apparently slow-cooled to allow formation of ferrite and carbides instead of martensite.

Ferritic Stainless Steel

A straight chromium steel in which the carbon is deliberately kept so low as to have a negligible hardening effect is classified as *ferritic*. The chromium content is usually above 14 weight-percent with some as high as 27 weight-percent. With low carbon and no nickel, these alloys are permanently ferritic. Figure 6-2 illustrates a ferritic stainless-steel microstructure.

Austenitic Stainless Steels

These alloys are sometimes referred to as *18–8 steels,* this being the nominal chromium–nickel composition of the first fully austenitic stainless alloys. They have the characteristics of being ductile, work-hardenable, and superbly corrosion-resistant. The presence of nickel at 8 weight-percent or higher makes them fully austenitic. Figure 6-3 illustrates a wrought austenitic microstructure.

Precipitation-Hardening Alloys

These alloys have the special property of being easy to form in the wrought condition but hardenable by low-temperature heat treatment. This hardening takes place through precipitation of submicroscopic phases that are normally dispersed throughout the microstructure. Additions of such elements as molybdenum, copper, aluminum, niobium, tantalum, or titanium are required for precipitation to occur. Figure 6-4 shows the microstructure of a precipitation-hardening alloy (17–4PH) that has been deliberately over-aged to produce phases large enough to be analyzed by microanalysis.

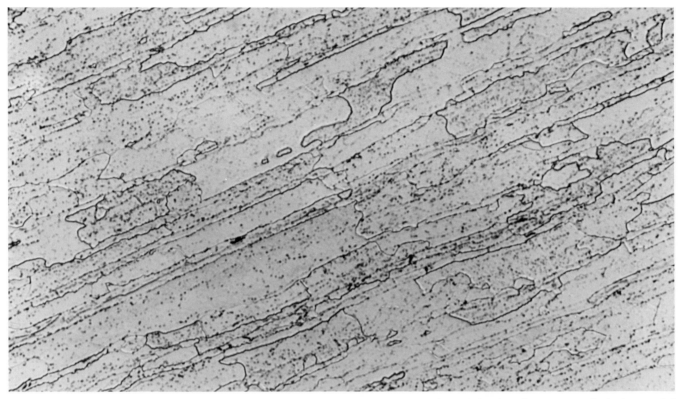

Fig. 6-1. Martensitic grade of type 410 stainless steel: This specimen was taken from as-received 1/4-in. plate that appears to be in the annealed condition. A martensitic structure is developed by rapid cooling and tempering. Etch: Villela's reagent. (250×.)

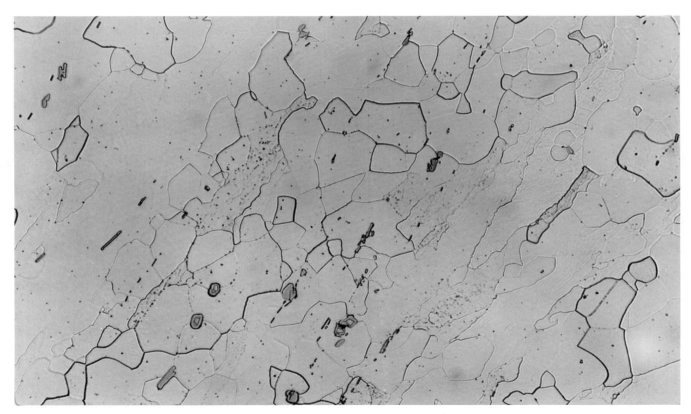

Fig. 6-2. Ferritic grade of type 409 stainless steel: This as-received 1/4-in. plate shows a generally recrystallized microstructure with an abundance of titanium carbides and carbonitrides. Etch: 25 ml HNO_3, 25 ml acetic, 5 ml HCl, 50 ml glycerol. (200×.)

74 APPLIED METALLOGRAPHY

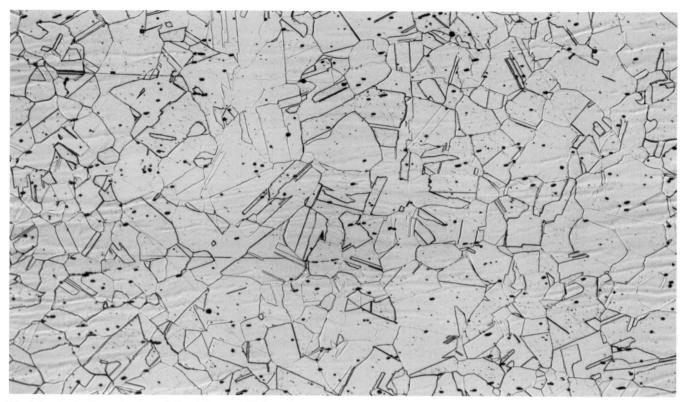

Fig. 6-3. Austenitic grade of type 316 stainless steel: This specimen, taken from mill-run, heavy-wall pipe, shows a fully recrystallized microstructure. Etch: HNO_3 + H_2O, electrolytic. (100×.)

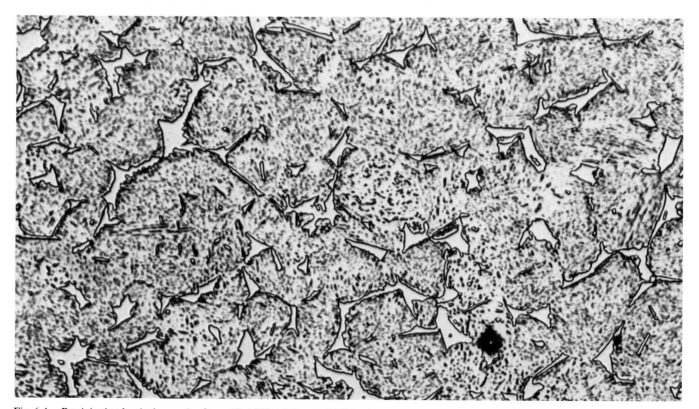

Fig. 6-4. Precipitation-hardening grade of type 17–4 PH stainless steel: This microstructure represents a somewhat over-aged condition. It underwent ½ hr at 620°C (1150°F) plus ½ hr at 400°C (750°F), and this produced massive phases enriched in Cr in a "salt-and-pepper" matrix. Etch: HF, HNO_3, glycerol. (500×.)

Because of the complexity of stainless alloys, it is practically impossible to depict a "typical" microstructure without first defining the history of all such alloys. Since such a presentation is beyond the scope of this chapter, Figs. 6-1 through 6-4 represent microstructures peculiar to those specimens and their thermomechanical treatments.

GENERAL CHARACTERISTICS OF METALLOGRAPHIC PREPARATION

The techniques used in the mounting, grinding, and polishing of stainless steels are those common to all metallographic operations, and, as such, need not be repeated here. Since the authors assume that the readers of this chapter are looking for information peculiar to stainless steel, they will attempt to attend to specifics, although some generalizations will, of necessity, occur.

Polishing Artifacts

Careless surface preparation, such as failure to fully remove all traces of prior grinding steps, can cause misinterpretation of a microstructure. This is a general statement that has a special relevance to stainless steels. In the wrought, fully annealed condition, stainless steels, especially the austenitics, are subject to considerable surface damage during the cutting and grinding steps of specimen preparation. Hence, these steps must be carefully controlled. Grinding carefully through 180-, 240-, 320-, 400-, and 600-grit silicon-carbide paper will usually assure success in the polishing step. A good rule of thumb to apply during grinding is to grind, in each step, for an additional 30 seconds after all traces of the previous grit scratches have disappeared.

Mechanical grinding and polishing procedures may produce strain-induced martensite in an austenitic alloy such as 304L; Fig. 6-5 is such a case. Often one finds it difficult, if not impossible, to remove this artifact by repeated etching and repolishing. In such cases, only electrolytic polishing seems to produce the proper microstructure quickly (see Fig. 6-6). If one has the time and vibratory polishing equipment, an artifact-free microstructure may be obtained by a prolonged (16- to 24-hr) final polishing on a nylon cloth with 0.5-μm diamond paste. This may seem an excessively long time, but since vibratory systems operate unattended, it may be only overnight, with less than one hour of operator time.

Fig. 6-5. Type 304L stainless steel: Two polishing artifacts that can occur with a low-carbon austenitic stainless steel. The dark, feathery feature is strain-induced martensite, and the mottled gray appearance of some of the grains is due to slightly disturbed metal. Etch: HNO_3, C_2H_5OH, and sodium metalisulfite stain after Beraha. (200×.)

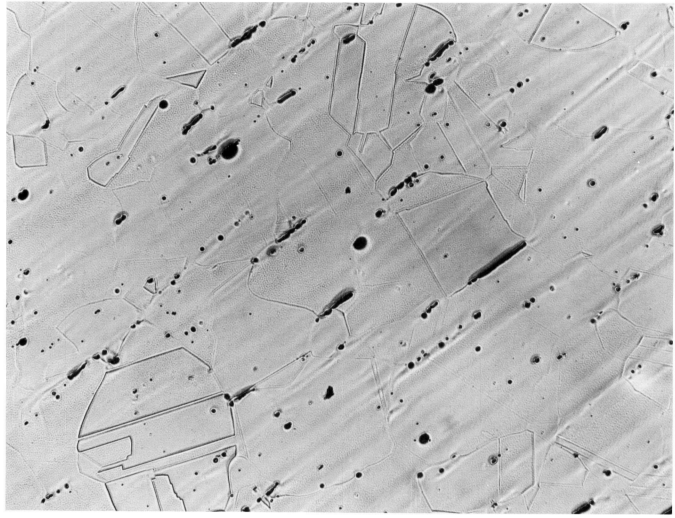

Fig. 6-6. Same specimen as in Fig. 6-5 but after electrolytic polish in CH_3OH and H_2SO_4 solution. Strain-induced martensite is removed. This figure illustrates one of the disadvantages of electrolytic polishing—the exaggeration of precipitates and phases. (100×.)

Chemical Polishing

Since metallographic preparation techniques at their most basic level are concerned with the removal of material from metal surfaces, one might think that chemical dissolution would be a viable technique. It does have an important role, especially with some high-purity metals and some clean solid-solution alloys; stainless alloys, however, do not meet these criteria. They frequently have inhomogeneities, phases, and precipitates that are attacked unevenly with respect to the matrix, resulting in unsightly pitting and loss of microstructural detail. This is not to say that chemical polishing should never be attempted. There are cases when, in the interest of saving time, one might find it quite useful. For example, chemical polishing makes it possible to produce a microstructure quickly in type 347 stainless steel from a machined surface for grain-size determination. Cain[14] describes a procedure that involves swabbing with an aqueous solution of hydrogen peroxide and hydrochloric and hydrofluoric acids (50 ml of H_2O, 25 ml of H_2O_2 (30 percent), 25 ml of HCl, and 5 ml of HF) for 10 sec to produce the microstructure shown in Fig. 6-7.

One may find other references in the literature to chemical polishing of stainless steels, but it has been the authors' experience that it has limited utility. One should not expect artifact-free microstructures from chemical polishing. What is more, a truly polished surface does not result, but rather an etched microstructure, because of the more rapid dissolution that takes place at grain boundaries and phase boundaries.

Etching

The "stainlessness" of stainless steels implies that one might have some difficulty in revealing their microstructures, and this is true to the extent that they do not respond to the etchants that one uses for carbon and

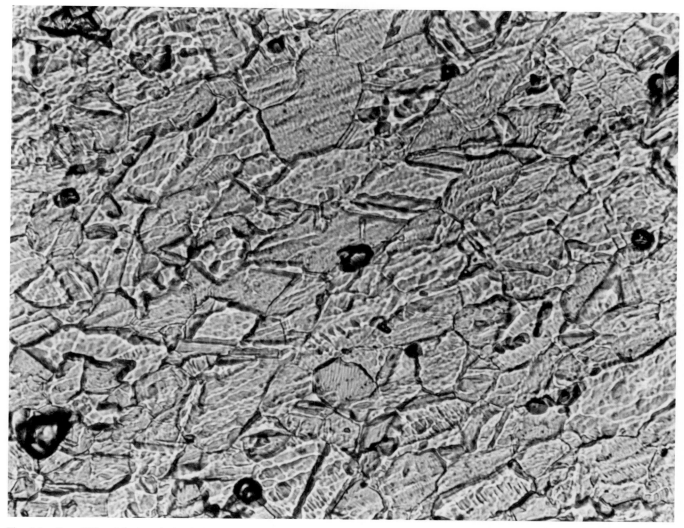

Fig. 6-7. Type 347 stainless steel chemically polished from a machined surface. This procedure is good for preliminary grain structure and size determination. (500×.)

alloy steels—i.e., nital and picral. The strong mineral acids (sulfuric, hydrochloric, and nitric) usually form the basis for the more successful stainless-steel chemical etchants. There are probably as many modifications of the basic mixed acid etchants as there are metallographers working with stainless steels. By experience, one tends to develop a list of favorite etchants as one encounters the variety of difficulties engendered by changing heat treatment, fabrication histories, and the inevitable alloy modifications that stainless alloys just seem to naturally invite.

Table 6-2 gives a listing of etchants that the authors have found effective. As mentioned above, metallographers will ultimately develop their own list of preferred etchants; this table merely reflects the preferences of the authors' laboratory.

The American Society for Metals (ASM) prints two very excellent publications that no metallography laboratory should be without.[15,16] They are the *Metals Reference Book* and volume 7 of the *Metals Handbook*. In these, one can find a collection of chemical and electrolytic etchants, similar to but more extensive than those in Table 6-2, and also of typical microstructures of stainless alloys.

It was previously mentioned that complications in etching may arise from deliberate alloy modifications. One such instance occurred at the Oak Ridge National Laboratory during the development of a radiation damage-resistant alloy for fusion-energy applications. The reference alloy was type 316 austenitic stainless steel (18 Cr–13 Ni–2 Mo–2 Mn–0.05 C–0.4 Si–balance Fe). The modification consisted of eliminating the phosphorous and sulfur, reducing the carbon and silicon, and fixing the nickel at 13 weight-percent.

The microstructure desired by the research program contained grain-boundary precipitates that had to be retained after etching to permit grain-boundary delineation for grain-size measurement. Unfortunately, the etchant normally preferred for grain-boundary delineation (nitric

Table 6-2. Chemical and Electrolytic Polishing and Etching Reagents.

CHEMICAL ETCHANTS	COMPOSITION		COMMENT
1. Ferric chloride and hydrochloric acid	$FeCl_3$ HCl H_2O	5 g 50 ml 100 ml	Swab or immerse for general structure
2. Chrome regia[1]	HCL CrO_3 solution (10% in H_2O)	25 ml 5.50 ml	Heat-treated austenitic grades
3. Ferric chloride and nitric acid[1]	Sat. solu. $FeCl_3$ in HCl plus a little HNO_3		Structures of stainless steels
4. Cupric chloride and hydrochloric acid (Kallings)[1]	$CuCl_2$ HCl C_2H_5OH H_2O	5 g 100 ml 100 ml 100 ml	Austenitic and ferritic stainless steels
5. Mixed acids and cupric chloride[1]	HCl HNO_3 Sat. with $CuCl_2$ and let stand 20 to 30 min	30 ml 10 ml	Swab for high Ni grades
6. Nitric and acetic acids[1]	HNO_3 Acetic	30 ml 20 ml	Swab for high Ni grades
7. Nitric and hydrofluoric acids[1]	HNO_3 HF (48%) H_2O	5 ml 1 ml 44 ml	Use cold for about 5 min for general structure of stainless steels
8. Ferricyanide solution (Modified Murakami's)[1]	$K_3Fe(CN)_6$ KOH H_2O	30 g 30 g 60 ml	Use fresh, boiling; colors sigma phase light blue; ferrite, yellow
9. Cupric sulfate (Marble's)[1]	$CuSO_4$ HCl H_2O	4 g 20 ml 20 ml	For general structures. Swab or immerse.
10. Aqua regia	HCl HNO_3	20 ml 10 ml	For general structures of austenitic grades. Swab with fresh solution.
11. Buffered aqua regia	$C_3H_6O_3$ (lactic acid) HCl HNO_3	12 ml 38 ml 10 ml	Swab with wet cottom swab before use; flush with water while swabbing with cotton saturated with etchant
12. Mixed acid in glycerol (Glyceregia)[1]	(A) HNO_3 HCl Glycerol (B) HNO_3 HCl Glycerol H_2O_2 (30%)	10 ml 20–30 ml 30–20 ml 10 ml 20 ml 20 ml 10 ml	A good basic etch for most stainless steel, especially austenitics May produce better results than 12(A); alternatively, etch and repolish for best results

ELECTROLYTIC POLISHING REAGENTS

COMPOSITION OF ELECTROLYTE		ELECTRICAL CONDITIONS	TIME	REMARKS
1. Perchloric acid* H_2O Ethanol[10]	50 ml 140 ml 750 ml	8–20 Vdc 0.6–2.1 A	20–60 sec	Use stainless steel cathode; keep below 24°C (75°F); agitate solution
2. Perchloric acid* Acetic anhydride[10]	335 ml 665 ml	6 A/dm²	3–5 min	Use stainless steel cathode; keep below 24°C (75°F); agitate solution
3. Acetic acid Perchloric acid*,[10]	1000 ml 50 ml	45 Vdc 11 Am/dm²	3–4 min	Use stainless steel cathode; keep below 24°C (75°F); agitate solution
4. H_2O H_2SO_4[11]	250–900 ml 750–100 ml	1.5–6 Vdc	1–10 min	Use stainless steel cathode; agitate solution, which may be buffered with varying amounts of glycerol
5. H_2O Chromic acid[11]	830 ml 620 g	1.5–9 Vdc	2–10 min	
6. H_2O H_3PO_4 H_2SO_4[11]	330 ml 550 ml 120 ml	0.05 A/dm²	1 min	

Table 6-2. (continued)

COMPOSITION OF ELECTROLYTE		ELECTRICAL CONDITIONS	TIME	REMARKS
7. H_2O	260 ml	0.6 A/dm²	30 min	Heat to 27–50°C (80° to 120°F)
Chromic acid	175 g			
H_3PO_4	175 ml			
H_2SO_4[11]	580 ml			
8. H_2O	210 ml	0.5 A/dm²	5 min	Heat to 21°–50°C (70° to 120°F)
HF	180 ml			
H_2SO_4[11]	610 ml			
ELECTROLYTIC REAGENTS FOR ETCHING				
1. HCl	10 ml	6 Vdc	10–30 sec	Keep water free; grain-boundary etch
Ethanol[1]	90 ml			
2. Lactic acid	45 ml	6 Vdc	10–30 sec	For delta ferrite in austenitic stainless steels
HCl	10 ml			
Ethanol[1]	45 ml			
3. KOH[1]	10 N	3 Vdc	0.2 sec	Colors sigma phase in austenitic grades
4. Oxalic acid	10 g	1.5 Vdc	20 sec	Attacks sigma phase and carbides
H_2O[10]	100 ml			
5. H_2SO_4	5 ml	2.5 Vdc	5–10 sec	Grain boundaries
HCl	25 ml			
HNO_3	5 ml			
H_2O	1000 ml			
6. Perchloric acid*	10 ml	40–70 Vdc	0.25–1 min	Use ice bath cooling, air agitation
Acetic acid	100 ml			
7. Ferric oxalate	10 g	6 Vdc	2 min	Reveals free carbides in all types of stainless steels
H_2O[1]	90 ml			
8. Chromic acid	60 g	6 Vdc	5–60 sec	Brings out carbides in stainless steels
H_2O	90 ml			
9. $(NH_4)_2S_2O_8$	35 g	6–10 Vdc	A few sec	For prior austenite grain boundaries in quenched and tempered martensitic stainless steels
HCl	10 ml			
H_2O[22]	10 ml			
10. Chromic acid	2 g	5 Vdc	35 sec	Attacks sigma phase in austenitic grades; barely attacks carbides
H_2O[13]	100 ml			
11. HNO_3	50 ml	>0.5 A	10–15 sec	For austenitic stainless steels; inspect frequently during etching
H_2O	50 ml			

* Perchloric acid has a reputation for explosiveness, and some laboratories prohibit its use categorically. The experience of metallographers who have used it indicates that by keeping the solution dilute, fresh, cool, and free from organics one is able to operate with safety.

acid and water—electrolytic) also removed the grain-boundary precipitates. This difficulty was resolved by very lightly etching the specimens with nitric acid and water electrolytically and photographing the microstructure in differential interference phase contrast (DIC). The resulting microstructure is seen in Fig. 6-8.

SENSITIZATION

Austenitic stainless steels are especially adaptable to use in highly corrosive environments because of their basic resistance to chemical attack. This resistance is imparted mostly by the presence of chromium in the 18-percent range and strongly augmented by nickel at about 8 percent. Nickel causes the alloy to be austenitic under practically all conditions of heat treatment. The resistance to chemical attack can be seriously compromised, however, if the alloy is held for some length of time in the temperature range 400° to 900°C (750° to 1650°F). As has been stated by Zappfe[7] (p. 43):

> While there is still dispute among experts over the exact nature of sensitization, there is complete agreement on its effect. Reports on corrosion rates in which 18–8, quenched from temperatures above 1000°C (1830°F), corroded at a rate of only a few mils per year in some strong solutions, but at a rate of thousands of mils per year in that same medium when the same steel was previously heat treated in the temperature range (maximum limits) of 400 to 900°C (750–1659°F). . . .

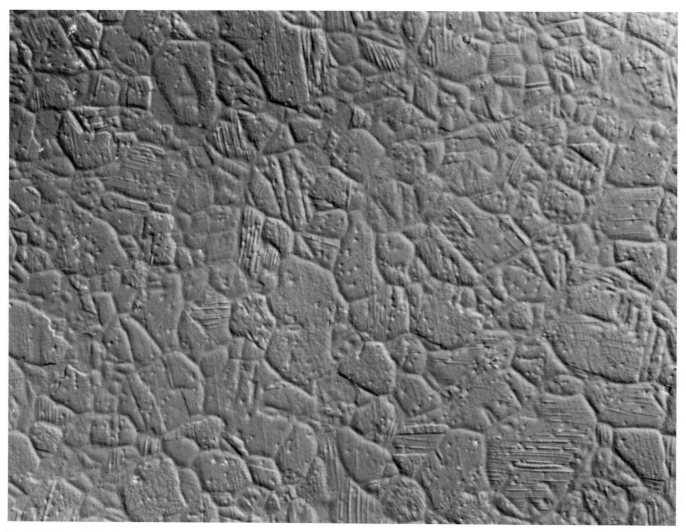

Fig. 6-8. Type 316 stainless steel of carefully controlled chemistry, specially treated to produce fine grains and grain-boundary precipitates. Etch: $HNO_3 + H_2O$, electrolytic, very lightly. (Photographed in DIC illumination; 500×.)

This sensitization heat treatment results in extensive precipitation of chromium carbides along grain boundaries with the result that in corroding environments the alloy is heavily attacked along these boundaries. The generally accepted theory concerning this phenomenon is that the carbon in the alloy combines with chromium in the ratio $M_{23}C_6$ and precipitates along the grain boundaries. The area along the boundary is denuded of chromium, causing it to lose its "stainlessness" and become susceptible to accelerated chemical attack.

A metallographer is often called upon to demonstrate whether or not a steel has become sensitized. The commonly accepted method for doing so is to apply the ASTM A262-75 standard test for sensitization. As with all ASTM Standard Recommended Practices, this one is set forth in great detail, with the metallographic aspect only part of the total practice. In summary, the metallographer should: (1) grind and polish a surface at least 1 cm square or at least 1 cm long, (2) electrolytically etch in 10-percent oxalic acid at $1A/cm^2$ for 1.5 min, and (3) examine the microstructure at 250X to 500X. Figures 6-9 and 6-10 show acceptable and unacceptable microstructures. If an alloy shows an unacceptable microstructure, further chemical testing should be pursued as prescribed. Acceptable microstructure precludes further testing.

Some extra-low-carbon stainless steels—such as 304L, 316L, and 317L—are tested after a sensitizing heat treatment of 1 hour at 675°C (1250°F). Since this is in the temperature range of maximum carbide precipitation, this treatment should reveal any tendency toward sensitization.

STRESS-CORROSION CRACKING

J. F. Mason, Jr.[17] defines stress-corrosion cracking (SCC) as "the combined action of static stress and corrosion

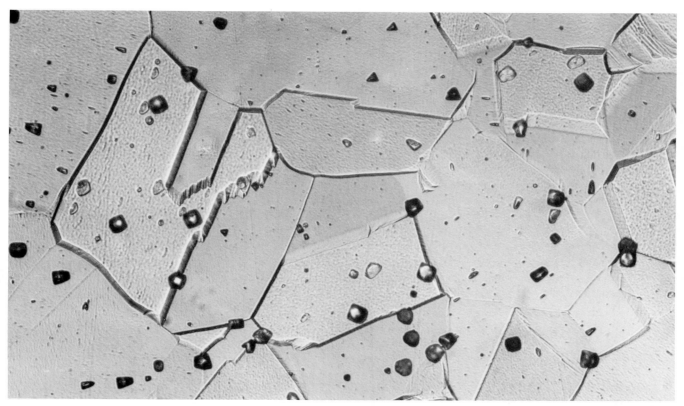

Fig. 6-9. Type 304LW stainless steel tested for sensitization according to ASTM Standard Recommended Practice A262–75. This microstructure illustrates a nonsensitized condition. Etch: 10-percent oxalic acid, electrolytic. (500×.)

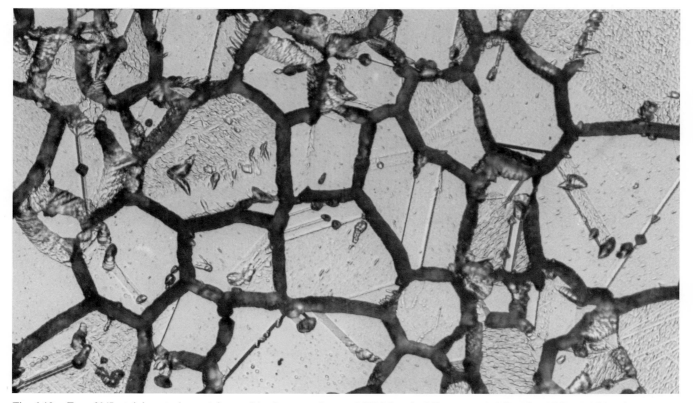

Fig. 6-10. Type 304L stainless steel tested for sensitization according to ASTM Standard Recommended Practice A262–75: This microstructure illustrates a sensitized condition; all grain boundaries are ditched. Etch: 10-percent oxalic acid, electrolytic. (500×.)

which leads to cracking or embrittlement of a metal." Of the four general types of stainless steels, only the ferritics do not appear to be susceptible to SCC.

The metallographer's usual role is to identify SCC when it is encountered in a microstructure. To equip oneself the better to do this, one needs to examine some of the more authoritative writings on the subject.[18,19] An understanding of the causes and mechanisms of crack development will aid in this identification.

The failures caused by SCC almost always occur in an aqueous environment, and most frequently seem to involve the chloride ion. The latter kind of failure is so prevalent that one frequently sees the phenomenon referred to as "chloride stress-corrosion cracking." It is well established that catastrophic failure of austenitic stainless steel can take place in environments where neither the corroding medium nor stress by themselves are sufficient to cause failure. It is important for the metallographer to have some information about the stress and corrosion environments of a sample to establish his analysis properly, and one should always insist on having such data when approaching a failure of this kind. The cracking may not always assume the classical pattern that may be identified as SCC by microscopic observation.

Chloride-induced SCC in austenitic stainless steels that have not been sensitized does seem to exhibit a characteristic branched, transgranular pattern. It is as though the crack proceeds in short bursts in directions dictated by submicroscopic conditions such as twin boundaries, preferential crystallographic attack planes, vacancy clusters, and stress concentrations. These may all play a part as well as other conditions. The "characteristic" crack pattern is shown in Fig. 6-11.

It is sometimes possible for a metallographer to demonstrate fairly conclusively the presence of chloride ions associated with cracking. The cracks revealed by metallography may be analyzed by electron-beam microanalysis for the presence of chlorine at the tips, which is where

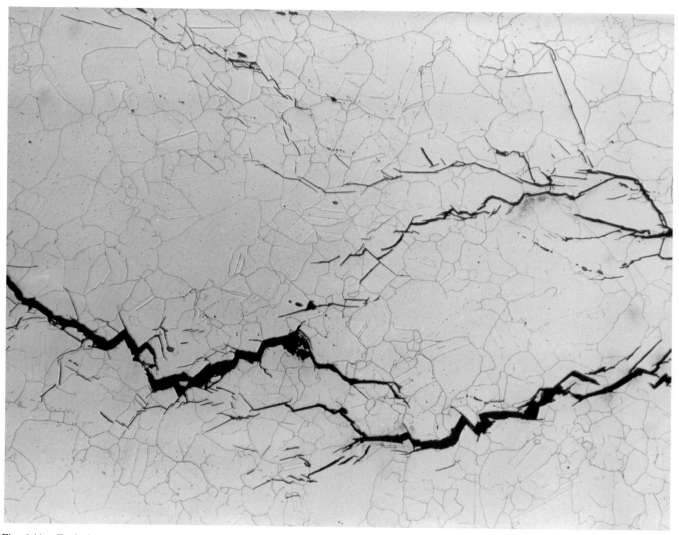

Fig. 6-11. Typical stress corrosion cracking in type 316 stainless steel: This material was taken from heat exchanger in a solvent refined coal plant. Etch: $HNO_3 + H_2O$, electrolytic. (100×.)

most theories indicate that the ions are concentrated. An example to illustrate this analysis will now be presented.

A failure occurred in a type 316 stainless-steel vent line in a solvent refined coal plant. This plant converts coal into petroleum products in a complex chemical process that involves heat, pressure, fluid flow, slurry transfer, and many other conditions that serve to create a very hostile environment. Large amounts of sulfur are known to be present as well as some chlorine.

A leaking line was removed from the system and submitted for metallographic examination. A large crack that was the obvious failure exhibited massive amounts of corrosion product and/or slurry from the process that effectively masked what was thought to be its true nature. There were, however, smaller secondary or branch cracks that had the appearance of SCC (see Fig. 6-12), and this was the area in which microanalysis was concentrated. Such microprobe analysis established that both sulfur and chlorine were present in the secondary cracks. By determining the ratio of chlorine to sulfur at several points, it was found that the smaller cracks contained far more chlorine than sulfur. These results are given in Fig. 6-13.

If one assumes that the small cracks typify the earliest cracks, then chlorine must be influential in their formation. Further analysis in the form of X-ray dot map displays showed that the small cracks contained chlorine but no detectable sulfur (Fig. 6-14). The conclusion was that chlorine contributed to crack initiation and growth and that sulfur corrosion ensued and enlarged the cracks.

Thus far, the stress-corrosion cracking discussed has been chloride-induced and described as being predominantly *transgranular* in nature. Sensitized stainless steels can also be susceptible to SCC with cracks that are exclusively *intergranular*. This is certainly not surprising in light of what was said about sensitization earlier, concerning the lack of corrosion resistance in sensitized grain boundaries. If one places stressed, sensitized stainless steel in a corrosive environment, grain-boundary attack should

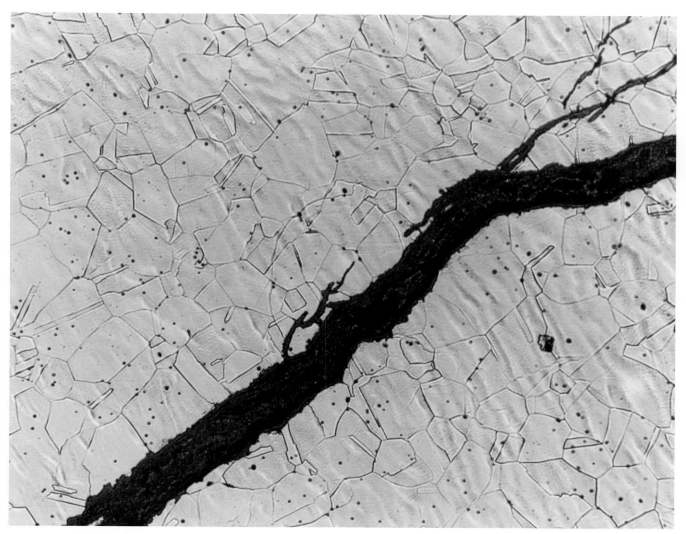

Fig. 6-12. Corrosion-filled crack in type 316 stainless steel from a solvent refined coal plant. Etch: $HNO_3 + H_2O$, electrolytic. (250×.)

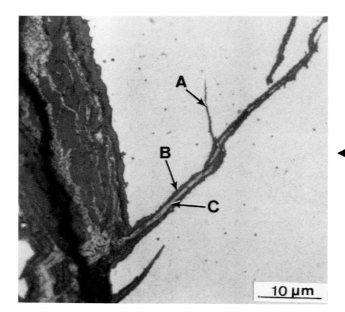

Fig. 6-13. Backscattered electron image of the specimen in Fig. 6-12: X-ray intensity ratios from different spots in the cracks showed a variation in both sulfur and chlorine, with chlorine more concentrated in the small crack tip.

POSITION	Cl/S X-RAY INTENSITY RATIO
A	4.2
B	0.5
C	0.3

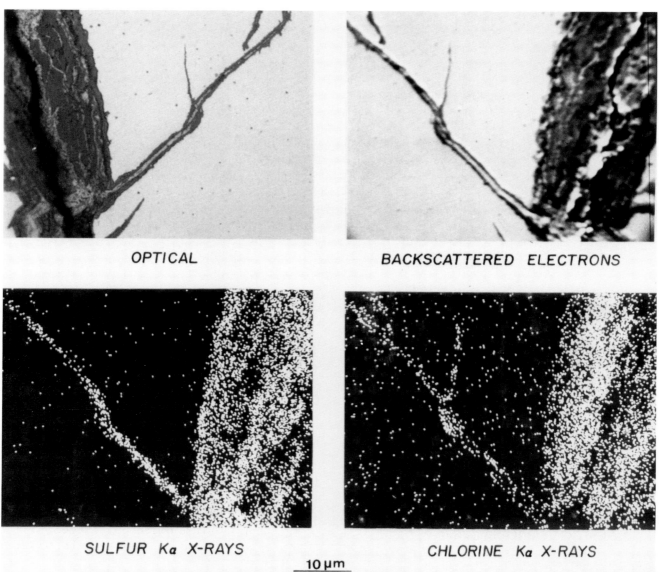

Fig. 6-14. Backscattered electron and X-ray dot map images of the specimen in Fig. 6-12. This type of presentation is graphic in its ability to visualize elemental distribution and confirms the presence of chlorine as an influence in stress-corrosion cracking.

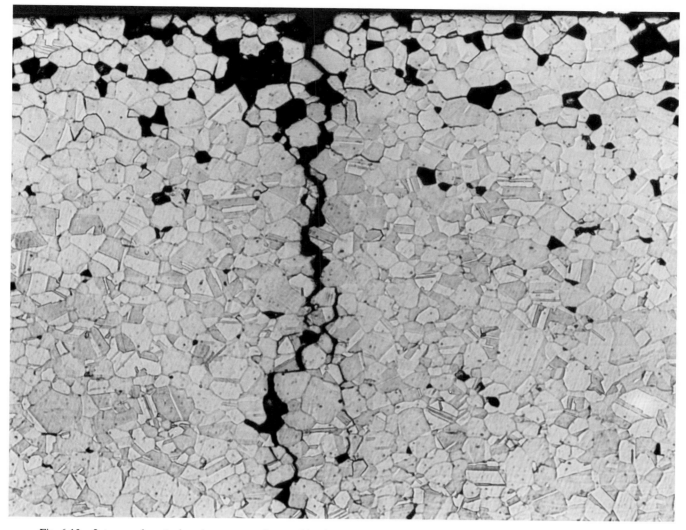

Fig. 6-15. Intergranular attack and stress-corrosion cracking in Type 304 stainless steel. Etch: $HNO_3 + H_2O$, electrolytic. (150×.)

follow, with the likelihood of eventual failure as a result of corrosion-weakened boundaries. This does indeed happen when the preceding conditions are met.

When one encounters an intergranular failure of an austentic stainless steel in a chemical environment, one should suspect sensitization and SCC. Figure 6-15 is a good example of this condition, showing a type 304 stainless-steel U-bend specimen that has been sensitized and tested in polythionic acid for 30 days at room temperature, according to ASTM standard G35–73. Although there is considerable intergranular attack, it is intergranular stress corrosion that has caused the failure.

SIGMA PHASE, FERRITE AND CARBIDES IN AUSTENITIC STAINLESS STEEL

None of the four general types of stainless steels (ferritic, martensitic, austenitic, and precipitation-hardened are full-time solid-solution alloys. They are all multiphasic in at least some conditions of fabrication or heat treatment. We chose to concentrate on the preceding phases in austenitics because special efforts seem to be required to identify and separate them. Metallographers are often called upon to identify them.

Sigma phase is known to have a definite deleterious effect on the mechanical properties of the alloys. A classic work was performed by Dulis and Smith[20] in 1950 on the identification of this phase, and collaterally, of ferrite and carbide. They investigated the effects of some 18 etching reagents on a variety of austenitic stainless steels, all containing one or all of the phases previously discussed. They tabulated the effects in a concise manner that allows one to choose the effect desired.

Sigma phase is a hard, brittle phase with the general composition 45 weight-percent Cr, 55 weight-percent Fe.[21] This composition may vary somewhat depending upon the alloy composition and heat-treatment history. It is a relatively low-temperature transformed product, its range of formation typically being from 500° to 900°C (935° to 1655°F). It will slowly transform from an austen-

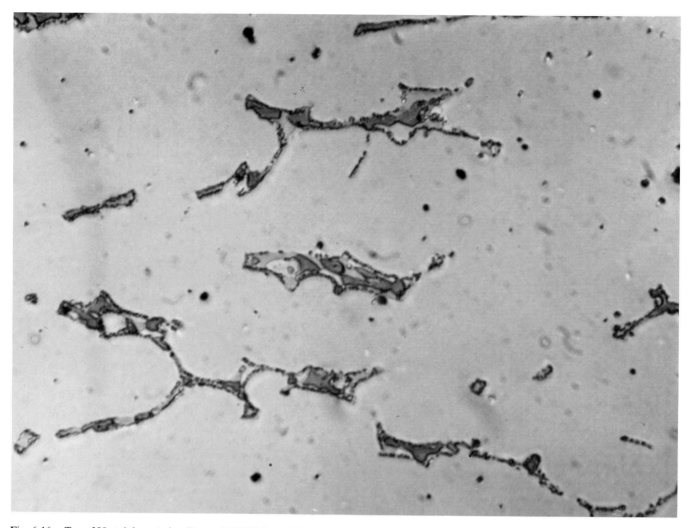

Fig. 6-16. Type 308 stainless steel weld, aged 20,000 hr at 593°C (1100°F): This specimen contained delta ferrite islands in an austenite matrix that partially transformed to sigma phase and austenite during aging. Sigma is the dark phase. Etch: Murakami's reagent [$K_3Fe(CN)_6$ + KOH + H_2O]. (1500×.)

ite matrix but occurs much more readily if ferrite is present. Microstructures showing sigma intermixed with islands of delta ferrite are often seen (Fig. 6-16).

If one were called upon to demonstrate the presence (or absence) of sigma phase in a stainless steel, one might proceed as follows. First, one would obtain the best possible mechanical polish on the sample. This is to assure that all phases are in the surface plane and not in relief or pitted as they would be after an electrolytic polish. Then, one would etch as recommended by White and LeMay,[22] first swabbing with a dilute aqua regia (15 ml hydrochloric acid, 5 ml nitric acid, 100 ml distilled water). This is a general reagent that attacks the austenite uniformly, leaving carbides, ferrite, and sigma phase clearly outlined and in relief. One would then immerse with Murakami's reagent (10 g potassium ferricyanide, 10 g potassium hydroxide, 100 ml distilled water). This is a color reagent and stains carbide dark, ferrite yellow or yellow-brown, and sigma phase blue. It is normally used at 80°C for up to 1 hour. Some sigma is resistant and may be stained with a more concentrated Murakami's. The color reagent follows the general reagent without repolishing. This technique was used very effectively by Bagnall and Witkowski[23] to identify phases in type 310 stainless steel.

PREPARATION FOR TRANSMISSION ELECTRON MICROSCOPY

Specimens to be examined by a transmission electron microscope must be thin enough for the accelerated electrons to pass completely through them with only a negligible loss of energy. The limiting thickness for transmission when 100-kV electrons are used varies from about 200 nm for aluminum to 50 nm for uranium.[24] Thicker specimens can be penetrated if higher accelerating voltages are used. Instruments exceeding 1 MV are in use today, albeit in limited quantity because of the enormous cost and increased difficulty of interpreting the microstructures. For stainless steels, the optimum accelerating voltage seems to be in the 200- to 400-kV range.

There are six ways of thinning specimens for transmission electron microscopy (TEM): chemical polishing, electrolytic polishing, ion bombardment (sputtering), mechanical abrasion, microtoming, and vacuum deposition.[24] Of these, mechanical abrasion combined with electro-polishing is the method most commonly used for stainless steels and will be the one discussed here. Further details can be found in References 24 through 27. Each of these publications lists, in turn, scores of additional references.

It is of utmost importance when preparing thin specimens not to introduce damage to the microstructure. From the time that the desired sample is removed from the bulk material, it should be handled as gently as possible. The need for gentleness becomes more acute as one nears the point at which the specimen goes into the microscope. One simply cannot afford to introduce artifacts into the microstructure by careless handling.

Cutting Methods

To obtain a sample of stainless steel from a parent piece, one should use the method that will produce the least damage commensurate with an acceptable cutting speed. Hacksawing or abrasive wheel cutting might be acceptable, but a fine jeweller's saw would be better since it will introduce far less damage. An abrasive slicing wheel is perhaps the most widely used device for cutting the thinnest possible samples for further thinning. Goodhew[25] presents an excellent and concise coverage of cutting methods.

The authors' laboratory uses an abrasive cutting machine* to obtain samples of stainless steel for electron microscopy. This machine has special capabilities that make it well suited for this application. The apparatus consists of a modified surface grinder with a thin (0.010-in.) alundum blade in place of a grinding wheel. The spindle speed is variable over a wide range, and the blade is liquid-cooled to minimize cutting damage. The spindle height above a horizontally movable table is also variable, thus allowing one to remove very small amounts of material per pass. It has been found that stainless steel wafers 0.020 to 0.025 in. thick are about optimum for the subsequent polishing and thinning required.

Grinding and Polishing

Standard metallographic procedures may be used to further thin and polish samples for TEM. Stainless steels, however, do not require any mechanical preparation other than grinding through 600-grit silicon-carbide abrasive paper. At this point, they can be thinned sufficiently by electrolytic jet polishing* such as described by Brammar and Dewey.[24]

Ion bombardment, or "ion milling," is now being used more often as a means of carefully controlled removal of material from TEM samples since it causes minimal damage to the internal structure of materials. The principle of ion bombardment is discussed in References 24 and 25. This technique is applied to stainless steel after the sample has been "dimpled" by electro-jet polishing. A number of different electrolytes may be used to thin stainless steel, but the authors' laboratory has adopted one that consists of seven parts methanol and one part H_2SO_4. It is used at a temperature of $-10°C$ and a current of 2 A/cm². This has been found to be quite useful for stainless alloys.

SUMMARY

As a class of alloys, stainless steels do not present metallographers with any great difficulty in preparation. The polished surfaces of these alloys are very easily scratched, however, and one's technique must include meticulous cleanliness and gentleness. Interpretation of microstructures requires appropriate etching, and a vast array of etchants is available. Most metallographers develop a preferred list that includes modifications.

The extremely widespread use of stainless steels means that they occupy a large portion of a metallographic laboratory's attention. It is hoped that this chapter will be of some aid to those whose job it is to prepare and interpret stainless steel microstructures.

REFERENCES

1. Kehl, George L. *The Principles of Metallographic Laboratory Practice.* New York: McGraw-Hill Book Co., 1949.
2. *Symposium on Methods of Metallographic Specimen Preparation.* Philadelphia: American Society for Testing Materials, 1960.
3. Williams, Robert S., and Homerberg, Victor O. *Principles of Metallography.* 5th ed. New York: McGraw-Hill Book Co., 1948.
4. Greaves, Richard H. *Practical Microscopical Metallography.* 4th ed. London: Chapman and Hall, Ltd., 1957.
5. Smith, Cyril Stanley. *A History of Metallography.* Chicago: The University of Chicago Press, 1960.
6. McCall, James L., and Mueller, William M., eds. *Metallographic Specimen Preparation: Optical and Electron Microscopy.* New York: Plenum Press, 1974.
7. Zapffe, Carl A., *Stainless Steels.* Cleveland, OH: The American Society for Metals, 1949.
8. Guillet, L. Chromium steels. *Revue de Metallurgie* V. 1 (1904).
9. Borchers, W., and P. Monnarty. German patent 246,015. Jan. 22, 1910.
10. Kehl, G. L., et al. Metallography. In *Metals Handbook, 1954 Supplement.* Cleveland, OH: ASM, 1954, pp. 164–177.
11. Methods of preparing metallographic specimens. *Standards of the ASTM,* Part 3. Philadelphia, PA, 1961.

---

* Micro-Matic Precision Slicing and Dicing Machine, Micromech Mfg. Co., Rahway, N.J.

* Using a TENUPOL thinning unit, Struers Scientific Instruments, 20102 Progress Dr., Cleveland, OH 44136.

12. Viswanathan, V. L. A new etchant to reveal prior austenite grain boundaries in martensitic stainless steels. *Metallography* 10:291–297 (1977).
13. Shehata, M. T. *et al.* A quantitative metallographic study of the ferrite to sigma transformation in type 316 stainless steel. *Microstructural Science* 11:89–99 (1983).
14. Cain, F. M., Jr. Simplified metallographic techniques for nuclear materials. *ASTM STP* 285:37–57 (1960).
15. *Metals Reference Book,* 2nd ed. Metals Park, OH: ASM, 1983.
16. *Metals Handbook.* 8th ed. *Atlas of Microstructures of Industrial Alloys.* Metals Park, OH: ASM, 1972.
17. Mason, J. F., Jr. Corrosion resistance of stainless steels in aqueous solutions. *Source Book on Stainless Steels.* Metals Park, OH: ASM, 1976.
18. Rhodin, Thor. N., ed. *Physical Metallurgy of Stress Corrosion Fracture.* AIME Metallurgical Society Conf. New York: Interscience Publishers, 1959.
19. Staehle, R. W.; Hochmann, J.; McCright, R. D.; and Slater, J. E., eds. *Stress Corrosion Cracking and Hydrogen Embrittlement of Iron Base Alloys.* Houston: NACE, 1977.
20. Dulis, E. J., and Smith, G. V. Identification and mode of formation and re-solution of sigma phase in austenitic chromium-nickel steels. *ASTM STP 110.* Philadelphia: ASTM, 1950.
21. Hansen, Max. *Constitution of Binary Alloys.* New York: McGraw-Hill, 1958.
22. White, W. E., and LeMay, I. Metallographic observations on the formation and occurrence of ferrite, sigma phase, and carbides in austenitic stainless steels. *Metallography.* 3:35–60 (March 1970).
23. Bagnall, C., and Witkowski, R. E. The influence of thermal aging and sodium corrosion on the microstructural stability of AISI type 310 'S' and alloy 330 at 700°C. *Microstructural Science* 6:301–319 (1978).
24. Brammar, I. S., and Dewey, M. A. P. *Specimen Preparation for Electron Metallography.* New York: American Elsevier Publishing Co., 1966.
25. Goodhew, P. J. *Specimen Preparation in Materials Science.* New York: American Elsevier Publishing Co., 1973.
26. Murr, Lawrence E., *Electron and Ion Microscopy and Microanalysis.* New York: Marcel Dekker, Inc., 1982.
27. Thomas, Gareth. *Transmission Electron Microscopy of Metals.* New York: John Wiley & Sons, Inc., 1962.

7
PROBLEM SOLVING USING QUANTITATIVE STEREOLOGY

R. T. DeHoff

*Department of Materials Science and Engineering,
University of Florida*

Stereology is a field that deals with the geometry of microstructures. During the past two decades, its mathematical foundations have been established, and a number of texts have appeared.[1-5] Since it would not be possible to present a comprehensive overview of this tool in a single chapter, this chapter will focus upon those stereological methods and relationships that are "general" in the sense of being valid for real microstructures of arbitrary complexity. It is fortunate, and perhaps remarkable, that these methods are also very straightforward and easy to implement.

From a geometric point of view, a microstructure consists of areas (phase particles, grains), lines (phase boundaries, grain boundaries), and points (triple points, particle centers). As observed under a microscope, these features represent sections through microstructural elements in the three-dimensional structure that the polished section samples. Features that are observed as areas are sections through particles or grains; lines on the plane of polish are sections through surfaces or interfaces; and points are sections through lines. Each of these three-dimensional features has geometric *properties;* for example, the particles or grains have *volume;* the interfaces have *area;* the lines have *length.* These properties have unambiguous meaning for real structures of arbitrary geometry.

Quantitative stereology provides the means for estimating these geometric properties of real three-dimensional microstructures (volume, area, length, etc.) from measurements that must necessarily be made on a two-dimensional section through the three-dimensional structure. These geometric properties may be easily and conveniently estimated from simple counting measurements made upon a microsection that is representative of the three-dimensional structure. Other properties, like size distributions and average sizes, can be estimated only if simplifying assumptions are made about the geometry of the structure (e.g., all the grains are spheres), and this restriction severely limits their range of validity. Still other properties can be estimated only by serial sectioning analysis.[6,7]

In this chapter, the basic stereological counting measurements made under the microscope will be described first. The fundamental relations of stereology will then be presented. These equations are used to calculate the geometric properties of the three-dimensional microstructure from the counting measurements. Next, the principles of operation of semiautomatic and automatic image-analyzing computers are reviewed, with emphasis on those kinds of situations for which they are useful and those for which they are not. The steps involved in designing and carrying out a particular stereological experiment are then outlined, and the chapter closes with some examples of applications to specific problems.

MAKING MEASUREMENTS

This section describes the kinds of measurements made in quantitative stereological analysis and then reviews the information about the geometry of the real three-dimensional microstructure that they provide. The measurements are described as if they had been made *manually,* i.e., without the assistance of image-analysis equipment. The assistance available from such equipment will be described in the next section.

Stereological measurements are fundamentally statistical in nature. A measurement (or series of measurements) is made on a field and recorded. Then, the stage is translated to reveal a new field of view; the measurements are made on that field, and so on. Since the individual measurements are very simple to make, repetition on a large number of fields may be justified for a single metallographic specimen. In a typical experiment, 25 to 100 fields may be examined. The mean value of each set of measurements is computed, along with the standard deviation; the latter quantity is the basis for characterizing the *precision* of the set of measurements.

The only equipment required to make stereological measurements *manually*, besides the microscope, is a *grid of lines* that may be superimposed on the mocrostructure. This may take the form of a reticle in the eyepiece, a plastic overlay on a ground-glass projection screen, or a plastic overlay on a photograph. Typically, the grid is square, though it need not be. A 6 × 6 grid is suitable for most purposes (i.e., six vertical lines crossing six horizontal lines). The design of the grid may influence the *efficiency* of the measurements, but not their validity. Figure 7-1 shows a microstructure with a 6 × 6 grid superimposed.

It is necessary to calibrate the dimensions of the grid with a stage micrometer so that the precise size of the field being examined is known. For a given instrument configuration (eyepiece, grid, and objective), the calibration, once established, is fixed. If photographs are used, their real magnification must be known with precision.

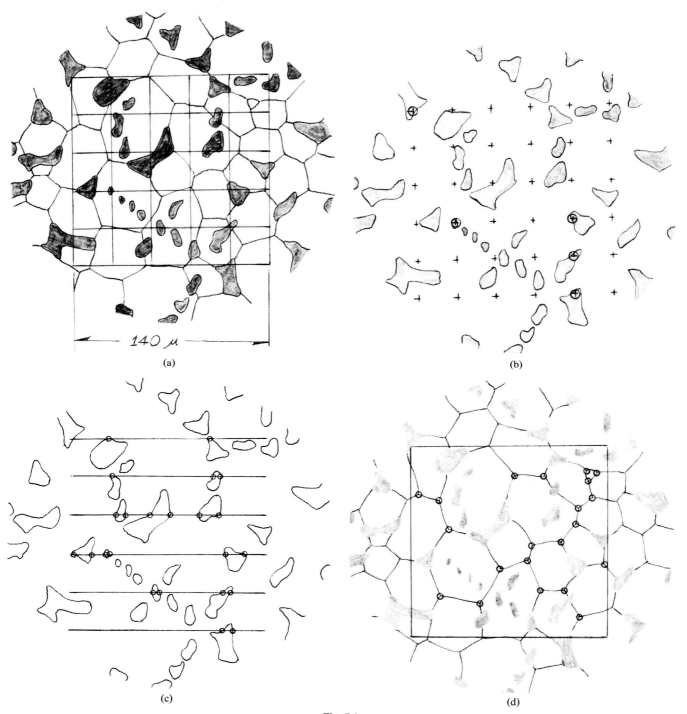

Fig. 7-1.

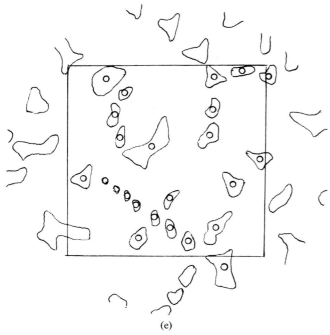

Fig. 7-1. Each of the stereological counting measurements is illustrated for a two-phase microstructure and a superimposed 6X6 grid (a), including: (b) the point count; (c) the line-intercept count; (d) the area point count; and (e) the feature count.

WHAT IS MEASURED

The basic stereological counting measurements, illustrated in Fig. 7-1, are as follows:

1. The *point count*, P_P: Focus on the 36 crossing points in the grid of Fig. 7-1(b). These points are scanned to make a count of the number of points lying in the feature of interest. If a grid point lies on a feature boundary, it is counted as ½. The *point count*, P_P, is the ratio of the number of points in the feature of interest to the total number of points in the grid. It is thus a unitless fraction. In Fig. 7-1(b), $P_P = 5/36 = 0.14$.

2. The *line intercept count*, P_L: Focus on the six horizontal lines in the grid of Fig. 7-1(c). The lines are scanned to make a count of the number of intersections they make with the boundaries of interest in the structure (grain boundaries, phase boundaries, etc.). The total length of the lines scanned for the field is known from the calibration of the grid [6×140 microns $= 840$ microns $= 8.4 \times 10^{-4}$ meters in Fig. 7-1 (c)]. The *line intercept count*, P_L, is the number of intersections counted, divided by the length of test lines sampled. It has inverse units of length. In Fig. 7-1(c), $P_L = 23/8.4 \times 10^{-4} = 2.7 \times 10^4$ counts/meter.

3. The *area point count*, P_A: Focus on the total square area delineated by the boundaries of the grid in Fig. 7-1(d). If the three-dimensional structure contains lineal features, they will appear on the sectioning plane as points. The most common kinds of lineal features are the *triple lines* that result when three cells meet in space (e.g., the grain edges in a single-phase structure) and *line imperfections* (dislocations). The former appear as triple points on the section where the three cells meet; the latter may be revealed as etch pits. A count is made of the number of such points within the area outlined by the grid. The grid divides this area into 25 subareas to help one keep track of the features as they are counted. The area scanned is known from the calibration of the grid: $140 \times 140 = 19,600$ square microns, or 1.96×10^{-8} square meters. The *area point count*, P_A, is the ratio of the number of points counted to the area scanned. It has inverse units of squared length. In Fig. 1(d), $P_A = 21/(1.4 \times 10^{-4} \text{ m})^2 = 1.1 \times 10^9$ counts/meter^2.

4. The *feature count*, N_A: Focus on the square area delineated by the boundaries of the grid in Fig. 7-(e). A count is made of the number of features in the class of interest contained in that area.* The area scanned is known from the calibration of the grid. The *feature count* is the number of features in the field divided by the area of the field. It has inverse units of squared length.† In Fig. 1(e), $N_A = 24/(1.4 \times 10^{-4} \text{ m})^2 = 1.2 \times 10^9$ counts/meter^2.

The four measurements—the *point count*, *line intercept count*, *area point count*, and *feature count*—are fundamental in stereology. They are *field* measurements, i.e., they have a value for each field scanned. Because they are simple counts of events that result from the interaction of the superimposed grid with features in the structure, they are relatively easy to make. In contrast, a variety of *feature specific measurements* in use require the measurement of some geometric aspect of the individual features of a structure, such as its area, perimeter, diameter, longest and shortest dimension, and so on. Such a measurement performed on all of the features in a field clearly

---

* Some features that intersect the boundaries of the area must carry a count of ½ since, if the total area of the sample were scanned by placing fields adjacent to each other, particles on the field boundaries would be counted twice. Alternatively, this correction for features on the boundary may be made by including in the count those features that intersect the top and left sides of the field boundary and excluding those that intersect the right and bottom sides.

† In some complex structures, the features of interest may contain internal holes. In this case, it is necessary to count the number of holes in the field and subtract this number from the feature count. An alternative procedure for obtaining the same information is provided by the *area tangent count*, T_A, obtained by sweeping a line across the field and counting separately the number of tangents made with convex segments of the feature boundary and those made with concave segments. The *area tangent count* is obtained by subtracting concave counts from convex counts and then dividing by the area of the field swept by the line. It is rigorously true that *the area tangent count is twice the feature count*.

demands significantly more effort than making simple counts with a grid. Feature specific measurements form the basis for estimating particle size distributions, average diameters, and extreme values of sizes and are usually made with the help of image-analysing instrumentation.

WHAT THE MEASUREMENTS MEAN

Each of the counting measurements reviewed in the previous section provides information about specific geometric properties of the features they sample. They are a part of the *fundamental relations of stereology* that will be presented in this section. These relations have the status of mean value theorems in statistics; that is, they are derived by assuming that the *entire population* of test probes that can be constructed in three-dimensional space (points for the point count, lines for the line intercept count, and areas for the area point and feature counts) are included in the sample to obtain the representative average value of the counting measurement. In practice, the set of probes scanned is a *random sample* from this population; if the sample is properly designed, then the sample mean provides an unbiased estimate of the population mean.

The strength and generality of these equations derives from the fact that *no geometric assumptions* are made about structure in their derivation. They may thus be applied to real microstructures to provide real estimates of the geometric properties with which they deal, as we shall now discuss.

The *point count*, P_P, provides an unbiased estimate of the *volume fraction*, V_V, of the feature counted. The relationship is as follows:

$$\langle P_P \rangle = V_V \qquad (7\text{-}1)$$

where the symbol $\langle \ \rangle$ means the average or expected value of the quantity contained. The volume fraction of a phase or constituent is the fraction of the volume of the structure that it occupies; it is the most straightforward measurement of the quantity of that phase in the microstructure. If the structure contains more than two phases, the point count may be applied separately to estimate the volume fraction of each phase. Alternatively, it may be applied to estimate the volume fraction of a constituent, such as pearlite, if it can be identified unambiguously on a microsection.

The *line intercept count*, P_L, is an unbiased estimate of the *surface area in unit volume*, S_V, of the boundary or interface whose intersections are counted on the microsection. The relationship is as follows:

$$\langle P_L \rangle = \frac{1}{2} S_V \qquad (7\text{-}2)$$

For a two-dimensional interface or boundary in a three-dimensional structure, *surface area* is the most straightforward measure of its *extent*. Other things being equal, a fine structure has a large, a coarse one a small, value of S_V. The line intercept count may be separately applied to traces of any identifiable class of interface in the structure, such as grain boundaries, twin boundaries, interphase interfaces, and the like.

The *area point count*, P_A, reports the *length in unit volume*, L_V, of the lineal feature that produces the observed points of emergence counted. The relationship is as follows:

$$\langle P_A \rangle = \frac{1}{2} L_V \qquad (7\text{-}3)$$

For lineal features in a three-dimensional microstructure, *length of line* is the most straightforward measure of *extent*. It is most commonly used to estimate the length of triple lines in the structure that produce triple points on the section, and it may be independently applied to estimate the total length of each identifiable class of lineal feature in the structure.

The *feature count*, N_A, of the number of particles in the unit area of a microsection provides an unbiased estimate of a geometric property of the boundaries of the particles in a three-dimensional structure that is called the *integral mean curvature per unit volume*, M_V. The relationship is as follows:

$$\langle N_A \rangle = \frac{1}{2\pi} M_V \qquad (7\text{-}4)$$

For the definition and geometric meaning of the integral mean curvature, consult References 8 and 9. This quantity is particularly useful in the description of plate structures, where it provides information about the average dimension of particles in the plane of the plate.[9]

Among the most used stereological parameters are measures of the *scale* of the grains or particles in the structure and of their *separation*. The most common measure of size or scale is the *mean lineal intercept*, $\bar{\lambda}$, which is illustrated in Fig. 7-2. This property of a particle or grain is the average surface-to-surface distance through the feature. Imagine drawing all possible lines that intersect the feature in three dimensions. Measure the length of each and compute the average value; this value represents the

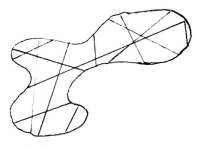

Fig. 7-2. Geometrical meaning of the mean lineal intercept of a feature.

mean lineal intercept of the particle. For a collection of particles or grains, imagine measuring the lengths of all intersections of lines with all of the features to obtain the mean lineal intercept, as follows:

$$\bar{\lambda} = \frac{4V_V}{S_V} \qquad (7\text{-}5)$$

Since the numerator in Eq. 7-5 is the volume fraction of the phase of interest, application of this formula to the *matrix phase* in a two-phase structure will give the *mean free path* between particles, which is simply the mean lineal intercept of the matrix. If V_V is the volume fraction of particles in the structure, then $(1 - V_V)$ is the volume fraction of the matrix, and its mean intercept is as follows:

$$\bar{\lambda}_{(MFP)} = \frac{4(1 - V_V)}{S_V} \qquad (7\text{-}6)$$

The surface area that must be employed in Eq. 7-5 is the total surface area of the particles of the phase of interest. In a single-phase polycrystal, each element of grain face is shared by *two* particles (grains) so that the total surface area is twice the value determined by applying Eq. 7-2. Since $V_V = 1.0$ for this case, the mean lineal intercept of a collection of grains is simply

$$\bar{\lambda}_{(GRAINS)} = \frac{4(1)}{2S_V} = \frac{2}{S_V} \qquad (7\text{-}7)$$

Measures of the *size of features* in the structure are often defined in terms of the mean lineal intercept. For example, a grain diameter, D, may be given by

$$D = \frac{4}{3}\bar{\lambda} = \frac{8}{3S_V} \qquad (7\text{-}8)$$

This equation is borrowed from the computed result that the diameter of a sphere is 4/3 of its mean intercept. In order to apply the equation to a real structure, however, it is necessary to assume, not only that all of the particles or grains are spheres, but that they are all the same size. Thus, while the mean intercept rigorously supplies information about the scale of the structure, as it is defined in Fig. 7-2, it has different geometric meaning for different particle shapes. For example, for a collection of plates, the mean intercept is related to their thickness and provides no information about their dimension in the plane of the plates.[9]

Feature-specific measurements provide the basis for estimating particle-size distributions of features in the three-dimensional structure. For a clear presentation of these procedures, consult Chap. 6 of Reference 1. The procedures for converting section measurements to estimates of the number of particles in each three-dimensional-size class require that the particles all be assumed to have the same (quantitative) particle shape. Since no two particles or grains are alike in real microstructures, there is no general procedure for assessing the validity of the shape approximations that are a necessary part of size-distribution estimating procedures.

Furthermore, it has been demonstrated that these procedures are statistically unstable, with small errors sometimes having important effects on the results.[10] For example, it is frequently found that estimates of the number of particles in small-size classes are *negative,* which clearly has no geometric significance. Since feature-specific measurements by their nature require more effort, they should be used only after the need for them has been justified and with full knowledge of their limitations.

INSTRUMENTATION FOR IMAGE ANALYSIS

During the past two decades, instrumentation designed to help accumulate measurements like those just described have evolved to a high level of sophistication. Semiautomatic instruments make use of a human operator to detect and discriminate the objects to be analyzed and to input the features of interest into a microcomputer, which then provides the geometric analysis and displays the numerical results. Fully automatic instruments view the image with a television scanner, digitize the image into a large number of picture points, or *pixels,* each with an associated grey shade, and transfers the digitized image into computer memory, where it can be processed by appropriate software to extract the geometrical information and display the results.

Semiautomatic instruments use the human operator as detector and discriminator of the image. Their primary advantage derives from this fact. Since certain microstructures cannot be properly discriminated by gray-level differences alone, an operator is required to perform the detection and discrimination. In such instances, semiautomatic instruments are more suitable than fully automatic ones. The features are input into a microcomputer by tracing them on a magnetic board or digitizing pad, either from a photomicrograph laid on the pad or from a TV image directly from a microscope. Software uses the input boundary-position coordinates to compute area, perimeter, selected dimensions, and other *feature-specific geometric properties.* The results can be reported as a features list in distribution histograms or can be summed to produce global or total properties for the structure.

Fully automatic instruments replace the human operator with automatic detection and discrimination of the features of interest. Early versions of these machines based their detection schemes entirely upon grey-shade discrimination, with the boundaries of the features located by setting a grey-level threshold. More recent instruments begin with grey-shade detection but may employ additional criteria—geometrically based criteria, for example. Since the digitized image information is stored in computer memory, virtually any detection scheme that may be formalized into an unambiguous algorithm may be applied. In addition, current instruments permit sophisticated image processing before detection, as well as editing of the detected image by the operator. Once the image has been satisfactorily detected, the geometric properties of individual features may be evaluated, as with semiautomatic equipment. The larger computers employed in fully automatic image analysers permit the evaluation of more sophisticated geometric properties of the features involved, including "erosions," "dilatations," "openings," "closings," and other operations derived from set algebra. For a more complete presentation of this class of operations, see Reference 4.

Image-analysis instrumentation brings speed and efficiency to quantitative metallography, while relieving most of the tedium associated with application of this tool. The availability of such instruments permits undertaking some characterization tasks that would otherwise be beyond human endurance. Routine quality control that contemplates analysis of a large number of similar samples on a daily basis would not be feasible, for example, without automatic instrumentation.

It is important that the decision to acquire and use image-analysis equipment be undertaken knowledgeably. Such equipment is expensive. Despite its great power in processing and editing an image, it is no match for a moderately experienced metallographer in discriminating what is to be measured. It must be remembered that *detection is always a compromise.* Some features that are easy for a human operator to detect, such as triple lines of various types, are almost impossible for automatic equipment. Grain structures are particularly difficult. If samples are etched to reveal grain boundaries as black lines, then incomplete lines will mislead the computer. (The most sophisticated equipment makes use of *skeletonization* software that handles this problem fairly well.) If the grains are etched to be seen by differential contrast—as is common with brass, for example, or aluminum under polarized light—detection is beyond the scope of image analyzers.

Image analyzers can provide reams of information that has no stereological meaning in the sense that it has no unambiguous relation to the three-dimensional geometry of the real microstructure that is being characterized. This information is presented in detail, with distributions, histograms, or even lists of properties for individual features. It is easy to obtain, requiring no more effort than knowing the global properties that are the equivalent of the stereological counting measurements. For these reasons it may get in the way of clear thinking about the real microstructural geometry.

In order to prepare samples that can be properly detected by an image analyser, it may be necessary to develop novel preparation routines. To make matters worse, there is sometimes a tendency to tailor the problem under study so that the image analyzer can conveniently see what it ought to see. It is thus important to maintain a proper perspective about the role of image-analysis instrumentation in microstructural characterization.

PLANNING AN EXPERIMENT

The basic steps in planning a stereological experiment are the following:

1. Obtain the samples or sequence of samples to be examined.
2. Carry out the requisite metallographic preparation.
3. Determine what you want to know and why you want to know it.
4. Select the stereological quantities to be measured.
5. Choose a realistic precision target.
6. Choose a magnification for the measurements.
7. Select a grid size if the measurement is manual.
8. Make the stereological measurements on a predetermined number of fields.
9. Compute the mean and standard deviation of the measurement series.
10. Compare the confidence interval with the precision target; continue if necessary or modify the target.
11. Interpret the results.

Each of these steps will be examined in this section.

Since the sequence of samples to be examined is specific to each problem undertaken, generalizations are accordingly not appropriate. It is important to note, however, that if the samples are intended to reflect the evolution of a microstructure during a process, it is essential that they lie along a continuous path. This requires that each sample be related to the previous one in the series, first by duplicating the conditions that produced the previous sample, then by adding new conditions. This is true, for example, for a series of samples prepared by isothermally annealing for a sequence of times. It is *not* true for a series of samples prepared *isochronally*—i.e., by heating for a fixed time, say one hour—at a series of increasing temperatures because such samples do not lie along a *path* and therefore do not experience the same history.

The specifics of metallographic preparation are beyond the scope of this chapter, since each material and each process examined will require the metallographer to draw upon personal experience for that unique combination. It is essential, of course, that metallographic preparation be of the highest quality for quantitative measurements, with the best definition and contrast that can be obtained. When possible, it is desirable to *calibrate the preparation* by measuring a property stereologically and comparing the result with an independently computed value. For example, in applying stereology to impregnated porous samples, it is possible to measure the density by water immersion methods and compare the volume fraction of porosity obtained from this measurement with that estimated from the point count.

Determining what you want to know about a structure, and why you want to know it, are crucial to the design of an experiment. This determination lies within the context of the overall research program at hand, and generalities do not apply. Nonetheless, it is important to think before you start, since stereological measurements are obviously effort-intensive. A frequent error made in this aspect of the design process arises from the assumption that size distribution information is essential to every application. The distribution is therefore measured, and the mean is calculated and used in all comparisons. Usually, however, a very simple and appropriate estimate of the mean may be obtained without determining the size distribution.

It is useful to make an exhaustive list of the features of the structure (cells, interfaces, triple lines, quadruple points) as well as of the geometric properties that each may exhibit. This list can then be narrowed by eliminating features or properties not appropriate to the problem at hand, thus identifying the things to be measured. The measures of extent previously discussed will usually provide a useful level of characterization.

As is true of any statistical variable, the precision of a stereological measurement is expressed in terms of the "95-percent confidence interval" around the mean value. This confidence interval may be estimated from the values of the mean and standard deviation computed for the set of measurements of the property. The standard error, S.E., of the mean is

$$\text{S.E.} = s_x/\sqrt{n} \quad (7\text{-}9)$$

where s_x is the standard deviation of the sample readings and n is the number of readings in the sample. The confidence interval, C.I., is a range about the sample mean given by

$$95\% \text{ C.I.} = \bar{x} \pm 2 \text{ S.E.} \quad (7\text{-}10)$$

A typical precision target for a stereological characterization is ±5 percent of the value of the quantity estimated. As a rough rule of thumb for manual measurements, if fields are examined until a total of 400 to 500 counts have been obtained, the precision will be about 5 percent. Thus, if a given field yields an average of about 20 counts, a total of 25 fields will yield 500 counts and a precision of about 5 percent. Bear in mind that such an estimate is *only a rough rule of thumb*. The value obtained in a given situation depends upon both the quantity being measured and the nature of the sample. It may be necessary to modify the precision target in the light of accumulating data.

The magnification is selected as a compromise between two limiting conditions. It must be high enough so that the features to be characterized are easily resolved. If the magnification is too low, then the number of undecided counts will be abnormally high. On the other hand, if the magnification is too high, each field may show only a few features. This scarcity will affect the efficiency of the measurements, since it is necessary to scan a large number of fields to obtain the precision target. As a rough rule of thumb, the magnification should be such that the size of a typical feature in the structure is equal to about half the spacing in the superimposed grid. In any practical application, implementation of these recommendations depends upon the scale of the structure, the sparseness of feature distribution, and the availability of the equipment needed to achieve the required magnification. Compromises are inevitable.

The grid reticle should be designed to cover most of the field of view. The number of lines in it should depend upon the density of the features to be sampled. For many purposes, a 5×5 grid is adequate. If the features are sparsely distributed, an 8×8 or 10×10 grid may be appropriate. Common sense is the primary criterion in this decision. For example, if the feature of interest is present at a volume fraction of 0.01 (1 percent), then one point out of 100 will exhibit an element of the feature. Since a 5×5 grid would reveal only one count in every four fields, a finer grid would be called for in this case. If the grid chosen is too fine, however, the result will just be confusing.

A combination of grid size and magnification should be chosen so that between 5 and 20 counts can be made in each field viewed (if this is possible for the features under examination). If fewer than 5 counts are made per field, then more than 100 fields will have to be scanned to produce reasonable precision. If more than 15 are made, counting errors will intrude no matter how careful the observer may be.

The fields selected for stereological examination must constitute an *isotropic-uniform-random* sample of the microstructure. Recall that the validity of stereological relationships rests on the assumption that the entire population of test probes (points, lines, or sectioning planes) is included in the computation. In practice, the set of probes examined is a small sample of the total population of probes in that category. Members of the set of points in space are described by their position; members of the set of lines and/or planes are described by both position and orientation. Fields to be viewed should be selected randomly from a uniform distribution of positions and an isotropic distribution of orientations.

If the structure itself is uniform, isotropic, and random, then any set of sample fields will be appropriately representative. If the structure exhibits gradients in the properties of the features (i.e., if they vary with *position*), then the gradients must be appropriately represented in the collection of sample fields. For example, if the volume fraction of a phase varies radially in a circular bar, this positional variation will be properly represented in a section transverse to the axis of the bar; longitudinal sections taken at different radii will each give different estimates of the volume fraction. If some of the feature properties of the structure exhibit anisotropies (i.e., if their values vary in accordance with their *orientation* in the structure), then sample fields must be chosen that are uniformly distributed over orientation. Since the sectioning operation is destructive, rigorous inclusion of anisotropies in the analysis may not be practical.

In some applications, it may be useful to view these sampling problems as an oppcrtunity rather than an obstacle. It is possible to *characterize the gradient* or *the anisotropy* in nonuniform structures. The gradient may be characterized by making a series of measurements as a function of position, e.g., by taking many fields at each position along the radius in the case just cited. Anisotropy (of surface area or line length) may be characterized by taking measurements in three mutually perpendicular directions. For detailed discussions of the characterization of anisotropy, consult References 1 and 2.

Stereological measurements may be made at first on 25 fields of view. At this point, compute the mean and standard deviation of the sample of 25 readings; calculate the confidence interval using Eq. 7-10. Compare the result with the precision target selected originally. Use the same formula to estimate the number of fields that will be necessary to attain the precision target. Use common sense either to revise the precision target, if the number of fields required is inordinately large, or to accept it as is and continue to examine new fields until it is approximately attained.

EXAMPLES

This section contains a variety of examples that demonstrate the strategies and methods of stereology. The data for these computations are presented in Table 7-1. In each case, the counts obtained on individual fields are listed, and the mean value, standard deviation, and standard error of the mean of each data set are computed.

Example 1: Volume Fraction

A sodium-modified aluminum–silicon alloy consists of irregularly shaped silicon-rich particles dispersed in an aluminum-rich matrix. Qualitative examination of the structure shows that the particles are fine and somewhat sparsely distributed. Accordingly, a high magnification is chosen: 1000 ×. A 5 × 5 eyepiece grid is available. Calibration with a stage micrometer yields 70 microns (7.0×10^{-5} m) for the length of each side of the grid at this magnification. Estimation of the volume fraction of silicon particles is an application of Eq. 7-1. A count

Table 7-1. Raw Counting Data for Examples of Applications of Stereology.

FIELD NUMBER	EXAMPLE NUMBER				
	1	2	3	4	5
1	1	8	15	3	9
2	1	7	12	5	11
3	3	7	8	6	8.5
4	0	9	18	4	12
5	2	6	20	2	10
6	1	5	16	1	13
7	2	8	10	2	12
8	1	8	11	4	10
9	1	7	17	7	9.5
10	0	10	17	0	10
11	3	12	12	3	11
12	4	9	20	4	11
13	0	4	19	4	13
14	0	5	13	3	10
15	1	7	14	5	12
16	1	7	16	6	10
17	2	8	17	2	11
18	1	9	17	2	11.5
19	1	8	14	1	9
20	3	8	19	5	12
21	0	7	11	4	10
22	1	6	23	3	10
23	0	10	12	4	12
24	1	9	16	2	13
25	2	7	15	3	9
x	1.28	7.64	15.28	3.40	10.78
s	1.10	1.75	3.60	1.70	1.35

is made of the number of grid points within the silicon particles in each of 25 fields (see Column 1 of Table 7-1).

The 95-percent confidence interval for these results is 1.28 ± 0.44 points per placement of the grid. The corresponding confidence interval for the point fraction, P_P, is obtained by dividing this by the number of points in each grid sample, or 25 in the case of a 5 × 5 grid. Thus,

$$P_P = 0.051 \pm 0.018$$

Since, by Eq. 7-1, the point fraction provides an unbiased estimate for the volume fraction,

$$V_V = 0.051 \pm 0.018$$

for this sample.

Note that the precision of this estimate is not very satisfactory: $0.018/0.051 = 0.345$, or 34.5 percent. This is not unexpected for this case, since only 32 hits (1.28×25) were made in 25 fields. In order to achieve a precision of ± 10 percent, it is necessary that ±2 S.E. = 0.10 $\bar{x}$ of the mean). Thus,

$$2\left(\frac{s_x}{\sqrt{n}}\right) = 0.10\,\bar{x}$$

$$n = \left[20\left(\frac{s_x}{\bar{x}}\right)\right]^2$$

Insert the value of s_x and solve for n, the number of fields required to achieve a precision of 10 percent. In the present example, $n = 298$ fields, which is too large a number for most practical applications. Thus, it is necessary to accept a lower level of precision of the estimate for the analysis of this structure. (Analysis of about 75 fields with a 10×10 grid will yield similar precision, but the grid spacing in this case might be too narrow for realistic counting.) In order to obtain high precision for this measurement, it may be necessary to use an automatic image analyser.

Example 2: Surface Area

In order to estimate the area of interface separating the silicon particles from the aluminum matrix, it is necessary to apply Eq. 7-2. The same grid and magnification are chosen for this exercise as were used in the volume fraction estimate. For each placement of the grid, a count is made of the number of intersections formed by the set of five horizontal lines in the grid and traces of the aluminum-silicon interface. This count is repeated on each of 25 fields, with the results displayed in Column 2 of Table 7-1.

The 95-percent confidence interval for the resulting count is 7.64 ± 0.70 intersections per field. Since the calibration showed that the length of one line is 7×10^{-5} m, the total length scanned in examining five lines in the field is 35×10^{-5} m. The corresponding confidence interval for the line intercept count is obtained by dividing the count by the total length scanned, or

$$P_L = (2.18 \pm 0.20) \times 10^4 \text{ m}^{-1}$$

Application of Eq. 7-2 yields the confidence interval on the following estimate of the surface area:

$$S_V = (4.36 \pm 0.40) \times 10^4 \text{ m}^2/\text{m}^3$$

Note that the precision of this estimate is $(0.40 \times 10^4)/(4.4 \times 10^4) = 0.09$, or about 9 percent. The fact that this is much better than that obtained for the estimate of the volume fraction follows from the observation that a total of 191 events (intersections) were counted on 25 fields, whereas only 32 events were observed in the point count.

Combination of these two results provides an estimate of the *scale* of the system of silicon particles expressed in terms of the mean lineal intercept of the particles. Applying Eq. 7-5 yields the following:

$$\bar{\lambda} = \frac{4(0.051)}{4.36 \times 10^4} = 4.7 \times 10^{-6} \text{ m} = 4.7 \text{ microns}$$

Thus, the silicon particles in this alloy occupy about 5 percent of the volume and are sized at about 4.7 microns, with an interfacial area of about 4.4×10^4 m^2/m^3.

Example 3: Mean Grain Intercept

A sample of polycrystalline iron is given a preliminary examination with a 5×5 grid at 250X. A line-intercept count is made, using the five horizontal lines in the grid. A total of 35 intersections is counted in the field at 250X. This number is too high for practical counting, and experience has shown that systematic errors will result if the number of counts in a field is too high. Accordingly, a new magnification of 500X is chosen for this study of grain size. In a preliminary experiment, 25 fields are examined, with the resulting line-intercept counts that are recorded in Column 3 of Table 1.

The 95-percent confidence interval for this set of readings is 15.3 ± 1.4 counts per field. Calibration shows that each side of the grid at 500X is 1.4×10^{-4} m long; the total length scanned in each placement of the grid is thus $5 \times 1.4 \times 10^{-4}$, or 7.0×10^{-4} m. The confidence interval on the line-intercept count is thus

$$P_L = (2.19 \pm 0.20) \times 10^4 \text{ counts/m}$$

The total surface area of grain boundary in unit volume may be computed from P_L by applying Eq. 7-2 as follows:

$$S_V = (4.38 \pm 0.40) \times 10^4 \text{ m}^2/\text{m}^3$$

For this example, a total of 382 intersections were counted, yielding a precision of (0.40/4.38), or 0.091, i.e., about 9 percent. Since the width of the confidence interval varies inversely with the square of the number of readings, it would be necessary to examine a total of four times as many fields (i.e., 100 fields) to achieve a level of precision under 5 percent.

The mean grain intercept is the average surface-to-surface distance through the grains of a structure. Application of Eq. 7-7 yields

$$\bar{\lambda} \text{ (grains)} = 2/S_V = 4.6 \times 10^{-5} \text{ m, or 46 microns}$$

for the mean grain intercept of this particular structure. An estimate of its average grain diameter using Eq. 7-8 gives 61 microns. Keep in mind that the mean grain intercept has unambiguous geometric meaning for the real three-dimensional grain structure. This computation of grain "diameter" assumes that all grains are spheres of the same size, an assumption that is clearly untrue, and potentially misleading, for real structures.

Example 4: Triple Line Length

A sample of sintered nickel powder has a pore volume fraction of 0.06 (6 percent). Some of the pores are distributed along grain boundaries in the nickel matrix. The intersection between a pore and two grains is a triple line that may be designated "ssp" (solid; solid; pore) for the three kinds of cells that meet to form the line. If the mechanism of densification involves grain-boundary diffusion, the source of vacancies lies along this triple line. Thus, it may be interesting to estimate the length of this lineal feature.

Preliminary examination of the sample demonstrated that a magnification of 1000× yielded a reasonable number of counts per field and provided sufficient resolution to observe the ssp triple points in the etched microstructure. The resulting counts are summarized in Column 4 of Table 1.

The confidence interval resulting from these triple-point counts is 3.40 ± 0.68. Calibration of the grid shows that its dimensions on a side are 7×10^{-5} m at this magnification (see Example 1); the area enclosed is the square of this value, or 4.9×10^{-9} m². The normalized triple-point count for ssp triple points is computed by dividing the number of counts by the area scanned, as follows:

$$P_A(\text{ssp}) = (6.94 \pm 1.39) \times 10^8 \text{ counts/m}^2$$

Application of Eq. 7-3 yields an estimate of the length of the ssp triple line in unit volume of structure, as follows:

$$L_V(\text{ssp}) = (1.39 \pm 0.28) \times 10^9 \text{ m/m}^3$$

It is interesting to contemplate the implication that this value represents about one million kilometers of ssp triple lines in a cubic meter.

Example 5: Integral Mean Curvature

An aged aluminum–copper alloy exhibits a dispersion of platelets of copper-rich phase in an aluminum-rich matrix. The integral mean curvature has been shown to be a useful parameter in the characterization of platelet structures. It is related to a count of the number of particle sections observed on the plane of polish. After a magnification of 750× has been chosen, a count is made of the number of platelets that lie within the area delineated by the boundary of the grid. Particles that cross the boundary of the field are counted at half value (0.5). The resulting counts are summarized in Column 5 of Table 7-1.

The confidence interval resulting from the feature count is found to be 10.78 ± 0.54, yielding a precision of (0.54/10.78), or 0.049, i.e., about 5 percent. A total of 269.5 features were counted in 25 fields. Calibration of the grid dimensions at 750× gave a boundary length of 9.3×10^{-5} m and hence an area equal to $(9.3 \times 10^{-5} \text{ m})^2$, or 8.7×10^{-9} m².

Normalization of the number of counts obtained gives a feature count of

$$N_A = (1.25 \pm 0.06) \times 10^9 \text{ counts/m}^2$$

Application of Eq. 7-4 yields the following integral mean curvature for the platelets in this structure:

$$M_V = (7.83 \pm 0.39) \times 10^9 \text{ m/m}^3$$

Additional stereological measurements on this system gave a volume fraction of plates of 0.013, a surface area of 3.8×10^4 m²/m³, and, by applying Eq. 7-5, a mean lineal intercept, λ, of $(4 \times 0.013)/(3.8 \times 10^4)$, or 1.4×10^{-6} m (1.4 microns). Reference 9 demonstrates that the mean intercept is equal to twice the average plate thickness, $\bar{\tau}$; hence, $\bar{\tau} = 0.7$ microns. The same reference shows that the mean lineal intercept *in the plane of the plate* is related to the integral mean curvature, as follows:

$$\bar{\lambda}(\text{plate}) = \left(\frac{\pi^2}{4}\right)\left(\frac{S_V}{M_V}\right)$$

$$= \left(\frac{\pi^2}{4}\right)\left(\frac{3.8 \times 10^4}{7.8 \times 10^9}\right) = 1.2 \times 10^{-5} \text{ m} = 12 \text{ microns}$$

Thus, the typical plate in this structure is about 0.7 microns thick and has a longitudinal dimension of about 12 microns.

SUMMARY

Stereology provides a powerful tool for the characterization of the geometry of real three-dimensional microstructures. The counts at the heart of experimental stereology are easy to make, and image analysing equipment is available for applications requiring large quantities of data or more exotic information. Relations of the counting measurements made on a section and the geometry of the microstructure in three dimensions are simple and free from limiting assumptions. The design of a stereological experiment is largely an exercise in common sense, salted with a mininal knowledge of statistics.

The information based upon counting measurements that is presented in this chapter does not exhaust the applicability of the tool. Additional rigorous information may be obtained about unusual properties (for example, torsion of space curves or topological measures that require serial sectioning). A large literature exists on approximate methods. Since these usually involve invalid geometric approximations, use the straightforward global information that is supplied by the methods reviewed here instead. They provide realistic numbers that encourage one to think realistically about three-dimensional microstructures.

REFERENCES

1. DeHoff, R. T. and Rhines, F. N. *Quantitative Microscopy*. New York: McGraw-Hill, 1968.
2. Underwood, E. E. *Quantitative Stereology*. Reading, MA: Addison-Wesley Press, 1970.
3. Weibel, E. R. Vol. 1 of *Stereological Methods. Practice Methods for Biological Morphometry*. New York: Academic Press, 1979.
4. Serra, J. *Image Analysis and Mathematical Morphometry*. New York: Academic Press, 1981.
5. H. Elias, and Hyde, D. M. *A Guide to Practical Stereology*. New York: S. Krager AG, 1983.
6. DeHoff, R. T.; Aigeltinger, E. H.; and Craig, K. R. Experimental determination of the topological properties of microstructures. *Journal of Microscopy*. 95, pt 1: 69–81 (1972).
7. DeHoff, R. T. Quantiative serial sectioning analysis: Preview. *Journal of Microscopy* 131: 259–266 (1983).
8. DeHoff, R. T. Geometrical meaning of the integral mean surface curvature. In *Microstructural Science,* vol. 5, ed. J. L. McCall. New York: Elsevier-North Holland Press, 1977, pp. 331–345.
9. DeHoff, R. T. Integral mean curvature and platelet growth *Metallurgical Transactions* 10A: 1948 (1979).
10. Exner, H. E. Analysis of grain and particle size distributions in metallic materials. *Int. Met. Rev.* 17: 25–128 (1972).

8
QUANTITATIVE FRACTOGRAPHY

Ervin E. Underwood

School of Materials Engineering
Georgia Institute of Technology

The goal of quantitative fractography is to express the features and important characteristics of a fracture surface in terms of true surface areas, lengths, sizes, numbers, shapes, orientations, and locations, as well as distributions of these quantities. Modern quantitative image analysis systems play an important part in furthering these objectives, not only to speed up the measurement process but also to perform operations that would not be feasible otherwise.

With an enhanced capability for quantifying the various features in the fracture surface, we can better perform failure analyses, determine the relationship of the fracture mode to the microstructure, develop new materials, and evaluate their response to mechanical, chemical, and thermal environments.

BACKGROUND

Over the past few years, there has been an increased interest in using more quantitative geometrical methods to characterize the nonplanar surfaces encountered in fractures.[1] A concerted effort has produced gratifying advances in our capabilities to quantify the fracture surface and its features.[2-10] Although experimental methods are still being explored and general theoretical treatments are still lacking, there have been enough results by now to justify a review and summary of progress.

The earliest attempts toward a quantitative fractography were concerned with individual features. Simple measurements of height or depth or separation could be performed, either by shadowing, stereoscopy, or interferometry.[11,12] Clearly, these methods were applicable only to the simplest of geometric configurations, such as a hillock raised above a flat background.

Optical examination of fracture surfaces contributed significantly to our qualitative knowledge of fracture geometry,[13] but low magnifications and an unsatisfactory depth of field limited the potentialities of this type of technique. With the advent of fracture replicas and the TEM, however, great strides were made toward the resolution of fine detail and higher magnifications.[14,15] Quantification techniques were still limited to the more primitive measurements, particularly to the projected image (or photograph) of the replica.[16]

The general availability of the SEM opened up new avenues toward the understanding of fracture structures in three dimensions. Researchers were increasingly encouraged to make quantitative measurements on SEM photomicrographs. Although direct measurements on these projected images do not yield correct spatial information, a very large step forward was being taken toward quantification.

The basic problem is a very general one. Briefly put, we must know the area of a fracture surface in order to put numbers on the components that comprise the nonplanar surface. Once the area is known, we can proceed hopefully by a fairly simple and direct procedure, to determine the other quantities of interest in the fracture surface.

The more prominent current techniques for studying the fracture surface can be classified in accordance with the following:

1. Stereoscopic methods
2. Projected images
3. Profile generation

Each method has its strong points and weaknesses. Of course, as metallurgists, we need a procedure that will handle the complex and irregular fracture surfaces found in metals and alloys. This overriding requirement immediately narrows down the selection of a suitable experimental technique.

We would also like to utilize the basic equations of stereology to the utmost.[17] These statistically exact, assumption-free relationships apply equally as well to the flat figures in the plane of polish as they do to the spatial objects in three-dimensional sample space. If we make a planar cut through the fracture surface, a profile is obtained. We can measure its length, the angular distribution of its segments, its orientation characteristics, etc., as if it were a line in a plane. On the other hand, we may also wish to determine the area of the fracture surface. Unfortunately, the facets of a fracture surface are usually preferentially, rather than randomly, oriented. This fact presents a difficult sampling problem if the stereological relationships based on random sampling and measurements are to be used. For example, to determine the spatial angular distribution of facets, a prohibitive number of sampling planes at many angles and locations are theoretically required. Of course, alternatives to the random-orientation equations are available, such as those that express the surface area in terms of roughness parameters, the degree of orientation, etc. Moreover, an investigation of sampling requirements on a computer-simulated fracture surface indicates that relatively simple serial sectioning may be adequate for determining the fracture surface area.[8]

Because of the interest in sectioning procedures, methods for characterizing their resulting profiles have proliferated. Several related profile parameters have emerged that show considerable promise.[9,10,18] These include the roughness parameters, R_S, R_L, and R_P, which are based on areal, lineal, and point measurements, respectively, and the fractal dimension, $\mathcal{D}$. A number of models have been proposed in order to relate the profile parameter to the fracture surface area.[2,3,4,8]

Another path is being pursued to exploit the profile characteristics. Mathematical relationships are available that transform the angular distribution of linear elements in a profile to the angular distribution of facets in three-dimensional sample space.[19,20,21] This type of method will be discussed later in terms of the Scriven and Williams analysis.[19]

At the same time, some proposed relationships between the profile parameter and the fracture surface area are obviously incorrect. Moreover, other relationships based on randomly oriented surface elements should be used with care unless the assumption of randomness can be justified. Such relationships represent only a trivial solution to an unrealistic condition.

To summarize the situation, considerable work must still be done before we can efficiently estimate facet and dimple areas, striation spacings, interparticle distances, crack path length, and other attributes of features in the fracture surface.

EXPERIMENTAL TECHNIQUES

As indicated above, we can classify the methods that are currently most frequently used under the three categories: stereoscopic methods, projected images, and profile generation. These techniques will be discussed briefly in terms of their advantages and disadvantages, accompanied by a tabulation of the quantitative relationships available for each.

Stereoscopic Methods

Under this category, we will consider conventional stereoscopic imaging and measurements, photogrammetric methods, and deep etching on the surface.

Stereoscopic Imaging.
Stereoscopic pictures can be taken by SEM and TEM quite readily.[22] In any SEM picture, there are two main kinds of distortion: (1) perspective error caused by tilt of the surface, and (2) magnification errors arising from surface irregularities. The first type of error may be minimized by keeping the beam close to perpendicular to the fracture surface. The second can be understood by referring to Fig. 8-1, which is the rectilinear optical equivalent of the SEM image.[23]

Magnification is defined as the ratio of the image distance to the object distance. In the case of an irregular surface, the object distance is not constant. Consequently, high points on the surface have higher magnification than low points. For example, at point p in Fig. 8-1, the magnification is proportional to ss'/sm, whereas at point q it is proportional to ss'/sn.

Usually, the coordinates of the points in the fracture surface are measured by stereo-SEM pairs, i.e., two photographs of the same field taken at small tilt angles with respect to the normal. The geometry of this is shown in Fig. 8-2, where the two points A and B appear at A', B' and A'', B'' in stereo pictures taken at tilt angles $\pm\alpha$. The lengths $A'B'$ and $A''B''$ can be measured either from the two pictures separately or from the stereo image directly.[24] The difference, $A'B' - A''B''$, is called the *parallax*, Δx.

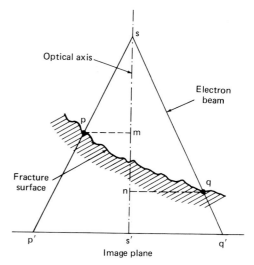

Fig. 8-1. Geometry of image formation in the SEM.

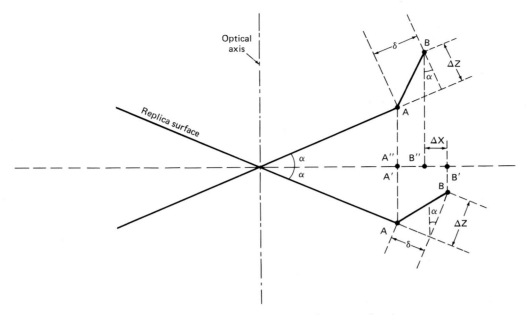

Fig. 8-2. Determination of parallax Δx from stereo imaging.

According to Fig. 8-2,

$$\Delta x = A'B' - A''B''$$
$$= (\delta \cos \alpha + \Delta z \sin \alpha) - (\delta \cos \alpha - \Delta z \sin \alpha) \quad (8\text{-}1)$$
$$= 2\Delta z \sin \alpha$$

in the scale of the micrograph. Rearranging gives the following:

$$\Delta z = \left[\frac{1}{2M \sin \alpha}\right] \Delta x \quad (8\text{-}2)$$

where the height difference, Δz, between two points is determined directly from the measured parallax, Δx. Dividing Δx by the average magnification, M, then gives Δz in terms of the actual height difference.[15,24]

Since the tilt angle is fixed for one pair of photographs, Eq. 8-2 may be rewritten as

$$\Delta z = K \Delta x \quad (8\text{-}3)$$

where K represents the constant terms. For a value of α of 10 degrees, the height value is approximately 2.88 times that of the corresponding measurement along the x or y direction.

Equation 8-2 is strictly correct for an orthogonal projection; that is, the point s in Fig. 8-1 is situated at infinity. This is a reasonable assumption at higher magnifications (>1000X), but at lower magnifications induced errors appear.[23] The x and y coordinate points can be measured directly with a superimposed grid or obtained automatically with suitable equipment.[25]

Once we have obtained the (x,y,z) coordinates at selected points on the fracture surface, elementary calculations can be made, such as the equation of a straight line or a planar surface, the length of a linear segment between two points in space, the angle between two lines or two surfaces (the dihedral angle), and so forth.[26] Some of these basic equations are collected for the reader's convenience in Table 8-1.

Photogrammetric Method. Another major procedure based on stereoscopic imaging utilizes modified photogrammetry equipment. Several reports from the Max-Planck Institut in Stuttgart have described the operation in detail.[5,25,27,28] Their instrument is a commercially available mirror stereometer with parallax measuring capability (adjustable light-point type). It is linked to a Videoplan and provides semiautomatic measurement of as many as 500 (x,y,z) coordinate points over the fracture surface. The output data generate height profiles or contours at selected locations, as well as the angular distribution of profile elements (see Fig. 8-3).[27] The accuracy of the z coordinates is better than 5 percent of the maximum height difference of a profile, whereas the measurement time is about 3 seconds per point.

Stereo-pair photographs with symmetrical tilts about the normal incident position are used to generate the stereoscopic model. The inbuilt "floating marker" (a point light source) of the stereometer is adjusted to lie at the chosen (x',y') position at the level of the fracture surface. Small changes in height are recognized when the marker appears to "float" out of contact with the fracture surface. The accuracy with which the operator can place the marker depends on his stereo acuity and amount of practice. When the floating marker is positioned on the surface, a push button sends all three spatial coordinates to the computer. A FORTRAN program then calculates

Table 8-1. Geometrical Relationships for Lines and Planes in Space.

- Distance l between two points, P_1 (at x_1,y_1,z_1) and P_2 (at x_2,y_2,z_2):

$$l = \sqrt{(x_2-x_1)^2 + (y_2-y_1)^2 + (z_2-z_1)^2}$$

- Direction angles of P_1P_2 between the line and the positive (x,y,z) axes, respectively: α, β, and γ

- Direction cosines of the two lines P_1P_2 and P_1P_3 with lengths l_1 and l_2:

$$\cos \alpha_1 = (x_2-x_1)/l_1 \quad \cos \alpha_2 = (x_3-x_1)/l_2$$
$$\cos \beta_1 = (y_2-y_1)/l_1 \quad \cos \beta_2 = (y_3-y_1)/l_2$$
$$\cos \gamma_1 = (z_2-z_1)/l_1 \quad \cos \gamma_2 = (z_3-z_1)/l_2$$

Note that $\cos^2 \alpha + \cos^2 \beta + \cos^2 \gamma = 1$.

- Angle θ between two intersecting lines in space:

$$\cos \theta = \cos \alpha_1 \cos \alpha_2 + \cos \beta_1 \cos \beta_2 + \cos \gamma_1 \cos \gamma_2$$

- In terms of coordinates rather than direction cosines, angle θ between the line segments from point P_1 (x_1,y_1,z_1) to points P_2 (x_2,y_2,z_2) and P_3 (x_3,y_3,z_3):

$$\cos \theta = [(x_2-x_1)(x_3-x_1) + (y_2-y_1)(y_3-y_1) + (z_2-z_1)(z_3-z_1)]/l_1 l_2$$

- Angle θ between the two planes, $A_1x + B_1y + C_1z + D_1 = 0$ and $A_2x + B_2y + C_2z + D_2 = 0$, is the angle between two intersecting lines, each perpendicular to one of the planes:

$$\cos \theta = \frac{A_1A_2 + B_1B_2 + C_1C_2}{\pm\sqrt{(A_1^2+B_1^2+C_1^2)(A_2^2+B_2^2+C_2^2)}}$$

where the quantities A, B, C, and D are constants whose values depend on the specific plane.

---

the three-dimensional coordinates and produces the corresponding profile or contour map on the plotter or screen.

This procedure is nondestructive of the fracture surface and also allows detailed scrutiny of some areas and wider spacings in regions of less interest. It is still a point-by-point method, however, that relies heavily on operator skill and training. No information is provided on any subsurface cracking or possible interactions between microstructure and fracture path. Overlaps and complex fracture surfaces are difficult to handle, if they can be handled at all. Moreover, smooth or structureless areas cannot be measured unless a marker is added.[28] This method appears best suited for relatively large facets and for samples that must be preserved.

Deep Surface Etching. Another procedure that uses stereoscopic viewing or single SEM images is based on heavy surface etching, such as bromine etching of Al–

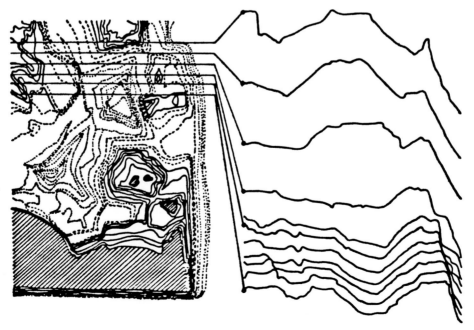

Fig. 8-3. Contour map and profiles obtained by photogrammetry.[27]

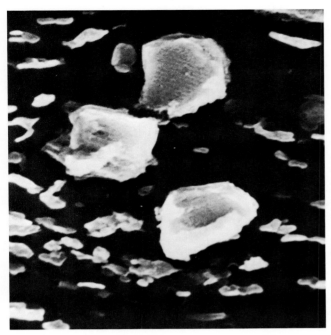

Fig. 8-4. Deep surface etching of Al-Li alloy using bromine etch.

Li alloys to reveal the surface particles (see Fig. 8-4).[29] Frequently it is possible to etch down about two to three layers of surface particles without removing them. In this case, an estimated value for N_V can be based on the general stereological equation,

$$N_V = N_A/\overline{D} \qquad (8\text{-}4)$$

for convex particles cut by a flat plane of polish.[17] Here N_V is the number of particles per unit volume; N_A, the number of particle sections per unit area; and $\overline{D}$, the mean tangent diameter. The latter can be determined readily from the SEM photographs, but the appropriate test volume is not immediately obvious.

The applicable equation for the surface-deep etching case is derived readily. We start with the equality, $N_3 = N'$, which reflects the fact that the number of particles in three-dimensions is the same as the number of particle images observed on the SEM photograph. Dividing by the general equation for a test volume, $V_T = A_T' t$, we get

$$\frac{N_3}{V_T} = \left(\frac{N'}{A_T'}\right)\frac{1}{t} \qquad (8\text{-}5)$$

or

$$N_V = N_A'/t \qquad (8\text{-}6)$$

The trick is to estimate the effective thickness of the test volume, t. Based on the number of particle layers observed, and their disposition with respect to one another, we can express t in terms of $\overline{D}$. For example, three layers of particles would correspond to a thickness t between $2\overline{D}$ and $3\overline{D}$, depending on their stacking. Thus, for this case, we could write

$$N_V = N_A'/2.5\overline{D} \qquad (8\text{-}7)$$

Other values of t for Eq. 8-6 would depend on the particular set of circumstances, including the number of layers, the particle size, and the type of stacking.

Projected Images

Although stereo-pairs are obtained readily with both SEM and TEM, single SEM images are also used extensively in fractographic studies. The vertical photographs permit some relative assessments to be made, and numerical (counting) results are quite common. The single SEM photographs are also used extensively for detailed qualitative interpretations concerning the fracture mechanisms and features involved in the fracture process.[14,15]

Other studies have attempted to extract more quantitative information from these planar projected images.[16,37] The analysis was largely two-dimensional since general stereological relationships bearing on the third dimension were not available.

In the absence of better procedures, these early attempts were laudatory. With the development of the relationships of projection stereology,[17,30] however, we now have additional tools at our disposal. Even so, these are not enough. Measurements restricted to SEM photomicrographs are limited in the amount of information they can provide about nonplanar surfaces. An additional degree of freedom, such as provided by stereometry or profile analysis, is necessary to generalize the three-dimensional spatial analysis.

Single SEM Image. What then is possible if we are restricted solely to the flat SEM or TEM picture? Basically, in a two-dimensional surface, we can measure only areal, lineal, or numerical quantities. These include projected areas, lengths, sizes, spacings, numbers, and more sophisticated parameters that deal with distributions, locational analyses, shapes, and orientational tendencies. No information is possible about the features in three-dimensional sample space except by making *a priori* assumptions.

Representative stereological relationships that apply solely to the projected images are listed in Table 8-2.[17,30] Part I pertains to individual projected features, and Part II applies to systems of projected features. Figure 8-5 portrays schematically the quantities involved in measurements on individual projected features. The equations in Part I of Table 8-2 that contain $\overline{d}'$, the mean tangent diameter, are valid only for convex figures, but there are alternative ways for obtaining the same quantity in some

Table 8-2. Stereological Relationships for Projected Images of Fracture Features in a Two-Dimensional Plane.*

PART I. FOR INDIVIDUAL PROJECTED FEATURES†

L'	Length of linear feature	$L' = (\pi/2)\,\overline{P_L'(\theta)}\,A_T'$
$\overline{L_P'}$	Perimeter length of closed figure	$\overline{L_P'} = (\pi/2)\,\overline{P_L'(\theta)}\,A_T'$
$\overline{L_2'}$	Mean intercept length of closed figure	$\overline{L_2'} = \pi A'/L_P' = 2\,\overline{P_P'(\theta)}/\overline{P_L'(\theta)}$
A'	Area of closed figure	$A' = \overline{L_2'}L_P'/\pi = \overline{P_P'(\theta)}\,A_T'$
Convex Figures		
L_p'	Perimeter length of convex figure	$L_p' = \pi\overline{d'}$
$\overline{L_2'}$	Mean intercept length of convex figure	$\overline{L_2'} = A'/\overline{d'}$
$\overline{k_L'}$	Mean curvature of convex figure perimeter	$\overline{k_L'} = 2/\overline{d'} = 1/\overline{r'}$
$\overline{d'}$	Mean tangent diameter	$\overline{d'} = 2\,\overline{r'}$

PART II. FOR SYSTEMS OF PROJECTED FEATURES‡

$\overline{L'}$	Mean length of discrete lineal features	$\overline{L'} = L_A'/N_A'$
$\overline{L_p'}$	Mean perimeter length of closed figures	$\overline{L_p'} = L_A'/N_A'$
$\overline{L_2'}$	Mean intercept length of closed figures	$\overline{L_2'} = L_L'/N_L'$
$\overline{A'}$	Mean area of closed figures	$\overline{A'} = A_A'/N_A'$
A_A', L_L', P_P'	Areal, lineal, and point fractions	$A_A' = L_L' = P_P'$
L_A'	Line length per unit area	$L_A' = (\pi/2)\,P_L'$
λ'	Mean free distance	$\lambda' = (1 - A_A')/N_L'$
Convex Figures		
$\overline{k_L'}$	Mean curvature of perimeters of convex figures	$\overline{k_L'} = 2\,N_A'/N_L'$
$\overline{d'}$	Mean tangent diameter of convex figures	$\overline{d'} = N_L'/N_A'$

* Primes denote projected quantities or measurements made on projection planes.
† A_T' is the selected test area in the projection plane.
  $\overline{P_L'(\theta)}$ is the number of intersections of a line per unit length of test line, averaged over angular placements θ of test line.
‡ P_L' is the number of intersections of a line per unit length of test line.
  N_L' is the number of interceptions of a feature per unit length of test line.
  N_A' is the number of sections of a feature per unit area of test plane.

---

cases. For example, $\overline{L_p'}$, the mean projected perimeter length, can be expressed in two ways, as follows:

$$\overline{L_p'} = \pi\overline{d'} \quad (8\text{-}8)$$

or

$$\overline{L_p'} = \frac{\pi}{2}\,\overline{P_L'(\theta)}\,A_T' \quad (8\text{-}8a)$$

Equation 8-8 is valid only for convex figures, whereas Eq. 8-8a is applicable for planar figures of any configuration (concave and convex perimeter elements), although a test area A_T' such as that shown in Fig. 8-5a is required. This is an arbitrary choice, however, and A_T' has a known value. The $\overline{P_L'(\theta)}$ quantity refers to the mean number of intersections of a linear grid with the perimeter at several angular placements, θ, per unit length of the linear grid. Details on these kinds of grid measurements can be found in Reference 17. Of course, if image analysis equipment with operator-interactive capability is available, most of the quantities listed in Table 8-2 can be measured directly. Note that the quantities in Part II can all be obtained by simple counting measurements.

When many individual measurements are required on a system of features (such as for a size distribution), more labor is involved. If appropriate or available, semiautomatic image-analysis equipment may be used in such cases. For example, to determine an areal size distribution of facets in Fig. 8-6, the perimeters of each facet in the SEM photomicrograph would be traced with the electronic pencil. With typical equipment, such as the Zeiss Videoplan, a printout would be produced with the data in the form of a histogram, along with other statistics.

It should be emphasized that the equations given in Table 8-2 are merely descriptive of features in projected images, not the true spatial quantities. Nevertheless, in lieu of anything better, the two-dimensional results are

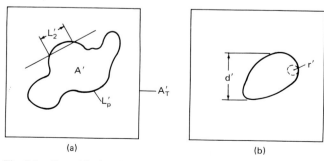

Fig. 8-5. Quantities involved in measurements on individual projected features of (a) closed figure of any configuration, and (b) convex figure where: L_P' = perimeter length; L_2' = intercept length; A' = area of figure; A_T' = test area, arbitrary size and shape; d' = tangent diameter; and r' = radius of curvature at a point.

QUANTITATIVE FRACTOGRAPHY 107

Fig. 8-6. SEM projection of facets in the fracture surface of an Al–4-percent Cu alloy.

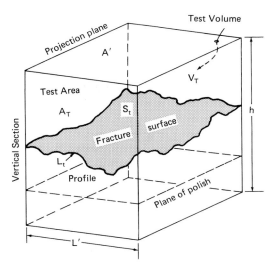

Fig. 8-7. Arbitrary test volume enclosing a fracture surface showing relationship to the projection plane and plane of polish and to the vertical section, where s_t = true fracture surface area; A' = projected area of fracture surface; L_t = true length of profile in vertical section; L' = projected length of profile; V_T = test volume, equal to $A'h$; and A_T = test area in vertical section, equal to $L'h$.

valid for that particular fracture surface and may even be compared with fracture surfaces of comparable roughness.

Assumption of Randomness. It would be better if the true magnitudes of areas, lengths, sizes, distances, etc., were always obtainable. If this is not possible, however, one can make order of magnitude calculations by assuming that the fracture surface is composed of randomly oriented elements. With this assumption, another set of stereological equations become available for the case when only one (or a limited number) of projection planes are available.[17] Figure 8-7 defines some terms and relationships involved in projections from a fracture surface.

Basic equations that can be used in this case are listed in Table 8-3. Note that the quantities on the left-hand side of the equations pertain to the fracture surface. Thus, instead of L_A, for example, we need L_S, which represents the length of the line per unit area of the (curved) fracture surface. Figure 8-8 points out graphically the differences between L and L_c and the relationship between S and A' for the fracture surface. Note also that L is a true line length and that $\overline{L}_3$ is the important mean intercept length through bounded regions in the fracture surface. Other, more specialized equations can be derived as required from the basic relationships in Table 8-3.

Assumption of No Correlation. Instead of assuming complete randomness of features in the fracture surface, we can assume that there is no correlation between the fracture surface configuration and the underlying microstructure. This, however, is usually a rather poor assumption. For example, we know that the fracture path frequently passes preferentially through particles, rather than along a random path through the material. However, if we can assume no correlation exists, the measurements of A_A', L_L', and P_P' on the SEM fractograph should yield the same result as from the conventional metallographic plane of polish.[23] That is, for a two-phase structure,

$$\begin{aligned} A_A' &= A_A \\ L_L' &= L_L \\ P_P' &= P_P \end{aligned} \quad (8\text{-}9)$$

and, of course,

$$A_A = L_L = P_P \quad (8\text{-}9\text{a})$$

Note that these equations are related to the bulk microstructural properties and are not necessarily the same for the fracture surface. Since the three projected quanti-

Table 8-3. Stereological Relationships between Spatial Features in a Random Fracture Surface and Their Projected Images.*

	PART I. FOR INDIVIDUAL FEATURES IN THE FRACTURE SURFACE	
L	Length of linear feature (fixed direction in projection plane)	$L = (\pi/2)L'$
L_c	Length of curved line (variable direction in projection plane)	$L_c = (4/\pi)L'$
L_p	Perimeter length of closed figure	$L_p = (4/\pi)L_p'$
$\overline{L_3}$	Mean intercept length of closed figure of area S	$\overline{L_3} = (\pi^2/2)\overline{A'}/\overline{L_p'}$
S	Area of curved figure (no overlap)	$S = 2\overline{A'}$
Flat Facets		
A	Area of flat figure in fracture surface	$A = 2\overline{A'}$
$\overline{d}$	Mean tangent diameter of flat figure	$\overline{d} = (4/\pi^2)\overline{L_p'}$
$\overline{k_L}$	Mean curvature of flat figure perimeter	$\overline{k_L} = \pi^2/2\overline{L_p'}$

	PART II. FOR SYSTEMS OF FEATURES IN THE FRACTURE SURFACE	
$\overline{L}$	Mean length of discrete linear features (each with fixed direction in projection plane)	$\overline{L} = (\pi/2)L_A'/N_A'$
$\overline{L_c}$	Mean length of curved lines (each with variable direction in projection plane)	$\overline{L_c} = (4/\pi)L_A'/N_A'$
$\overline{L_p}$	Mean perimeter length of closed figures	$\overline{L_p} = (4/\pi)L_A'/N_A'$
$\overline{L_3}$	Mean intercept length for closed figures (fixed direction of test line in projection plane)	$\overline{L_3} = (\pi/2)L_L'/N_L'$
$\overline{S}$	Mean area of curved closed figures (no overlap)	$\overline{S} = 2A_A'/N_A'$
λ	Mean free distance (fixed direction of test line in projection plane)	$\lambda = (\pi/2)(1 - A_A')/N_L'$
S_S	Area fraction of areal features (no overlap)	$S_S = A_A' = L_L' = P_P'$
L_S	Length of line per unit area of fracture surface (variable direction in projection plane)	$L_S = (2/\pi)L_A'$
P_L	Number of intersections per unit length of test line in the fracture surface (fixed direction in projection plane)	$P_L = (2/\pi)P_L'$
N_L	Number of interceptions of features per unit length of test line in the fracture surface (fixed direction in projection plane)	$N_L = (2/\pi)N_L'$
P_S	Number of point features per unit area of fracture surface	$P_S = P_A'/2$
N_S	Number of figures per unit area of fracture surface	$N_S = N_A'/2$

* Primes denote projected quantities or measurements made on projection planes. (Also see footnotes to Table 8-2.)

ties in Eq. 8-9 are dimensionless ratios, they are thereby independent of magnification and distortion in the image.

From the preceding discussion, it can be seen that considerable quantification is possible from SEM photomicrographs in two ways. First, calculations can be made in the plane of projection alone, without any attempt to convert to three-dimensional quantities (Table 8-2); and second, spatial quantities in the fracture surface can be calculated provided we can reasonably assume random orientation of surface elements (Table 8-3).

This second approach is admittedly on rather shaky ground when one considers the strongly oriented nature of a fracture surface. The next section, whereby partially oriented surfaces can be treated in a more quantitative manner than heretofore, addresses this problem to some extent.

Profile Generation

The analysis of the fracture surface by means of profiles appears more amenable to quantitative treatment than by other methods. Profiles are essentially linear in nature, as opposed to the two-dimensional sampling of SEM pictures and the point sampling involving photogrammetry procedures. Several types of profiles can be generated either directly or indirectly, but here we will discuss only three: those selected by metallographic sectioning, by nondestructive methods, and by sectioning of fracture-surface replicas.

Metallographic Sectioning Methods. Although many kinds of sections have been explored—e.g., vertical,[31,32] slanted,[33,34,35] horizontal,[36,37] conical,[38] etc.—we

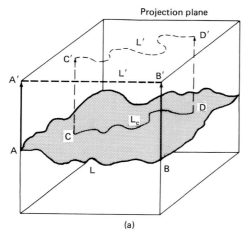 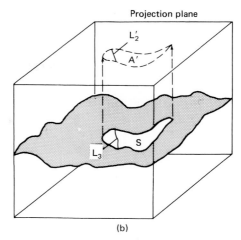

Fig. 8-8. Clarification of the difference between related features in the fracture surface and projection plane: (a) Line L in a fixed direction in the projection plane and curved line L_c with variable direction in the projection plane. (b) Feature of area S in the fracture plane and projected area A', where L_3 is the intercept length in the fracture surface and L_2' is the projected length.

will confine our discussion to planar sections. The major experimental advantages accruing to planar sections are that they are obtained from ordinary metallographic mounts and that any degree of complexity or overlap of the fracture profile is accurately reproduced. Moreover, serial sectioning is quite simple and direct; planar sections reveal the underlying microstructure and its relation to the fracture surface; the standard equations of stereology are rigorously applicable on the flat section; and the angular characteristics of the fracture trace (or profile) can be related mathematically to those of the surface facets.

Some objections have been raised to planar sectioning on the grounds that (1) it is destructive of the fracture surface and sample; (2) the fracture surface must be coated with a protective layer before sectioning in order to preserve the trace; and (3) in retrospect, additional detailed scrutiny of special areas on the fracture surface is not possible once it is cut. Bearing these objections in mind, however, if one carefully inspects the fracture surface by SEM in advance, areas of interest can be recorded photographically before coating and cutting takes place.

The preliminary experimental procedures are straightforward. When the specimen is fractured, two (ostensibly matching) nonplanar surfaces are produced. After inspection and photography by SEM, one or both surfaces can be coated electrolytically to preserve the edge upon subsequent sectioning.[39] The coated specimen is then mounted metallographically and prepared according to conventional metallographic procedures.[39] One fracture surface can be sectioned in one direction and the other at 90 degrees to the first direction, if desired.

Once the profile has been clearly revealed and the microstructure underlying the crack path properly polished and etched, measurements can begin. Photographs of the trace or microstructure can be taken in the conventional way for subsequent stereological measurements. With presently available commercial image-analysis equipment, one can by-pass photography, however, and work directly from the specimen, thereby greatly improving sampling statistics, costs, and efficiency.

Measurements that are possible with a typical semiautomatic image analysis system, such as the Zeiss Videoplan, include coordinates at preselected intervals along the trace, angular distributions, true profile length, and fractal data.[18] If the unit is interfaced with a large central computer, printouts and graphs are also readily available.[10] After the desired basic data have been acquired, the analysis of the profile and calculation of three-dimensional properties can proceed.[9]

Nondestructive Profiles. The sectioning methods described above are basically destructive of the fracture surface. What we would like is a nondestructive way to generate profiles representative of the fracture surface. Then the wealth of detail available in the flat SEM photomicrograph could be supplemented by the three-dimensional information inherent in the profile.

Comparatively simple procedures for generating profiles nondestructively across the fracture surface are available. One attractive method provides profiles of light using the Leitz Light-Section Microscope.[40] A narrow illuminated line is projected onto the fracture surface and reveals the profile of the surface in a specified direction. The line can be observed through the light microscope and photographed for subsequent analysis. Profile "roughness" is reported as the mean peak-to-trough distance, with values quoted of around 50 ± 5 μm for a 1080 steel.[40] The resolution obtained depends on the objective lens optics. Unfortunately, the other quantitative profile roughness parameters were not investigated.

In a second method, a narrow contamination line is deposited on the specimen across the area of interest, using the SEM linescan mode.[41] Then the specimen is tilted, preferably around an axis parallel to the linescan direction, by an angle α. The contamination line then appears as an oblique projection of the profile that can be evaluated (point by point) by the relationship

$$\Delta z = \left[\frac{1}{M \cos(90° - \alpha)}\right] \Delta k \tag{8-10}$$

where M is the magnification; Δk, the displacement of the contamination line at the point of interest; and Δz, the corresponding height of the profile at that point.

The tilt angle should be as large as possible.[42] The major source of error comes from broadened contamination lines with diffuse contours.[27] (Fortunately, the lines appear considerably sharper in the tilted position.) Another drawback to this method is that it cannot be used on surfaces that are too irregular. Other than for these problems, however, it appears that both the contamination line and light-beam profiles possess useful attributes that should be investigated more thoroughly for the study of metallic fractures.

Profiles from Replicas. Another way to generate profiles nondestructively is to section replicas of the fracture surface.[15] The major concern here is to minimize distortion of the slices during cutting of the replica. Recent experimental studies have established suitable foil material, as well as procedures for stripping and coating the replicas.[27] After the replica has been embedded, parallel cuts are made with an ultramicrotome. In one investigation, a slice about 2 μm thick was found to be optimum as far as resolution was concerned.[27] Although thicker slices deformed less, the resolution was also less.

The slices obtained by serial sectioning can be analyzed to give the spatial coordinates of a fracture surface. Coordinates along several profiles were obtained using stereogrammetry, as described above. The right hand side of Fig. 8-3 depicts six vertical sections spaced 25 μm apart (those with dots) and seven vertical sections that are spaced 5 μm apart. On the left-hand side of Fig. 8-3, the locations of the more widely spaced sections are shown with respect to a contour map of the same surface.

This method of obtaining profiles nondestructively from a fracture surface looks rather useful. It is limited, of course, to relatively smooth surfaces. If only R_L is sought, however, the replica sectioning method appears to be a viable way to quantify the usual measurements made from a flat SEM photomicrograph.

ANALYTICAL PROCEDURES

Roughness Parameters

Since profiles are easily obtained and experimentally accessible, it is natural that considerable attention has centered on their properties. Several types of roughness measurements, quantities, and parameters have been evaluated as to their suitability for characterizing an irregular planar curve.[4,43] From these evaluations, a few parameters have emerged that possess potential usefulness for quantitative fractography.[10,18] We particularly favor those profile parameters that express "roughness" well, relate readily to the physical situation, and equate to spatial quantities.

Profile and Surface Roughness Parameters. Basically, the profile parameters consist of ratios of lengths or points of intersection. They are consequently dimensionless and do not vary with size for the same shape curve. We select here for definition and further discussion four profile parameters and one surface-roughness parameter.

1. R_L, the lineal roughness parameter,[31] is equal to the true profile length divided by the projected length, or

$$R_L = L_t/L' \tag{8-11}$$

(See Fig. 8-9.) For a profile consisting of randomly oriented linear segments, $R_L = \pi/2$.

2. R_P, a profile roughness parameter due to Behrens,[44] is equal to the average height divided by the average spacing of peaks, or

$$R_P = \frac{1}{2 L_T} \int_{y_1}^{y_2} P(y) \, dy. \tag{8-12}$$

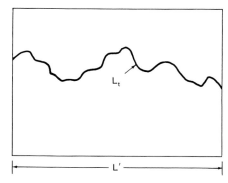

Fig. 8-9. $R_L = L_t/L'$ for an irregular planar curve, where R_L is the lineal roughness parameter, L_t is the true length and L' is the projected length.

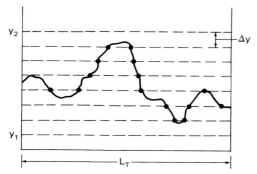

Fig. 8-10. Profile roughness parameter R_P is equal to the average height divided by the average spacing of peaks. Here, Δy is the displacement distance of the test line, and y_1 and y_2 are positions of the test line where there are no intersections.

where y_1 and y_2 give positions of the test line below and above the profile where no intersections are possible. The working equation can be expressed by

$$R_P = \frac{\Delta y}{2 L_T} \sum_i P_i \qquad (8\text{-}12a)$$

where Δy is the constant displacement of the test line of length L_T, and ΣP_i is the total number of intersections of the test line with the profile (see Fig. 8-10).

3. R_V, the vertical roughness parameter as defined by Wright and Karlsson,[4] is the sum of the projected hill heights on the vertical projection line ($\perp$), divided by the projected length of the full curve on the horizontal projection line ($\parallel$), or

$$R_V = \Sigma(L_i')^\perp / (L')^\parallel \qquad (8\text{-}13)$$

(see Fig. 8-11a). The equivalent working equation is the ratio of intersection points projected from the profile to the vertical projection line ($\perp$), to the intersection points projected from the profile to the horizontal projection line ($\parallel$). A parallel line grid is used in both the vertical and horizontal directions. The result is

$$R_V = \Sigma(P_i')^\perp / \Sigma(P_i')^\parallel \qquad (8\text{-}13a)$$

(see Fig. 8-11b and c). Note that the length of the intersection grid is immaterial as long as it covers the test area in both positions.

4. $\mathscr{D}$ is the fractal dimension of an irregular planar curve as described by Mandelbrot.[45] Its value is obtained from the slope of the log-log plot of the expression

$$L(\eta) = L_o \eta^{-(\mathscr{D}-1)} \qquad (8\text{-}14)$$

where $L(\eta)$ is the apparent length of the profile and L_o is a constant with dimensions of length. The apparent profile length varies inversely with the size of the measurement unit, η, and $\mathscr{D}$ can have fractional values equal to 1 or greater. If both sides are divided by the projected length of the $L(\eta)$ curve (a constant), Eq. 8-14 can be simplified to

$$R_L(\eta) = (L_o/L')\eta^{-(\mathscr{D}-1)} \qquad (8\text{-}14a)$$

or to the linear form,

$$\log R_L(\eta) = \log(L_o/L') - (\mathscr{D}-1)\log \eta \qquad (8\text{-}14b)$$

where the (negative) slope is equal to $-(\mathscr{D}-1)$. Figure 8-12 portrays the essential relationship of these quantities.

These parameters are interrelated to some extent. For example, R_L and $\mathscr{D}$ give qualitatively similar experimental curves, whereas R_L is equal to πR_P for a profile composed of randomly oriented linear elements. Under some circumstances, Wright and Karlsson's stepped fracture surface model[4] yields the relation, $R_V = R_L - 1$.

The major interest in these profile roughness parameters lies in their possible relationships to the fracture sur-

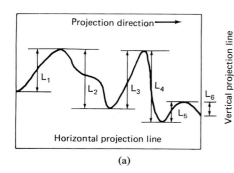
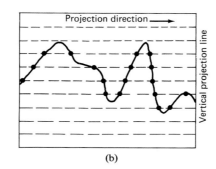
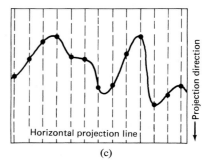

Fig. 8-11. Vertical roughness parameter R_V equal to the total projection of the curve to a vertical projection line, divided by the projection of the curve to a horizontal projection line: (a) R_V in terms of lengths, and (b) and (c) procedure to obtain R_V by projected points of intersection.

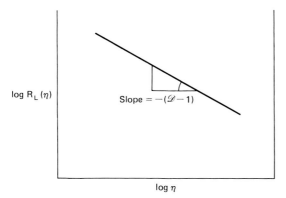

Fig. 8-12. Fractal dimension $\mathscr{D}$ and its relationship to the slope of a log-log plot of $R_L(\eta)$ vs. η.

face area. A surface roughness parameter, R_S,* can be defined in terms of true surface area, S_t, divided by its projected area, A', as follows:

$$R_S = S_t/A' \qquad (8\text{-}15)$$

or, in an alternate form, as

$$R_S = S_V/A_V' \qquad (8\text{-}15a)$$

(see Fig. 8-7). For a randomly oriented fracture surface with no overlap, $R_S = 2$.

The profile and fracture surface roughness parameters are related by the coefficient $4/\pi$ in the equation,

$$R_S = (4/\pi) R_L \qquad (8\text{-}16)$$

provided completely random angular sampling of the fracture surface elements is achieved. This result comes about readily from the general stereological equation,

$$S_V = (4/\pi) L_A \qquad (8\text{-}17)$$

Since $S_V = S_t/V_T$ and $L_A = L_t/A_T$, by substituting $V_T = A'h$ and $A_T = L'h$, we obtain Eq. 8-16. Actually, Eq. 8-17 and other equations based on the assumption of a randomly oriented fracture surface, or on complete randomness of sampling, should not be used unless justified.

At this point, it may be appropriate to summarize the more useful expressions between quantities in the fracture plane and the roughness parameters, R_L and R_S. A listing for partially oriented structures is given in Table 8-4. It will be recalled that Table 8-3 tabulates relationships for randomly oriented structures. Here, for the partially oriented fracture surfaces, directionality is very important. Thus, the linear quantities in Table 8-4—which are defined for a fixed direction ϕ in the projection plane, e.g., $[R_L]_\phi$—would have different values for different directions of ϕ. Other quantities are averages of measurements made at several angles of ϕ in the projection plane, e.g., $\overline{P_P{}'(\phi)}$. Note also that some expressions in Table 8-4 are stated in terms of R_S. If R_L is known, R_S can be calculated (see Table 8-5).

Parametric Relationships between R_L and R_S

A more difficult problem is that of relating surface roughness and profile roughness if they are not random but are oriented or partially oriented instead. Several attempts have been made to relate R_S and R_L for nonrandom surface configurations, but they do not agree too well. Table 8-5 lists the principle parametric equations proposed to date. Parametric Eq. No. 1 in Table 8-5 represents a rather complicated solution based on a "stepped surface" model. Other equations by the same author[2] were proposed for specified surface geometries—i.e., flat, random, stepped, and other shapes—with different solutions depending on whether $R_S \gtreqless 2$.

Parametric Eq. No. 2 in Table 8-5 was developed on a model-free basis for all configurations extending from a completely oriented surface to surfaces with any degree of roughness, including randomly oriented surface elements.[8] It is based on the very general Eq. 8-17 and the equation for a completely oriented surface,

$$(S_V)_{or} = (L_A)_{or} \qquad (8\text{-}18)$$

Note that both R_S and R_L equal 1 in this case.

Parametric equations Nos. 3, 4, and 5 in Table 8-5 were derived by Wright and Karlsson.[4] Equations Nos. 3 and 5 are based on a linear model and quadratic model, respectively, and represent upper and lower limits to R_S. Parametric Eq. No. 4 represents a linear relationship between the limits of completely oriented and completely random surfaces, with the intermediate surfaces composed of various area fractions of random and flat components. It is identical to parametric Eq. No. 6, which was derived by Coster and Chermant[3] on the same basis as Eq. No. 4. The major shortcoming of parametric Eqs. No. 4 and 6 is that they are valid only for values up to $R_S = 2$, a condition similar to that imposed on parametric Eq. No. 1.

Parametric Eqs. Nos. 2, 3, and 5, on the other hand, extend indefinitely. This characteristic is correct, of course, because R_L and R_S can increase without limit. The general stereological equation, Eq. 8-17, moreover, does not represent a cut-off point (or an upper limit) because it applies to any surface configuration, provided each element of the surface is sampled randomly. The completely random fracture surface is accordingly only

---

* The earlier literature used other symbols for R_S—e.g., K_A, R_A, and S_A. We prefer R_S because of its position with regard to R_L and R_P.

Table 8-4. Parametric Relationships Between Spatial Features in a Partially Oriented Fracture Surface and Their Projected Images.*

PART I. FOR INDIVIDUAL FEATURES IN THE FRACTURE SURFACE†

L	Length of discrete linear feature (fixed direction ϕ in projection plane)	$L = [R_L]_\phi L' = \left(\dfrac{\pi}{2}\right)[R_L]_\phi \overline{P_L'(\phi)} A_T'$
S_f	Area of curved closed figure with no overlap (ϕ represents different angular positions of grid in projection plane)	$S_f = (R_S)_f A_f' = (R_S)_f \overline{P_P'(\phi)} A_T'$
A_f	Area of flat closed figure	$A_f = (R_S)_f A_f' = (R_S)_f \overline{P_P'(\phi)} A_T'$

PART II. FOR SYSTEMS OF FEATURES IN THE FRACTURE SURFACE‡

$\bar{L}$	Mean length of discrete linear features (all oriented in same direction ϕ in projection plane)	$\bar{L} = [R_L]_\phi \bar{L}' = [R_L]_\phi L_A'/N_A'$
$\bar{S_f}$	Mean area of curved closed figures (no overlap)	$\bar{S_f} = R_S \bar{A_f'} = R_S (A_A')_f/(N_A')_f$
λ	Mean free distance between features (fixed direction ϕ in projection plane)	$\lambda = \dfrac{[R_L]_\phi - [(R_L)_f]_\phi (L_L')_f}{N_L'}$
$(L_L)_f$	Intercept length fraction of features (fixed direction ϕ in projection plane)	$(L_L)_f = [(R_L)_f]_\phi (L_L')_f/[R_L]_\phi$
$(S_S)_f$	Area fraction of areal features (no overlap)	$(S_S)_f = (R_S)_f (A_A')_f/R_S$
N_L	Number of interceptions of features per unit length of test line in fracture surface (fixed direction ϕ in projection plane)	$N_L = [N_L']_\phi/[R_L]_\phi$
N_S	Number of features per unit area of fracture surface	$N_S = N_A'/R_S$
L_S	Length of linear features per unit area of fracture surface (fixed direction ϕ in projection plane)	$L_S = [(R_L)_f]_\phi (L_A')_f/R_S$

* Primes denote projected quantities or measurements made on projection planes. (Also, see footnotes to Table 8-2.)
† $R_S = S_{\text{total}}/A'_{\text{total}}$ and $(R_S)_f = S_f/A_f'$.
‡ $R_L = L_{\text{profile}}/L'_{\text{profile}}$ and $(R_L)_f = L_f/L_f'$.

an intermediate condition between the perfectly flat surface and the extremely elongated surface (large surface area).

The curves for the six parametric equations in Table 8-5 are shown in Fig. 8-13, with all known pairs of data points (R_S and R_L) entered on the graph. There are six points for 4340 steels,[10] four points for Al–4-percent Cu alloys,[9] and eight points* for a ceramic consisting of Al_2O_3 and 3-percent glass.[5] The data points appear to favor curve No. 2 since they fall to both sides of it; however, more data are needed to make any clear-cut decision. The credibility of parametric equation No. 2 was enhanced using a computer-simulated fracture surface of known surface area.[8] The surface area calculated according to Eq. No. 2 agreed within about 1 percent. Thus, it is felt that the approach adopted in deriving Eq. No. 2 is fundamentally sound.

Estimation of Fracture Surface Area

There are other ways to determine the actual fracture surface area than by using a roughness parameter equation. A real surface can be approximated with arbitrary precision by triangular elements of any desired size. Another method is based on the relationship between the angular distribution of linear elements along a profile and the angular distribution in space of the surface elements of the fracture surface. These procedures will be discussed next.

Triangular Elements. One approach utilizes serial sections and a triangular network constructed between adjacent profiles.[5] The individual triangular facets are formed by two consecutive points in one profile and the "closest point" in the adjacent profile, giving about 50 triangular facets per profile. The estimated total facet surface area is given by

$$S_\Delta = \sum_{i,j} a_{i,j} \qquad (8\text{-}19)$$

A similar procedure is being evaluated elsewhere in which the triangular facets between adjacent profiles are formed automatically at 500 regularly spaced coordinate points along each profile.[46] The accuracy of these triangle approximation methods depends, of course, on the closeness

---
* These eight points are the original unaveraged data points, without an applied correction factor.

Table 8-5. Surface Roughness R_S of Partially Oriented Fracture Surfaces in Terms of the Profile Roughness R_L*

	RANGE OF VALIDITY	MODEL
EL-SOUDANI[2] No. 1: $R_L = \left(\dfrac{\pi}{4}\right) \dfrac{[R_S(2-R_S)]^{1/2}}{\tan^{-1}[(2-R_S)/R_S]^{1/2}}$	$1 \leq R_S \leq 2$	Based on "stepped" surface†
UNDERWOOD & BANERJI[8] No. 2: $R_S = (4/\pi)(R_L - 1) + 1$	$R_S \geq 1$	Linear relationship between oriented surface and any randomly sampled surface
WRIGHT & KARLSSON[4] No. 3: $R_S = \left(\dfrac{\pi}{2}\right)(R_L - 1) + 1$ (lower limit)	$R_S \geq 1$	Based on "stepped" surface†
No. 4: $R_S = \left(\dfrac{2}{\pi - 2}\right)(R_L - 1) + 1$ (intermediate curve)	$1 \leq R_S \leq 2$	Based on linear approximation between RNO‡ surface and flat surface
No. 5: $R_S^2 = \left(\dfrac{\pi}{2}\right)^2 (R_L^2 - 1) + 1$ (upper limit)	$R_S \geq 1$	Based on surface of triangular elements
COSTER & CHERMANT[3] No. 6: $R_S = \dfrac{R_L + \left(\dfrac{\pi}{2}\right) - 2}{\left(\dfrac{\pi}{2}\right) - 1}$	$1 \leq R_S \leq 2$	Based on linear approximation between an ideally flat and a random surface

* No overlap.
† A "stepped" surface has a profile with square corners on its hills and valleys, for example:
‡ An "RNO" surface is defined as having a uniform angular distribution of surface element normals in space.

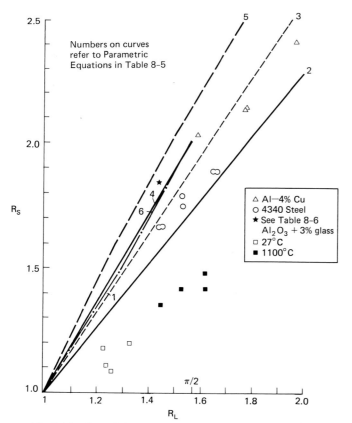

Fig. 8-13. Plot of six parametric equations relating R_S and R_L.

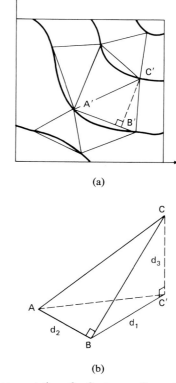

Fig. 8-14. Representation of a fracture surface using triangular elements between contour lines: (a) projected view of contours and triangles, and (b) three-dimensional view of triangle ABC.

of the serial sections and the number of coordinate points in relation to the complexity of the profile.

Wright and Karlsson[4] also propose an approximate surface composed of triangular elements but use a contour map rather than serial sections to form their triangles. The triangles are connected to adjacent contour lines, then divided into two right-angle subtriangles as portrayed in Fig. 8-14. The area of the triangular element ABC can be calculated since the sides $\overline{AB}$ and $\overline{BC}$ are determinable. The area of the entire model surface can then be obtained by summation.

The roughness parameter, R_S, for the triangular element ABC is given by

$$R_S = \sqrt{1 + c^2} \qquad (8\text{-}20)$$

where $c = d_3/d_1$. Thus, the overall R_S is easily determined by a computer program knowing only the coordinates of the triangular nodes. This method is simple and direct, once the coordinates of the contours are known. Experimentally, however, we are warned that contour sections ("horizontal" cuts) are prone to yield wide lines that cannot be evaluated accurately, especially with flat fracture surfaces.[27]

Profile Angular Distributions.

A rather promising analytical procedure, although not one specifically developed for fracture surfaces, provides a methodology for transforming the angular distribution of linear elements along a trace through the surface into an angular distribution of surface elements in space. From this spatial distribution we can obtain the fracture surface area.

Three versions of this procedure have been proposed for grain boundaries.[19,20,21] There are problems, however, in applying these analyses directly to the case of fracture surfaces. Transformation procedures require an axis of symmetry in the angular distribution of facets, as well as facets that are planar, equiaxed, and finite. Unfortunately, most fracture surfaces do not possess the angular randomness exhibited by grain boundaries. It does appear, however, that there may be an axis of symmetry. This possibility will be discussed later.

One of the first analyses for the angular distribution of grain-boundary facets was offered by Scriven and Williams.[19] They too require an axis of symmetry, so that all sections cut through the axis would have statistically identical properties. The authors provide a recursive formula and tables of coefficients, and from these the three-dimensional facet angular distribution, Q_r, can be obtained from the experimental profile angular distribution, G_s. The essential equations are, first,

$$Q_r = Q(rh) - Q[(r-1)h] \qquad (8\text{-}21)$$

where $r = 1, 2, 3, \ldots 18$ (for 0 to 90 degrees); h is a constant interval value of 5 degrees; and $Q(0) = 0$. Moreover,

$$Q(rh) = \frac{1}{N} \sum_{s=1}^{r} b_{rs} G_s \qquad (8\text{-}22)$$

where the coefficients b_{rs} are tabulated for each successive value of r; the values of G_s are obtained experimentally from $s = r$; and N is a constant defined by the equation,

$$N = \sum_{s=1}^{18} b_{18,s} G_s \qquad (8\text{-}23)$$

where the coefficients $b_{18,s}$ are tabulated in Scriven and Williams' tables under the heading, "r-18." The values of Q_r give the distribution of elevation angle θ—per unit angular interval in θ—in the form of a histogram. The total surface area is obtained simply by multiplying the Q_r values by the cosine in each class interval and summing over all class intervals, the sum equaling $1/R_s$.[9]

In order to illustrate their procedure, we select two profiles representative of two important types of fracture surfaces.[18] The first profile, shown in Fig. 8-15(a), comes from an actual dimpled fracture surface for 4340 steel. The other, in Fig. 8-15(b), is a prototype profile of a fracture surface with 100-percent cleavage or intergranular facets. The experimental data for these two extreme cases were obtained simply by tracing the profile image with a cursor on a digitizing tablet. The measuring distance η between coordinates on the dimple fracture profile was 0.68 μm. A very large value of η (= 123.6 μm) was selected to generate the prototype facet profile. The relationship of η to the fractal dimension has been discussed previously (see Eqs. 8-14a and 8-14b).

Two histograms showing the angular distribution of segment normals along the two profiles are given in Figs. 8-16a and 8-16b. As would be expected, there is a much broader angular distribution in (a) than in (b). In fact, the distribution for the dimple fracture profile is extremely broad, which might seem to justify the assumption of angular randomness. The surface areas estimated from the two profiles were calculated according to Scriven and Williams' procedure,[19] as well as parametric Eq. No. 2 in Table 8-5. The results are given in Table 8-6. Again, the Scriven and Williams' values for R_s are higher than those calculated by Eq. No. 2. Deviations of less than 6 percent would appear to be satisfactory for most purposes, however. In order to check the assumption of complete angular randomness for the dimple fracture profile, we use Eq. 8-16 to calculate R_s. The values obtained with this equation are also entered in Table 8-6, where a deviation of plus 10.7 percent is found for the dimpled profile and over 23 percent for the prototype faceted profile. Thus, we might conclude that the assumption of complete angular randomness in the dimple profile is not a good one.

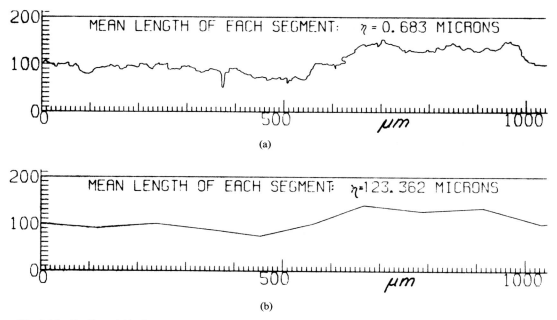

Fig. 8-15. Profiles of (a) dimpled fracture surface of 4340 steel, and (b) prototype faceted fracture surface.

Another major requirement in the Scriven and Williams' treatment is that there should be an axis of symmetry in the facet orientations. Surprisingly enough, an axis of symmetry has been observed in both two-dimensional (profile)[18,47] and three-dimensional (facet) angular distributions[47]; for real fracture surfaces[9,18,47] and a computer-simulated fracture surface[47]; and in compact tensile specimens[47] and round tensile specimens.[9,18]

The type of angular distribution observed is essentially bimodal, showing symmetry about the normal to the "macroscopic" fracture plane. The experimental angular distribution curves for two Ti–V alloys (compact tensile specimens) reveal this bimodal behavior quite clearly, as shown in Fig. 8-17. Similar distributions were obtained for both longitudinal and transverse sections, which establishes the three-dimensional angular symmetry about the macroscopic normal. This three-dimensional bimodality was also clearly confirmed with a computer-simulated fracture surface.[47] Bimodal behavior also appears to a greater or lesser extent in other published papers.[5,28] Histograms similar to those in Fig. 8-17 for the two-dimensional angular distribution reveal a relative decrease in

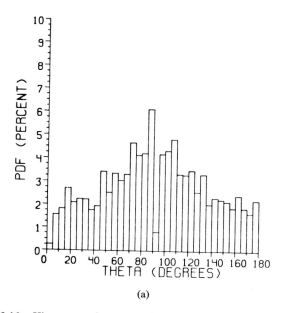

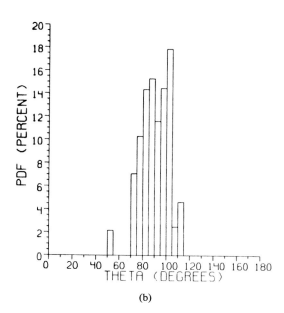

Fig. 8-16. Histograms of angular distribution of normals along profiles of (a) dimpled fracture surface of 4340 steel, and (b) prototype faceted fracture surface (see Fig. 8-15).

Table 8-6. Calculations of R_S for Dimpled and Prototype Faceted 4340 Steel Fracture Surfaces.

	R_S (SCRIVEN AND WILLIAMS' ANALYSIS)	R_S (PARAMETRIC EQ. NO. 2 IN TABLE 8-5)	R_S (EQ. 8-16 IN TEXT)
Dimpled rupture R_L ($\eta = 0.68$ μm) = 1.4466	1.6639	1.5686 ($\Delta = -5.7\%$)*	1.8419† ($\Delta = 10.7\%$)
Prototype facets R_L ($\eta = 123.6$ μm) = 1.0163	1.0482	1.0207 ($\Delta = -3.1\%$)	1.2940 ($\Delta = 23.5\%$)

* Deviations (Δ) are calculated with respect to the Scriven and Williams' values of R_S.[19]
† This value, representing a completely random fracture surface, is entered in Fig. 8-13 as a star.

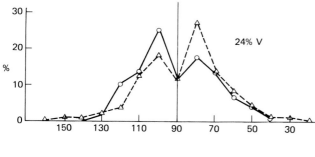
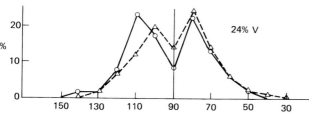
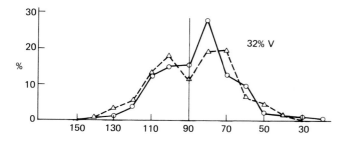
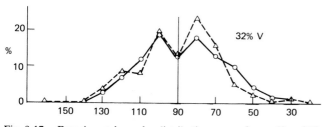

Fig. 8-17. Experimental angular distribution curves for profiles of Ti-24V and Ti-32V alloys fractured in fatigue. Longitudinal sections made in compact tensile specimens along edge are indicated by —○— and those made through center by ---△---.

the density function at or near 90 degrees. The conclusion to be drawn from this combined evidence is that there is a symmetry axis about the normal to the macroscopic fracture plane. Thus the mathematical transform procedures that require an axis of symmetry may be more applicable to metallic fracture surfaces than originally thought.

It is interesting to speculate briefly on the physical reasons for such an unexpected type of angular distribution. One of the simplest explanations appears to be that the fracture surface (and profile) consists *statistically* of alternating up- and down-facets (or lineal segments) rather than a basic horizontal component with slight tilts to the left and right (which would give a unimodal distribution curve centered about the normal direction). Passoja[48] confirms this explanation of a zig-zag profile in his studies of Charpy fractures. Using a discrete Fourier transform method, his results were rationalized on the basis that the profile was built up from a random shifting of a basic triangular element. This type of configuration leads automatically to a bimodal angular distribution.

APPLICATIONS

The methods and equations presented in this chapter are still incomplete, but the available results permit some problems to be solved. To illustrate a few of the approaches that can be taken, some specific examples will be examined here in depth.

Example 1: Fatigue Striation Spacings

The problem of determining the actual striation spacing in the nonplanar fracture surface has been previously examined.[6] Corrections for orientation and roughness effects were formulated, but with separate equations. Reference to Fig. 8-18 indicates the geometrical relationships involved. We measure striation spacings in the SEM picture in the Crack Propagation Direction (CPD) and according to the relation,

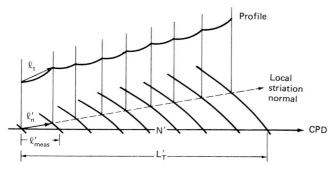

Fig. 8-18. Geometrical relationships involved in fatigue striation spacings measurements, where CPD is the crack propagation direction, l_t is the true striation spacing and l_n' its projection, and l'_{meas} is the apparent spacing measured in the projection plane along the CPD.

$$\overline{l'_{meas}} = L_T'/N' = 1/N_L' \qquad (8\text{-}24)$$

where L_T' is the distance over which measurements are made and N' is the number of striations within this distance. Since the directions of the striations vary widely from the CPD, an angular correction is required. From basic stereology,[17] we have the relation,

$$\overline{l_n'} = (2/\pi)\,\overline{l'_{meas}} \qquad (8\text{-}25)$$

where $\overline{l_n'}$ is the mean normal striation spacing.

There is also a roughness factor to consider, and this can be evaluated if the lineal roughness parameter, R_L, is known. We assume that R_L over the entire profile length can be equated to the local ratio of mean true striation spacing, $\overline{l_t}$, to its mean projected length, $\overline{l_n'}$. That is,

$$R_L = \frac{L_{\text{profile}}}{L'_{\text{profile}}} = \frac{\overline{l_t}}{\overline{l_n'}} \qquad (8\text{-}26)$$

We can now combine Eqs. 8-24, 8-25, and 8-26 to yield an overall equation that combines corrections for both orientation and roughness, namely

$$\overline{l_t} = R_L\,\overline{l_n'} = R_L\,(2/\pi)\,\overline{l'_{meas}} = (2/\pi) R_L/N_L' \qquad (8\text{-}27)$$

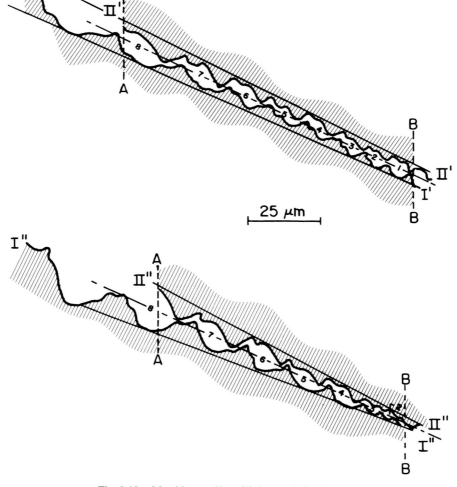

Fig. 8-19. Matching profiles of fatigue striations in nickel.[49]

This relationship gives the mean true striation spacing in terms of measurable quantities.

Example 2: A Check on Precision Matching

In an unusually careful and well-documented fractographic investigation, Krasowsky and Stepanenko[49] reported on a precision-matching study of fatigue striations in nickel. Two pairs of matching profiles were obtained using a stereoscopic method. These four curves are reproduced in Fig. 8-19, with the pairs of matching profiles identified as I'II' and I"II".

In order to assess the degree of matching, R_L and R_P were measured on all profiles. The results are tabulated in Table 8-7, including values for duplicate curves that appeared in the article. Of major interest is the correspondence between (ostensibly) matching profile curves. For R_L, the maximum difference between these pairs is about 8 percent (for I'II') and about 2 percent (for I"II"). Based on R_P, the corresponding values are about 20 and 8 percent, respectively. The relative insensitivity of R_L is probably due to the fact that only the total line length of the profile is involved, whereas R_P depends on both its tortuosity and the heights of the peaks. Thus, the results indicate that these pairs of curves are not "matching," at least not in a strict geometrical sense.

Table 8-7. Analysis of Matching Profiles of Fatigue Striations in Nickel.[43]

PARAMETER	PROFILE I'II'		PROFILE I"II"	
R_L				
Fig. 8-19:	1.164	1.251	1.251	1.231
Duplicate:	1.215	1.242	1.265	1.240
R_P				
Fig. 8-19:	0.330	0.392	0.347	0.340
Duplicate:	0.305	0.367	0.366	0.338

Example 3: Quantitative Property-Microstructure Models

A common procedure in the metallurgical literature is to devise a microstructural-property model based on features with assumed simple geometrical shapes. Calculations are then made relating microstructural characteristics to the property of interest. An illuminating example of such a procedure is afforded by the following analysis of microroughness in the fracture surface.[50] The published analysis is recapitulated in the left-hand column; the quantitative stereological and fractographic alternatives, in the right-hand column:

A simple shape parameter M is proposed of the form,

$$M = h/w \qquad (8\text{-}28)$$

where h is the height of the (half) dimple (or microvoid) in the fracture surface and w is the mean dimple "diameter." It is assumed that the dimples are circular and fill the fracture surface completely. (Actually, however, for a square array of circular dimples, over 21 percent of the surface would be uncovered.)

The number of dimples per unit area is given as

$$N_A = 4/\pi w^2 \qquad (8\text{-}29)$$

based on circular dimples with mean diameter w.

Next, an equation is given for the number of uniform size particles (or nuclei) per unit volume:

$$N_V = 6(V_V)_p / \pi d_p^3 \qquad (8\text{-}30)$$

where d_p is the mean diameter for spherical particles. This equation is based on measurements in the metallographic plane of polish.

The intent of Eq. 8-30 is to equate N_V for particles to N_V for dimples, based on the assumption of only one particle per dimple. This assumption is not

This mean height-to-width ratio is the same as the R_P profile roughness parameter defined by Eqs. 8-12 and 8-12a. Although a vertical section across the fracture surface would intersect the dimples at different locations, it is interesting to note that the h/w ratio for a spherical segment is a constant regardless of where it is intersected.

Apparently, the authors are referring in Eq. 8-29 to an SEM photomicrograph and a relationship for N_A' in the projection plane. We prefer to use the more precise equation,

$$N_S = N_A'/R_S \qquad (8\text{-}29\text{a})$$

since the correction provided by R_S amounts to over 66 percent in some cases.

The assumption of spherical particles of one size is unduly restrictive and not necessary. The actual particle "diameter" (d_p) can be measured in several ways, depending on the method of observation.

1. In a thin foil picture, $\overline{d_p'}$ is measurable directly, provided particle overlap is not excessive.
2. In the plane of polish, the mean diameter, $\overline{d_2}$, of the particle sections is related to $\overline{d_p}$ by

$$\overline{d_p} = (4/\pi)\overline{d_2} \qquad (8\text{-}30\text{a})$$

valid, however. It is obvious that there cannot be one particle per dimple in a fracture surface. Simple probabilities indicate that, on the average, only one-half the total number of particles would be associated with one of the fracture surfaces.

An equivalent equation for dimples is written next:

$$N_A = N_V \, 2h \qquad (8\text{-}31)$$

This equation was derived originally for random sections through randomly located particles.[17] For Eq. 8-31 to be valid here, the fracture surface must intersect the dimples within a slice of thickness $2h$, and all the dimple centers must lie in a common plane.

Assuming N_V in Eqs. 8-30 and 8-31 are equal, Eqs. 8-28 through 8-31 are combined, giving

$$\frac{h}{d_p} = \left[\frac{M^2}{3(V_V)_p}\right]^{1/3} \qquad (8\text{-}32)$$

The true local fracture strain is expressed as

$$\epsilon_f \approx \ln(h/d_p) \qquad (8\text{-}33)$$

which is evaluated by substituting Eq. 8-32.

for spherical particles. For convex particles, $\overline{d_2}$ is obtained from the mean perimeter length, $\overline{L_p}$, by

$$\overline{d_2} = \overline{L_p}/\pi \qquad (8\text{-}30b)$$

3. In the SEM photograph, we can measure the size of the particles in the fracture surface according to

$$\overline{d_p{'}} = N_L{'}/N_A{'} \qquad (8\text{-}30c)$$

(see Table 8-2). This may or may not be the true equatorial diameter, depending on how much the particle is exposed.

Actually, the N_A in Eq. 8-31 is not the same as in Eq. 8-29, which should be $N_A{'}$. A more serious problem is the determination of the test volume, V_T, in N_V in both Eqs. 8-30 and 8-31, where N_V is defined by $\Sigma N_i/V_T$. The number of particles and the number of dimples (ΣN_i) are assumed to be the same, but the two test volumes (V_T) are not the same.

Since ϵ_f depends on the height to which the void grows, however, it seems more appropriate to define the true local strain as

$$\epsilon_f \approx \ln(2h/d_p) \qquad (8\text{-}33a)$$

since the void grows in two directions with respect to the mean fracture plane.

The result obtained in Eq. 8-32 is based on the model developed by the authors.[50] An alternative to this treatment is not to make a priori assumptions, but to express $\overline{h}$ and $\overline{d_p}$ directly in terms of measurable quantities. To do so, we need the digitized profile and the SEM photomicrograph of the fracture surface. The mean height of dimples, $\overline{h}$, and the mean dimple width, $\overline{L_2{'}}$, are related by

$$\overline{h} = R_P \, \overline{L_2{'}} \qquad (8\text{-}34)$$

where R_P is given by Eq. 8-12a and $\overline{L_2{'}} = (L_T{'}/N'){_{\text{dimples}}}$. The mean particle diameter, $\overline{d_p}$, may be measurable di-

rectly from the SEM photomicrograph as $\overline{d_p{'}}$, the mean projected diameter, using

$$\overline{d_p{'}} = (N_L{'}/N_A{'})_p \qquad (8\text{-}35)$$

If a roughness correction is needed, we would use the relationship

$$\overline{d_p} = R_L \, \overline{d_p{'}} \qquad (8\text{-}36)$$

from Table 8-4. Combining results, we can write

$$\frac{\bar{h}}{d_p} = \frac{R_P \, (N_A{}')_p}{R_L \, (N_L{}')_p \, (N_L{}')_{\text{dimples}}} \quad (8\text{-}37)$$

or

$$\frac{\bar{h}}{d_p} = \frac{R_P \, \overline{L_2{}'}}{d_p} = \frac{R_P}{d_p{}' \, (N_L{}')_{\text{dimples}}} \quad (8\text{-}38)$$

if the mean particle diameter can be measured directly. For a profile with randomly oriented elements, $R_L = \pi R_P$, and further simplification is possible.

This type of approach appears to be capable of more meaningful quantification than to make model assumptions (size, shape, and configuration) in advance. In any event, it is best to use measured quantities as much as possible when developing microstructural property relationships.

SUMMARY

Perhaps the most obvious conclusion to come out of this survey is that quantitative fractography is undergoing considerable development and still has far to go. There is no one "best"way to proceed since techniques are still being proposed and assessed, measurement methods are yet to be optimized, and relationships of general validity are still lacking.

Nevertheless, considerable progress has been achieved to date and several viable paths have been partially explored. An assessment of the current status of quantitative fractography indicates that sectioning in one form or other predominates. Thus the profile and its characteristics assume considerable importance, and serial sectioning also emerges as an important adjunct to experimental methods and analysis.

There is still need for a quantification technique that can supplement the wealth of information available in the SEM photomicrograph. Perhaps a nondestructive profile-generating procedure is best suited for this role. If three-dimensional parameters can be provided readily and with sufficient accuracy, then a complete spatial characterization of the fracture surface may be possible on a routine basis.

Acknowledgments. This work was performed under the auspices of the National Science Foundation, Metallurgy, Polymers and Ceramics Section, Division of Materials Research: Grant No. DMR-8204018. Thanks are due Mr. K. Banerji, who provided some experimental data and calculations.

REFERENCES

1. Recent sessions include the following:
   November, 1979. Williamsburg, VA, ASTM. Fractography and Materials Science.
   October, 1980. Pittsburgh, PA. AIME-TMS Fall Meeting. Session on Quantitative Fractography.
   June, 1981. Ljubljana, Yugoslavia. 3d European Symp. on Stereology. Session entitled "Quantification of Nonplanar Surfaces."
   September, 1982. Leoben, Austria. 4th Europ. Conf. on Fracture. *Fracture and the Role of Microstructure.*
   December, 1983. Gainesville, FL. 6th Int. Congr. for Stereology. Session entitled "Fractures and Other Nonplanar Surfaces."
   July, 1984. Philadelphia, PA. 17th Ann. Meeting I.M.S. Session entitled "Fractography, Failure Analysis and Microstructural Studies."
2. El-Soudani, S. M. Profilometric Analysis of Fractures. *Metallography* 11:247–336 (1978).
3. Coster, M., and Chermant, J. L. Recent developments in quantitative fractography. *Int. Metals Reviews* 28, No. 4:228–250 (1983).
4. Wright, K., and Karlsson, B. Topographic quantification of nonplanar localized surfaces. *J. Micros.* 130:37–51 (1983).
5. Exner, H. E., and Fripan, M. Quantitative assessment of three-dimensional roughness, anisotropy, and angular distributions of fracture surfaces by stereometry. *J. Micros.* 139, Pt. 2:161–178 (1985).
6. Underwood, E. E., and Starke, E. A., Jr. Quantitative stereological methods for analyzing important microstructural features in fatigue of metals and alloys. In *Fatigue Mechanisms,* ed. Fong, J. T. ASTM STP 675:633–682 (1979).
7. Underwood, E. E., and Chakrabortty, S. B. Quantitative fractography of a fatigued Ti-28V alloy. In *Fractography and Materials Science,* eds. Gilbertson, L. N., and Zipp, R. D. ASTM STP 733:337–354 (1981).
8. Underwood, E. E., and Banerji, K. Statistical analysis of facet characteristics in computer simulated fracture surface. *Proc. 6th Int. Congr. Stereology,* Gainesville, Florida. *Acta Stereologica,* ed. Kališnik, M. Ljubljana, Yugoslavia:75–80 (1983).
9. Banerji, K., and Underwood, E. E. On estimating the fracture surface area of Al-4% Cu alloys. In *Microstructural Science,* eds. Shiels, S. A.; Bagnall, C.; Witkowski, R. E.; and Vander Voort, G. F. 13:537–551 (1985).
10. Banerji, K., and Underwood, E. E. Fracture profile analysis of heat treated 4340 steel. *Proc. 6th Int. Conf. on Fracture.* New Delhi, India, 1984. In *Advances in Fracture Research,* eds. Valluri, S. R.; Taplin, D. M.; Rao, P. R.; Knott, J. F.; and Dubey, R. 2:1371–1378 (1984).
11. Brandis, E. K. Comparison of height and depth measurements with the SEM and TEM using a shadow casting technique. *Scan. Electr. Micros.* 1:241 (1972).
12. Gifkins, R. C. *Optical Microscopy of Metals.* London: Pitman, 1970.
13. Zapffe, C. A., and Moore, G. A. A micrographic study of the cleavage of hydrogenized ferrite. *Trans. AIME* 154:335–359 (1943).
14. Beachem, C. D. Microscopic fracture processes. In *Fracture,* vol. 1, ed. Liebowitz, H. New York: Academic Press, 1969, pp. 243–349.
15. Broek, D. Some contributions of electron fractography to the theory of fracture. *Int. Metall. Reviews* 19:135–182 (1974).
16. Broek, D. A study on ductile fracture. *Nat. Lucht. Ruimteraartlab.* NLR (TR71021):98–108 (1971).
17. Underwood, E. E. *Quantitative Stereology.* Reading, MA: Addison-Wesley, 1970.
18. Banerji, K., and Underwood, E. E. Quantitative analysis of fractographic features in a 4340 steel. *Proc. 6th Int. Congr. Stereology.* Gainesville, Florida. *Acta Stereologica,* ed. Kališnik, M. Ljubljana, Yugoslavia:65–70 (1983).
19. Scriven, R. A., and Williams, H. D. The derivation of angular distributions of planes by sectioning methods. *Trans. AIME* 233:1593–1602 (1965).
20. Morton, V. M. The determination of angular distributions of planes in space. *Proc. Roy. Soc.* A302:51–68 (1967).

21. Hilliard, J. E. Specification and measurement of microstructural anisotropy. *Trans. AIME* 224:1201–1211 (1962).
22. Goldstein, J. I.; Newbury, D. E.; Echlin, P.; Joy, D. C.; Fiori, C.; and Lifshin, E. *Scanning Electron Microscopy and X-Ray Microanalysis*. New York: Plenum Press, 1981. (See pp. 143–146.)
23. Hilliard, J. E. Quantitative analysis of scanning electron micrographs. *J. Micros.* 95, Pt. 1:45–58 (1972).
24. Boyde, A. Quantitative photogrammetric analysis and quantitative stereoscopic analysis of SEM images. *J. Micros.* 98, Pt. 3:452–471 (1973).
25. Howell, P. G. T. Stereometry as an aid to stereological analysis. *J. Micros.* 118, Pt. 2:217–220 (1980).
26. Eisenhart, L. P. *Coordinate Geometry*. New York: Dover Publ., 1960.
27. Bauer, B., and Haller, A. Determining the three dimensional geometry of fracture surfaces. *Pract. Metallog.* 18:327–341 (1981).
28. Bauer, B.; Fripan, M.; and Smolej, V. Three dimensional fractography. *Proc. 4th Europ. Conf. on Fracture.* Leoben, Austria. *Fracture and the Role of Microstructure*, eds. Maurer, K. L., and Matzer, F. E. 591–598 (1982).
29. Bresnahan, K. Personal communication (1984).
30. Underwood, E. E. The Stereology of Projected Images. *J. Micros.* 95, Pt. 1:25–44 (1972).
31. Pickens, J. R., and Gurland, J. Metallographic characterization of fracture surface profiles on sectioning planes. *Proc. 4th Int. Congr. Stereology*, eds. Underwood, E. E.; deWit, R.; and Moore, G. A. NBS Spec. Publ. 431, Gaithersburg, MD: 269–272 (1976).
32. Shieh, W. T. The relation of microstructure and fracture properties of electron beam melted, modified SAE 4620 steels. *Met. Trans.* 5:1069–1085 (1974).
33. Shechtman, D. Fracture-microstructure observations in the SEM. *Met. Trans.* 7A:151 (1976).
34. Kerr, W. R.; Eylon, D.; and Hall, J. A. On the correlation of specific fracture surface and metallographic features by precision sectioning in titanium alloys. *Met. Trans.* 7A:1477–1480 (1976).
35. Almond, E. A.; King, J. T.; and Embury, J. D. Interpretation of SEM fracture surface detail using a sectioning technique. *Metallography* 3:379–382 (1970).
36. Passoja, D. E., and Hill, D. C. Comparison of inclusion distributions on fracture surfaces and in the bulk of carbon-manganese weldments. In *Fractography-Microscopic Cracking Processes*, eds. Beachem, C. D. and Warke, W. R. ASTM STP 600:30–46 (1976).
37. Van Stone, R. H., and Cox, T. B. Use of fractography and sectioning techniques to study fracture mechanisms. In *Fractography-Microscopic Cracking Processes*, eds. Beachem, C. D., and Warke, W. R. ASTM STP 600:5–29 (1976).
38. Chestnutt, J. C., and Spurling, R. A. Fracture topography-microstructure correlations in the SEM. *Met. Trans.* 8A:216 (1977).
39. Vander Voort, G. F. *Metallography: Principles and Practice*. New York: McGraw-Hill Book Co., 1984, pp. 86–90, 538–540.
40. Gray, G. T., III; Williams, J. G.; and Thompson, A. W. Roughness-induced crack closure: An explanation for microstructurally sensitive fatigue crack growth. *Met. Trans.* 14:421–433 (1983).
41. Swift, J. A. Measuring surface variations with the SEM using lines of evaporated metal. *J. Phys. E.: Sci. Instrum.* 9:803 (1976).
42. Wang, R.: Bauer, B.; and Mughrabi, H. The study of surface roughness profiles of fatigued metals by scanning electron microscopy. *Z. Metallk.* 73:30–34 (1982).
43. Underwood, E. E. Practical solutions to stereological problems. *Practical Applications of Quantitative Metallography*. eds. McCall, J. L., and Steele, J. H. ASTM STP 839:160–179 (1984).
44. Behrens, E. W. Personal communication (1977).
45. Mandelbrot, B. B. *Fractals: Form, Chance and Dimension*. San Francisco: W. H. Freeman and Co., 1977.
46. Underwood, E. E., and Banerji, K. Unpublished research at Georgia Inst. of Techn., Atlanta, GA. (1984).
47. Underwood, E. E., and Underwood, E. S. Quantitative fractography by computer simulation. *Proc. 3d Eur. Symp. for Stereology.* Ljubljana, Yugoslavia. *Acta Stereologica*, ed. Kališnik, M. 2d Part: 89–101 (1982).
48. Passoja, D. E., and Amborski, D. J. Fracture profile analysis by Fourier transform methods. *Microstructural Science*, eds. Bennett, J. E.; Cornwell, L. R.; and McCall, J. L. New York: Elsevier, 1978, pp. 143–158.
49. Krasowsky, A. J. and Stepanenko, V. A. A quantitative stereoscopic fractographic study of the mechanism of fatigue crack propagation in nickel. *Int. J. of Fracture* 15, No. 3:203–215 (1979).
50. Thompson, A. W., and Ashby, M. F. Fracture surface micro-roughness. *Scripta Met.* 18:127–130 (1984).

9
METHODS AND APPLICATIONS OF MICROINDENTATION HARDNESS TESTING

Peter J. Blau

National Bureau of Standards

HARDNESS, MICROHARDNESS, AND MICROINDENTATION HARDNESS

There is a great deal of confusion in conceptions about penetration hardness. Part of the problem results from a popular misconception that "hardness" is a basic property of a material. This is not the case. Hardness numbers are, more accurately, measures of the response of given materials to various types of indentation processes. Microindentation hardness numbers depend on a combination of surface and bulk material properties. As indenter geometry and test loads are changed (particularly at loads below about 100 g), different combinations of surface and bulk properties may contribute to the net response of the material. Issues of impression size versus microstructural feature size ("scale effects") are important. Also, at lower loads and with blunter indenters (e.g., Knoop), material is penetrated less deeply, bringing about an increased sensitivity to surface preparation technique. The metallographer should recognize this fact and take particular care in preparing polished samples that may be used for low-load microindentation studies.

Taken literally, the term "microhardness" means a very small degree of hardness. In the present context, one should more properly use the term "microindentation hardness," which means hardness that is obtained using very small indentations. For the purpose of this chapter, however, the term "microhardness" will be used on occasion because it is so thoroughly ingrained in the engineering literature and linked to common usage. Adoption of the more correct term is advocated whenever possible, and hopefully a complete transition in usage will be forthcoming.

Figure 9-1 shows the interrelated chain of factors that produces microindentation hardness numbers. One selects a type of test and uses an imperfect machine to produce an indentation. The material responds to the action of the real indenter and test machine. Then, some predetermined method of measuring the impression and using the value in a calculation is applied to obtain a final number. Sources of inaccuracy or error are possible in any factor in the chain. For example, perhaps the type of test method was improperly selected; perhaps the manufacturer's load was not accurate; perhaps the machine was vibrated by an external source during the test; perhaps the sample possessed preparation artifacts; or perhaps the optical measuring microscope was out of calibration or read improperly. These and other sources of error and hardness number variation will be discussed later. The central problem in microindentation-hardness number utilization is to separate real property test-to-test variations from those introduced in making the measurements.

This chapter is intended to provide a background guide for the selection of microindentation-hardness testing techniques, selection of testing machines, preparation of samples, and understanding something of the nature and limitations of the resulting data. It is not meant to be a detailed treatise on the subject, for several outstanding ones already exist,[1-4] but was rather designed to alert users regarding the strengths and limitations of several established microindentation-hardness test methods. References have been provided for those readers requiring more information in specific areas.

SELECTING A MICROINDENTATION HARDNESS TEST METHOD

In a laboratory where more than one option for microindentation hardness testing is available, the choice of the proper procedure(s) can be critical to maximizing the quality and meaningfulness of data. This section will describe how one may go about choosing the best procedure(s) for their needs and specimen materials. The factors bearing on this decision are the following:

1. Form, size, and thickness of samples (e.g., bulk sample, metal powder sample, layered coating, etc.)

124 APPLIED METALLOGRAPHY

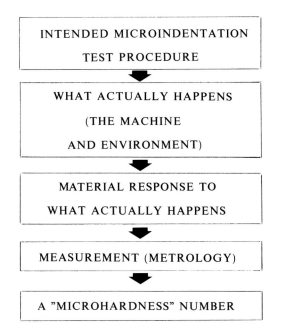

Fig. 9-1. Evolution of a "microhardness" number.

2. "Hardness" and ductility of the material (e.g., brittleness, likely range of hardness)
3. Sample homogeneity (very important in terms of the size of indentations compared with microstructural features and the amount of scatter in readings taken from one place to another on the sample)
4. The desired statistical confidence limits of the results
5. Availability of indentation measuring equipment beyond just the tester itself (e.g., a calibrated scanning electron microscope)
6. Reason for which the microindentation hardness data is needed. For example, for abrasion resistance, a scratch technique may be more useful than a quasi-static technique.

Two broad categories of microindentation hardness testing can be identified: (1) quality control and process development, and (2) research, both basic and applied. The selection of test method could be influenced strongly by which of these two categories the testing has to address. In the first category, generally larger numbers of tests must be performed on a continuous basis. The statistical variation of "microhardness numbers" and their range on a given product may be of immediate economic importance in product acceptability. The use of automated machines with frequent recalibration becomes very important in view of productivity and operator fatigue. Standard fixturing and set sampling techniques for production samples may have to be developed.

In research uses, testing may require a greater flexibility of specimen mounting and sampling procedures. Amounts of material could be limited and the cost of preparing each final sample could be very high. In both cases, it is important that both the quality of testing equipment and the skill of the operators be kept high. The investment made in using more careful, highly trained personnel in performing critical tests will be rewarded by improved data quality.

The remainder of this section will review common microindentation hardness scales, types of indenters, and standards developed for microindentation hardness; it finally provides examples of various applications, giving bibliographic references from which to obtain details.

Microindenter Types

Several microindenter geometries have been used in microindentation testing. Table 9-1 summarizes the characteristics of a number of these. The most commonly used microindenter in the U.S. and Europe is the Vickers, the Knoop being second. Still other types have been developed for special applications, and the reader may wish to consult the table references for further information on these. It should be noted that the Vickers indenter is somewhat less sensitive to microstructural orientation effects than the Knoop and penetrates about three times deeper for a given load and the same hardness number. The Knoop was developed for shallow penetration and to minimize cracking in brittle materials.[5] It tends to be more sensitive to low-load effects and surface preparation than the Vickers, but its elongated impression shape is often an advantage in placing indentations in narrow bands of material. Besides this, the shallow surface penetration of the Knoop can offer advantages when thin coatings are to be tested.

Several other characteristics of the Knoop indenter, such as anisotropic elastic shape recovery effects and sensitivity to crystallographic grain orientation, have made it useful in studying directional microstructural properties.[6-7] The anisotropic recovery effects can sometimes produce large errors in the following standard equation for Knoop microhardness number:*

$$\text{KHN} = 14229 \left(\frac{P}{D^2}\right) \qquad (9\text{-}1)$$

where P = load (in g) and D = long diagonal length (in μm). To obtain actual load per unit-projected-area on elastically relaxed Knoop impressions, a more accurate equation that uses the short diagonal (d) of the elongated impression has been suggested as a way to overcome anisotropy:[8]

$$\text{PAH} = 2000 \left(\frac{P}{Dd}\right) \qquad (9\text{-}2)$$

---
* The most commonly used units are kg/mm^2; however, MPa equivalents may be obtained by multiplying by 9.81.

Table 9-1. Microindenter Types and Characteristics.

NAME	GEOMETRY	REFERENCE	COMMENTS
Vickers	Equiaxed diamond pyramid; four faces; 136° facet apex angle	10, 11	Common commercially available indenter; ratio of length of diagonal to depth about 7:1
Knoop	Elongated diamond pyramid; four faces; 172°30' major-edge apex angle; 130° minor-edge apex angle	10	Common commercially available indenter; shallower than Vickers at given hardness and load; ratio of length of long diagonal to depth typically 30:1
Brinell	10-mm diameter ball	4	Usually used for higher loads (> 1 kg), as well as microhardness; a very well established technique in macrohardness testing
Berkovich	Diamond indenter; three facets; 142° edge-to-opposite facet	1, 12	Triangular impression
Grodzinski	Diamond; edge of bases of two base-to-base 66° cones; 2-mm base radius	1	Highly elongated indentation; length-to-depth can exceed 130:1 ratio
Pfund	Hemisphere of sapphire	13	Quartz indenter also used
Brookes	Pentagonal	14	Less sensitive to crystallographic orientation effects than Vickers, Berkovich, and Knoop

where PAH (projected area Knoop hardness) is also given in kg/mm^2. Other schemes for correcting for impressions that do not replicate the true indenter shape have been suggested by Tate,[9] and newer methods that use indenter displacement under load to calculate microhardness numbers have been appearing in the literature.[15,16] A major problem in implementing new displacement measurement methods and attempting to compare resulting data with previous values is similar to the problem of converting one scale to another in that different combinations of microindentation properties may be sampled by the new displacement techniques. Numerical equivalence of displacement-type measurements with traditional optical measurements is not to be expected, and new sets of baseline comparison data would be required for hardness numbers obtained using the displacement technique.

Standards Related to Microindentation Hardness Testing

A considerable number of international standards for conducting microindentation hardness testing exist. Most are found in three sources: DIN (W. German), ISO, and ASTM. The DIN standards for Vickers and Knoop testing are 50133 and 52333, respectively. The ISO lists three Vickers related standards: R81, R192, and R399. ASTM* is an internationally recognized voluntary standards organization and publishes a number of microindentation hardness standards; these are listed in Table 9-2.

Procedures for testing are one matter, but obtaining accurate standards to calibrate hardness testing equipment is quite another. Many equipment manufacturers advertise "microhardness standards." These commercial standards should be rigorously evaluated under carefully controlled testing conditions to certify that their range of numerical reproducibility is sufficient for the user's purposes. It is always best to use well-characterized standards that are similar in hardness to the test materials of interest and have been "certified" at the test loads to be used. It may not be wise initially to project the blame for "bad data" on the machine or the operator when calibration standards could vary from one place to another. Improper "touching up" of the surface of standards by rough regrinding and repolishing may lead to errors. Certifications could be invalidated by repolishing.

Recently, a series of low-load (25-, 50-, and 100-g) "microhardness" standards have been developed at the National Bureau of Standards. These are numbered Standard Reference Materials as follows: SRM 1893 (Copper–Knoop), SRM 1894 (Copper–Vickers), SRM 1895 (Nickel–Knoop), and SRM 1896 (Nickel–Vickers).† More low-load standards of higher microindentation hardness than copper and nickel are being developed.

Sources of Inaccuracy and Errors

There are three definable sources of inaccuracy and error in conducting microindentation hardness tests: the machine, the operator, and the material. This trilateral responsibility—which the author finds convenient to remember by using its initials M.O.M.—advocates that one should not be too quick in blaming any one factor for problems in microindentation hardness data. In standardization work, the frequent lack of a clear source of data scatter is particularly acute. For example, one might ask:

---

* 1916 Race Street, Philadelphia, PA 19103.

† Purchasable from the Office of Standard Reference Materials, National Bureau of Standards, Gaithersburg, MD 20899.

126 APPLIED METALLOGRAPHY

Table 9-2. ASTM Standards Related to Microindentation Hardness Testing.

MATERIALS	FORM	INDENTER TYPE	ASTM NO.	COMMENTS
Solids	—	Knoop Vickers Brinell Rockwell	E 6	Subsection of a general specification for mechanical testing
	—	Knoop Vickers	E 384	General microhardness
Metallics	—	Vickers	E 92	1–120 kgf load range
	—	Knoop Vickers Brinell Rockwell Rockwell Superficial	E 140	Approximate scale conversions
	Electroplate	Knoop	B 578	
	Electrical contacts	Knoop Vickers Rockwell Brinell	B 277	
Ceramics	Whiteware	Knoop	C 849	
	Glass	Knoop	C 730	Similar to C 849
Organics	Films	Pencil points	D 3363	Scratch test
	Films, paint, lacquer	Knoop Pfund	D 1474	A Pfund indenter is a hemisphere of quartz or sapphire

Is our machine out of calibration? Is the reader not measuring correctly? or, Is the material really varying *that much* from place to place? Knowing the "microhardness" number at one location on a sample with great accuracy does not guarantee that the same number will obtain a few micrometers away even if the microstructure *appears* to be identical.

Table 9-3 lists a number of possible sources of inaccuracy and error in microindentation data. When the need to use low loads arises, second-order effects that are negligible under higher loads can become very important. The proportion of surface-versus-bulk properties that are responsible for the material's response to the indenter is different in such cases.

Impression Measurement Methods

There are three principal methods currently employed to measure microindentations: (1) optical, (2) electro-optical, and (3) indenter-displacement. The first method involves both traditional light microscopy, which has comprised the vast majority of historical usage, and scanning electron microscopy (SEM) which must be used to measure reliably the dimensions of the tiniest impressions commonly produced at very small loads and on hard materials. As Buckle has discussed,[1] the quality of the optics can effect the reproducibility of the data. For example, as the objective numerical aperture increases, the points at the tip of indentations have narrower fringes and the indentations appear smaller, giving higher microindentation hardness numbers. Another critical factor in light-optical measurement is the calibration of the observers. Each optical microscope observer should be periodically required to recalibrate his own eye on standard *impressions* of known size (obtained by calibrated SEM measurement). Doing so will maintain consistency in the laboratory.

Electro-optical methods involve both television-operator interactive methods that reduce operator fatigue and image digitizing methods that use preset electronic detectors to measure and calculate impression sizes, areas, and shapes. Coupling television cameras and/or image-analysis equipment to microindentation instruments can lessen operator fatigue, but this procedure can also result in some additional sources of measuring errors if not properly applied. For example, rastering electronic images is limited in resolution by picture element (pixel) size. The finite scan-line widths and direction can also have subtle effects on measurements, which may be illustrated by the following experiment on a commercial image analyzer that used 525 lines × 1082 pixels per scan line. A Knoop indentation shape was carefully drafted, measured with a toolmarkers' microscope, and placed on a camera stand beneath the image-analyzer camera on an easel. The analyzer screen provided a 2.23X enlargement of the drawn image. The image analyzer was calibrated using a black on white series of rulings and set to measure both area (in mm^2) and longest figure dimension (in mm). Figure

Table 9-3. Some Possible Sources of Inaccuracy in Microindentation Hardness Testing.

SOURCE OF INACCURACY	NATURE OF INACCURACY
Testing Machine	Insufficient vibration isolation
	Mounting not rigid
	Indenter "bounce"
	Mechanical interference with lowering of the indenter
	Loose indenter mount (wobble)
	Load overshoot
	Loads not exactly what are printed on dial
	Standards specimens used to calibrate the machine are not correct
Operator	Sample/indenter not clean
	Improper microscope calibration factor
	Reading wrong impressions
	Wrong microscope objective in place
	Testing too near the edge of the sample
	Poor metallographic sample preparation
	Rough handling of the machine
	Eye fatigue or operator in a hurry
	Improper training
	Specimen loosely mounted
	Extra weights not removed
	Bowed substrate on thin sample (not mounted correctly)
Material	Indentation distortions on ductile samples
	Brittle phase cracking
	Subsurface, unobservable defects
	Crystallographic effects, elastic recovery anisotropy
	Grain size similar to indentation size
	Lamellar sample; "anvil" effects
	Wavy test surface (sample must be tested as-received)
	Directional surface residual stresses due to machining or grinding
	Indentation creep
	Microstructural arrangement of phases and grain boundaries

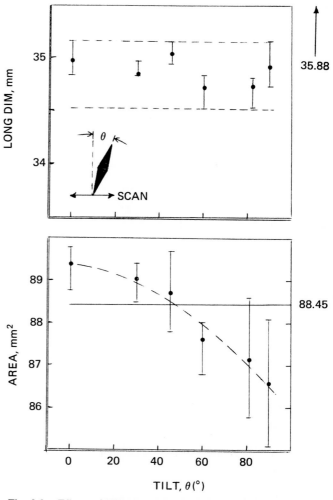

Fig. 9-2. Effects of TV scanning line direction with respect to the orientation of a Knoop impression shape on length and area measurements in an automated image analyzer.

9-2 shows the effect of image orientation on these two output quantities. Six readings were taken for each tilt angle of the Knoop-shaped figure with respect to the scan-line direction. As shown, long-dimension values were much less sensitive to figure orientation than was area, which decreased as the figure was laid over on its side. Also, long dimension, while suffering little directional bias, was consistently low compared with the actual length, indicating the additional care necessary for calibration of such systems. The effective scan-line width of about 0.34 mm at the plane of the drawn figure limited the ability to detect the tips of the figure image.

On real samples, indentation edge defects and other contrast-producing artifacts can greatly complicate automated image measurement of impressions. User-interactive TV-based systems can alleviate part of the problem, but much further work on instrumentation is needed before fully automated electro-optical impression measurement can be applied broadly.

As displacement measurement techniques have improved, the feasibility of using them to measure impressions without observer interpretation has greatly improved. The Russian PMT-3 tester has been used in this way in several investigations (e.g., Ref. 15), and other research machines using this principle have been built (e.g., Ref. 16). Data obtained by measuring impressions under load must not be compared directly with data from residual impression measurements since the elastic recovery of the "relaxed" indentation shape is usually not accounted for in the former case.

Scratch Testing

Scratch hardness testing is one of the oldest forms of hardness testing. Despite this, the combination of properties resulting in a scratch are so complex (e.g., strain rate sensitivity, friction forces, machinability parameters) that it is more difficult to quantify such tests to the extent that quasi-static (i.e., vertical) methods have been successfully quantified. For example, scratch widths, particularly in brittle materials that fracture along the scratch edges, are often difficult to measure. Also, since various indenter shapes from cones to pyramids to fine fibers have been used to generate the scratches, there is a great lack of standard methodology. Equations based on assumed contact geometry during indenter movement have been developed to provide scratch hardness numbers, but the validity of the assumptions used in their development must be evaluated in terms of the differences in the micromechanical responses of each material to a traveling indenter. For this reason, average scratch widths, not a derived number, have been reported in the following section, "Interconversion of Microindentation Hardness Scales." Further discussions of scratch testing may be found in Refs. 17–18.

Scratch testing geometry has also been used to observe critical loads for thin coating disruption. In one case, optical sensors detect the penetration of metallic coatings on transparent substrates,[19] whereas in another, bursts of acoustic emissions signal coating failure.[20] Not a scratch hardness number, but rather a critical load for failure, is the parameter of interest.

INTERCONVERSION OF MICROINDENTATION HARDNESS SCALES

There are a substantial number of published tables for interconversion of one hardness scale to another; great care should be used, however, in trying to apply these tables to situations where the compositions, heat treatments, and/or product forms of interest are different from those used to generate the original tables. The following list of references may be helpful for those who require approximate conversions:

1. ASTM *Annual Book of Standards* (E 140): (a) 1983, Sec. 3, vol. 03.01, pg. 284; (b) 1982, Part 10, pg. 359; and (c) 1981, Part 10, pg. 355.
2. ASM *Metal Progress,* 1981, Materials and Processing Databook, pg. 164.
3. ASM *Metals Handbook 1961,* 8th ed., vol. 1, pg. 1234.
4. SME *Tool and Manufacturing Engineers Handbook,* 1976, 3rd ed., McGraw-Hill Book Co., Inc., pg. 36-20.
5. G. F. Vander Voort, *Metallography: Principles and Practice,* 1984, McGraw-Hill Book Co., pp. 382–383.
6. L. Emond, *Metal Progress,* vol. 74, Sept. 1958, pg. 97; vol. 76, Aug. 1959, pg. 114.

A study was recently completed to obtain the microindentation hardness and scratch widths for a series of met-

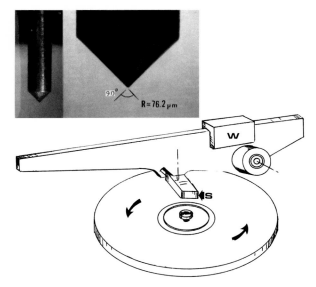

Fig. 9-3. Geometry of a stylus-on-flat scatch-hardness tester. Inset is an enlargement of the hemispherically tipped, conical diamond stylus.

als and alloys at a series of low loads (10 to 100 g). The same samples were used to obtain Vickers microhardness numbers (VHN); Knoop microhardness numbers (KHN); Projected Area Hardness numbers (PAH)— these being a modified Knoop microhardness that uses both long and short diagonal measurements; and scratch widths (using a commercial rotary scratch tester, as shown in Fig. 9-3). The average values at various loads for these measurements are given in Table 9-4. If KHN and PAH differ greatly, it means that the ratio of the long to short diagonal lengths of the Knoop indentations varied from the value that was assumed in computing KHN from the standard formula (i.e., elastic recovery and surface upset effects on indentation shapes were significant). In such cases, KHN would not truly reflect the load per unit projected area, and PAH would be a more accurate quantity to use.

Examples of Applications

Microindentation techniques have been used for hundreds of applications throughout the whole spectrum of materials science and engineering. From routine quality control and nondestructive testing* to highly specialized basic research and materials development, a range of micropenetration techniques have become available. Sometimes the desired output is a "microhardness" number, whereas in other cases, it is the nature of impressions and material response. This chapter, of course, can only hint at the broad possibilities of applications. Table 9-5 has been provided as a representative sample to provide the reader with a feeling for the creative approaches taken in the

---

*Whether a technique is considered "nondestructive" or not depends on the application.

Table 9-4. Comparison of Four Microindentation Hardness Testing Methods for Selected Metals and Alloys.*

Conditions: Mechanically polished, unetched specimens with same surface used for all four methods; data for 10 g and 100 g loads.

MATERIAL	KHN (kg/mm²)		PAH (kg/mm²)		VHN (kg/mm²)		SCRATCH WIDTH (μm)	
	10 g	100 g	10 g	100 g	10 g	100 g	10 g	100 g
Sn	6.8	7.1	6.3	7.1	7.8	6.4	35	130
Cd	20.3	23.1	17.3	23.1	21.0	19.5	29.5	71.2
Cu	36.3	27.3	47.8	31.0	—	—	22.5	56.7
Ni	125.	116.	102.	98.2	124.0	103.	16.6	43.4
Fe	285.	281.	212.	198.	197.	249.	11.6	32.1
Co	324.	279.	227.	250.	261.	264.	11.3	27.5
Mo	391.	281.	341.	224.	306.	278.	11.3	27.5
CDA 683 bronze	150.	160.	121.	135.	112.	148.	13.5	38.1
CDA 688 bronze	149.	151.	127.	120.	109.	146.	13.9	42.5
AISI 1010 steel	143.	123.	122.	105.	—	—	15.7	48.2
AISI 52100 steel	—	956.	—	908.	—	868.	—	17.1
Nitinol	237.	120.	371.	306.	196.	258.	9.7	25.8

* Use only as an approximate guide. Relationships may vary based on product form, heat treatment, and microstructural condition. Note that MPa equivalents can be obtained by multiplying table values by 9.81.

past. Given the proper technical awareness, the strengths and weaknesses of specific microindentation procedures should become increasingly apparent and lead to the selection of a test method judiciously suited to the problem at hand.

Thin Coatings

Coatings represent a special class of problems in microindentation hardness testing. A great deal of care must be taken in making and interpreting measurements on coatings, but, treated in an informed manner, such data can be very useful in evaluating their mechanical properties. Sample preparation, load selection, indenter selection, and appropriate calibrations accounting for lamellar test pieces should all be a part of the data-collection process. Because developing technological processes incorporate increasing numbers and types of coatings, special techniques and interpretations for coating tests are continually being developed.

The effect of a coating on the Vickers microindentation hardness number is demonstrated in Fig. 9-4. The test specimen was Cu-electroplated with 1.2 μm of Cr. The indentation depth (Z) in micrometers was calculated from indenter geometry (no vertical elastic recovery assumed) using the equation,

$$Z = 6.15 \sqrt{(P/VHN_p)} \quad (9-3)$$

where P = the load (in g) and VHN_p = the measured Vickers microindentation hardness number of the plated surface (in kg/mm²). The average Vickers microindentation hardness number of the unplated Cu substrate (VHN_s) was 50 kg/mm². At penetrations less than about 12 times the plating thickness, the Cr plating began to influence the VHN. The dashed curve shows a reasonable exponential fit to the given data; however, it should not be used to extrapolate data to very near the surface or to below the ratio, (VHN_p/VHN_s) = 1.0. If one plots the log of the penetration-to-thickness ratio (Z/t) versus log (VHN_p/VHN_s), then Fig. 9-5 results. Extrapolating the low-load (< 50 g) portion of the plot to the ratio (Z/t) = 0.1 (ASTM E384 criterion for valid testing of coatings), then (VHN_p/VHN_s) becomes 6.6, making $VHN_p \cong 330$ kg/mm². This value is somewhat low for Cr platings, but it suggests a possible approach to overcoming such substrate effects. Other work in understanding such "anvil" effects is underway at this writing (e.g., Jonssen and Vingsbo, Ref. 21).

SELECTING A MICROINDENTATION HARDNESS TESTER

After the appropriate hardness test methods and load range(s) have been selected for the materials of interest, criteria for the tester should be established. Not the least important of these is the ease of impression measurement and operator comfort since many hours must often be spent to obtain microindentation hardness numbers by traditional optical methods. Although process automation continues to improve testing productivity, it is important to recognize that valid, accurate microindentation hardness numbers are not automatically a consequence of automation and digitization of tester outputs. This section

Table 9-5. Selected Applications of Microindentation Methods.

APPLICATION	CHOSEN TEST METHOD	REFERENCES
Electrical contacts	Knoop; 25-g load	Hain, et al. *IEEE Trans on Components, Hybrids and Manu. Tech.* CHMT-5(1):16 (1982).
Ion implanted surfaces	Special apparatus; Vickers indenter; capacitance depth measurement; loads up to 2g	Pethica. Conference, "Modification of the Surface Prop. of Metals by Ion Implantation," Manchester, UK, 1981.
Plastic zone mapping near fatigue cracks	Knoop	Saxena and Antolovich. *Met. Trans.* 6A:1809 (1975).
Deformation and fracture of dual-phase steel	Vickers; 10-g load; hardness versus load curves	Szewezyk and Gurland. *Met. Trans.* 13A:1821 (1982).
Temper embrittlement	Vickers; acoustic emission during indentation	Clough and Wadley. *Met Trans.* 13A:1965 (1982).
Electrical current effects on semiconductor hardness	Vickers; 10-g load; coupled with electrical resistivity measurements	Westbrook and Gilman. *J. Appl. Phys.* 33(7):2360 (1967).
Grain-boundary solute migration	Vickers; 5-g load	Lozinsky and Ferenetz. *Acta Met.* 12:1255 (1964)
Thin films of LiF on glass	Triangular tipped diamond; special apparatus; 1 to 100-mg loads	Nishibori and Kinosita. *Japan J. of Appl. Phys.* 11:758 (1972).
Dislocation cell structure microhardness correlations in Cu	Knoop; two-diagonal method	Wey, et al. *Wear of Materials,* ASME Conf. Proceedings, 1983.
Deformation of separate constituents in a two-phase Cu–Al alloy	Vickers inferred but not stated explicitly; 5-g load	Yegneswaran and Tangri. *Met. Trans.* 14A:2407 (1983).
Wear correlations with plated surface hardness	Spherical indenter; function of load and penetration depth versus wear	Clinton, et al. *Wear* 7:354 (1964).
Electrical contacts, Au	Knoop; various depths of penetration	Antler and Drozdowicz. *Bell Sys. Tech. Journal* 58(2):323 (1979).
Creep; hot hardness; adhesion	Review of previous work theories	Maugis. *Wear* 62:349 (1980).
Degree of recrystallization	Vickers; combined micro-macrohardness techniques	Kaspar. *Praktische Metallog.* 11:671 (1974).
Wear of recording heads	Knoop indentation dimensional changes	Begelinger and deGee. *Wear* 43:259 (1977).
Anodic coating adhesion on Al alloys	Knoop; 1 to 50-g load	Thomas. *J. Am. Electropl. Soc.* 70(7):53 (1983).
Sliding wear deformation studies	Knoop; 10-g load; aspect ratios; two-diagonal measurement method	Blau. *Scripta Met.* 14:719 (1980).
Fracture toughness in ceramics	Vickers; crack lengths near indentations used to estimate toughness	Anstis, et al. *Am. Ceramic Soc.* 64(9):533 (1981).
Electroplates of Au	Scratch tests using Vickers; 5 to 25-g load and down to 0.1 mg with scratching-mode Vickers indenter	Lo. *Plating,* March, 1973, p. 249.
Thin film adhesion of metal on glass	Special apparatus; various loads on a spherical-tipped, chrome-steel indenter	Benjamin and Weaver. *Proc. Roy. Soc. A.* 254:163 (1960).
Abrasive wear studies	Vickers indenter	Torrence. *Wear* 67:233 (1981).
High temperature indentation; rate-controllable testing of steels	Special apparatus; Vickers sapphire pyramid indenter; 0.2 to 250-g load; up to 1600°C	Taoka, et al. *Trans. Japan Inst. Metals* 21(12):773 (1980).
Rate effects on indentation behavior	Special apparatus; Vickers indenter	C. Fairbanks, et al. *J. of Matl. Science Letters* 1:391 (1982).

lists a number of machine characteristics that should be considered in addition to merely cosmetic factors and lowest costs.

The following checklist includes the most desirable attributes of a high-quality microindentation-hardness testing machine and some comments on these based on the author's experience:

- The base should be solid, stable, and free from wobble, with a low center of gravity. High heavy indenter/microscope heads could lead to increased vibration.
- Built-in leveling features should be provided.
- A vibration-isolated table for bench-model testers is very desirable, especially if low (less than 100-g) loads are to be used.
- Once the test cycle is initiated, it should proceed to complete indenter withdrawal without further need for the operator to touch the machine. Ideally, a delay to allow for vibration damping between the time the operator presses a button on the base and the cycle start-up occurs should be included (unless remote electronics are used).
- The indentation process should be smooth and free from indenter bounce.
- Adequate clearance for specialized fixturing or hold-

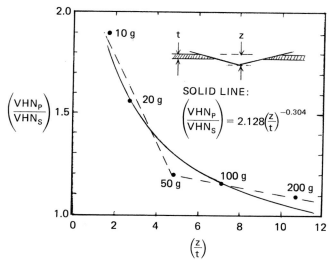

Fig. 9-4. Ratio of apparent Vickers microindentation hardness of a Cr-plated Cu sample to that of the substrate for various relative indentation depths through the electroplate. Exponential fit to the data is shown by the solid line. The two-segment dashed curve seemed to fit better.

- Indenter fixturing, changing, and targeting alignment should be easily and quickly achievable.
- A calibrated-precision, button-tipped compression-load cell can be used in place of a specimen to verify the accuracy of selected test loads. Tests in the load range of the expected application should be performed. Extrapolations of mechanical performance, especially to low load ranges, should be avoided.
- Microscope optics and measuring scales should be sharp and focusable for individual operators. Not only light intensity, but also adjustable diaphragms, often help greatly in defining the edges of impressions. Sometimes a calibrated high-quality metallograph can be used to make up for a mechanically good but optically mediocre microhardness tester. Scanning electron microscopes (SEM) with calibrated magnifications* have also been used for critical applications or tiny impressions.
- A fixed cross hair or other indicator for targeting indentations is very desirable. Some machines do not have provisions for accurate targeting but require

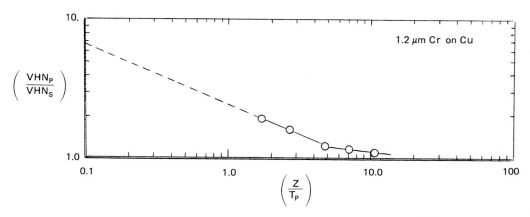

Fig. 9-5. Log-vs-log plot of Fig. 9-4 data showing extrapolation of low-load data (below 50 g) to $(z/t) = 0.1$.

ers and provision for easy mounting of the fixtures should be provided, but too much clearance can lead to a top-heavy machine, with increased vibration sensitivity and reduction in stiffness.
- Internal switches to prevent accidental cycle start-up if the indenter is not in position are desirable.
- Adequate clearance should be allowed between the specimen or specimen holder and the indenter tip during positioning and stage translation.
- Several attempts to target a given microstructural feature using the micrometer stage controls will give indications of drive train backlash, alignment accuracy, and ease of reading stage markings. Rotation, as well as X–Y stage motion, is often an advantage.
- Controls should be easy to find and manipulate, but the main cycle start switch should be at a point where accidental activation is less likely.

one to remember where a movable cursor line should be, etc.
- Several selectable magnifications on the microscope will permit easier test-area location and measuring accuracy when impressions vary greatly in size because of load and/or hardness differences. Each objective lens should be calibrated for measurement by specific operators since judgments of indentation edge positions can vary. ASTM Test Method E384 provides guidelines in this area.
- Company field recalibration and field maintenance services are worthy of consideration, as are warranty provisions.

---

* National Bureau of Standards, Standard Reference Material No. 484 is a magnification calibration standard designed for the SEM.

Commercial Machines

A wide variety of commercial microindentation-hardness testing equipment and accessories is available; several examples are shown in Figs. 9-6 to 9-10. Their inclusion here does not constitute an endorsement or certification by the U.S. National Bureau of Standards, but only a means of depicting the range and styles of equipment in use at this writing. Readers should consult various manufacturers for detailed specifications and available accessories. Once the choice of equipment type has been narrowed, there is no substitute for a demonstration that involves the type(s) of samples on which the equipment will be used. Advice of technicians who must use the testers on a day-to-day basis should be sought.

Often manufacturers will offer special fixtures, such as self-leveling mounts, thin sample vises, or other odd-shaped sample holders, as accessories. Television monitors, digital filar oculars, and data printers are also widely available.

An ultra-low-load Vickers-type microindentation-hardness accessory for *in situ* use with the scanning electron microscope (SEM) is made by Anton Paar Co. in Austria. It applies loads up to 2 grams by a leaf-spring arrangement. Finding the indentations in some cases requires skill and practice on the part of the SEM operator.

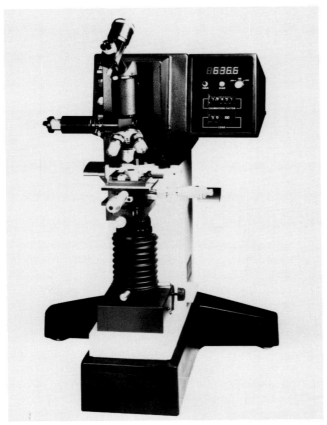

Fig. 9-6. Example of a dash-pot mechanism-based machine (Courtesy, Page-Wilson Corp.).

Research Machines

Numerous microindentation producing machines have been built for specialized research on materials. Some of these are capable of applying μN loads and measuring micro-stress/strain behavior.[16] Others have been used to study loading rate effects.[22] Still others have been used to study abrasion behavior.[23] Machines have been developed to study cracking of transparent ceramics and glasses *in situ* during the indentation process by optical microscopy through the back of the specimen.[24] Many of the same criteria for selecting a tester (see the previous checklist) can be applied in designing and building research machines for specialized studies. A suitably chosen, commercially manufactured machine can be modified for specialized research in certain cases, thereby saving time and "reinvention" engineering problems.

Microindentation hardness numbers obtained at room temperatures may not be representative of those in either elevated- or low-temperature service, and relationships between microhardness numbers and temperature may be complex. Figure 9-11 demonstrates such effects for Fe, Cr, Co, and Au.[25] Such complex behavior is not unexpected because of the influence of temperature on microstructure and mechanical properties such as flow stress and elastic modulus. Specialized, high-temperature microindentation-hardness measurement instruments are commercially available (e.g., see Fig. 9-12), and others have been constructed for individual research investigations (e.g., see Ref. 26).

SPECIMEN PREPARATION AND FIXTURING

Some kinds of microindentation-hardness test samples, such as smooth platings on flat coupons, can simply be clamped in place on the testing machine and tested without a need for special mounting or further preparation, whereas other kinds of samples, such as those cut from larger pieces or those requiring indentations in particular orientations, require special techniques in preparation and fixturing. Sometimes the mounting procedures can involve special metallographic techniques, especially when materials with cross-sectioned thin coatings, lamellar microstructures, stepped surfaces, or other complications are involved. This section will describe several methods of specimen preparation and fixturing that may be helpful. Obviously, many other approaches besides those given here may be effective, given the needs of specific samples.

Metallographic Considerations

In order to obtain valid representative microindentation hardness numbers, especially at the low range of test loads, no effects of sample grinding or polishing damage can be tolerated. If indenter loads are over about 100 to 200 g (0.98 to 1.96 N), careful mechanical metallo-

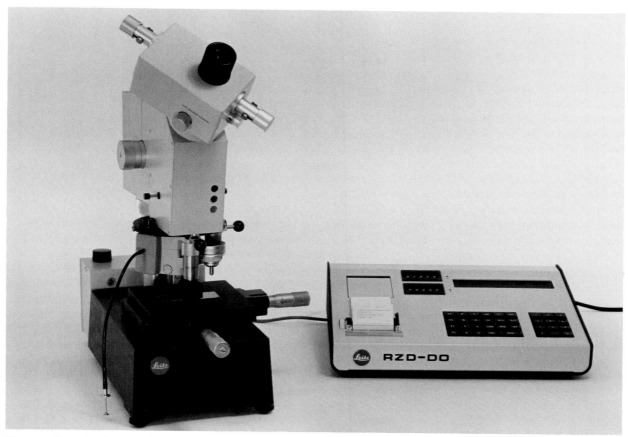

Fig. 9-7. Example of a digital measurement testing system with revolving nosepiece to allow targeting precision (Courtesy, E. Leitz).

Fig. 9-8. Example of an automated programmable stage accessory for producing a prespecified test matrix of microindentations (Courtesy, Leco Corp.).

134 APPLIED METALLOGRAPHY

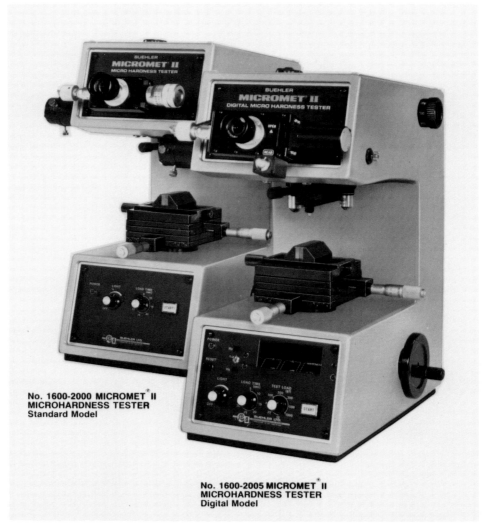

Fig. 9-9. Two compact digital and manual reading machines (Courtesy Buehler Ltd).

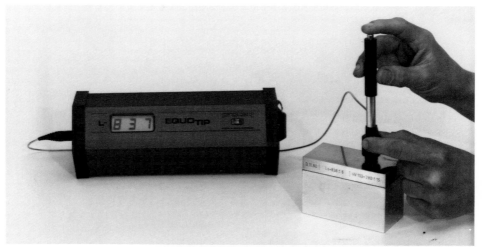

Fig. 9-10. Portable, spring-loaded hardness tester for field use. The instrument is supplied with conversion tables for various traditional scales. (Courtesy, Hentschel Instruments, Inc.)

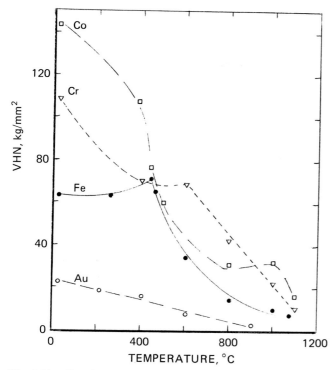

Fig. 9-11. Complex microindentation hardness behavior of several metals over a range of temperatures. Changes in elasto-plastic properties as well as crystal structure transformations contribute to such behavior.

too much damage (as it might under low loads), electrolytic or chemical polishing is suggested.[27] Occasionally, etching of indented samples may be required to observe the extent of lateral microindentation damage and to make sure that individual tests are spaced widely enough apart so as not to interfere with one another. The reader should be cautioned that excessive chemical or electrolytic polishing can also produce undesirable surface irregularities as a result of different polishing rates from one phase to another or one crystallographic orientation to another.

A particularly challenging problem involves composite materials whose constituents differ greatly in microindentation hardness and/or chemical activity. Final polishing invariably produces surface relief in such cases. Use of finer grades of abrasive and diamond for final polishing may help reduce the relief effects.[27] It may be necessary to choose a different indenter geometry (e.g., more acute facet angles or elongated shape) to fit between upstanding constituents. Changing the orientation of the polished section in an aligned composite may also help.

As discussed above, good metallographic technique may be combined with more favorable sample orientation in the mount to permit a sufficient volume of material to be presented to the indenter. Taper sectioning, as discussed by Moore,[28] has proven to be extremely useful in this respect. Figure 9-13 demonstrates several advantages of this technique as applied to near surface microindentation hardness-versus-depth profiling. Instead of using an orthogonal cross section of a plated surface, the sample is mounted at a selected tilt angle (α) so as to "magnify" the microstructural features lying in the "downhill direction" of the original surface by a factor of M ($M = 1/\sin \alpha$). For the given example, this is shown to greatly increase the number of indentations pos-

graphic procedures usually suffice. Etching is not advisable before testing or reading indentations since it can create irregular surfaces. If the phases to be tested cannot otherwise be located, however, they can be etched, their locations marked with a pattern of surrounding microindentations, and the sample repolished before indentation.

If it is suspected that mechanical polishing will produce

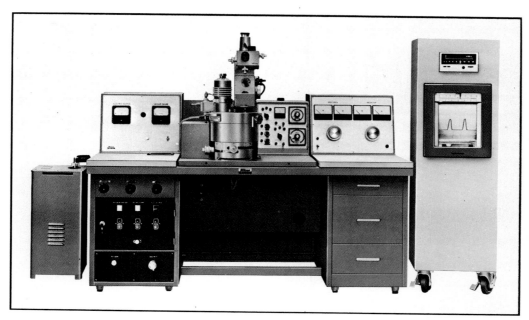

Fig. 9-12. A specialized high-temperature (specifications indicate up to 1600°C) commercial microindentation tester (Courtesy, Nikon).

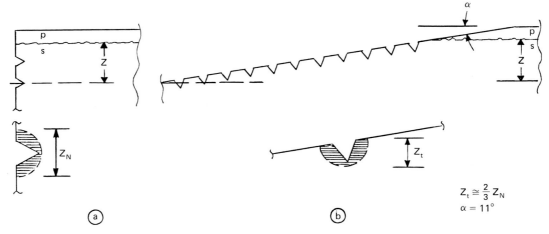

Fig. 9-13. Comparison of a taper-polished section with a normal cross-section of a plated sample. More indentations are possible in the near-surface regions, but angled relative loading direction of indentations and changing location below the surface are possible drawbacks to the technique.

sible within the near surface region and to alleviate part of the problem of testing near a free surface. Moreover, the direction of indenting is aimed more at the bulk sample than parallel to the original free surface. As Moore has pointed out,[28] one drawback to the technique is that, unlike the case with normal cross sections, taper polishing exposes material below a range of points on the original surface rather than below the same line of cut on the free surface. Therefore, the microindentation hardness must be assumed to vary in the same way with depth below every point on the original surface. Another drawback is that tapered layers are wedge-shaped where they intersect the plane of polish. Taper sectioning, however, has been used successfully for coatings,[29] worn surface/subsurface studies,[30] and machining damage assessments.

Polishing tangentially to a shaft or other cylindrical component is a form of pseudo taper sectioning. It can provide the same kind of effect, and the depth of a given location is easily obtained. Figure 9-14 displays an example of pseudo taper sectioning on a cylinder. Plating thicknesses can be measured at the same time as the survey of microindentation hardness versus depth is taken.

Microindentation hardness testing of particles usually involves embedding them in a metallographic mounting medium and polishing the mount to display a series of cross sections. Loads should be chosen to avoid free-surface effects. Incorporation of metal particles in electrodeposits has also proven to be an effective mounting technique.[31]

Fixturing Methods

The proper fixturing method for microindentation hardness test specimens should be determined by the size, shape, and number of replicate specimens to be tested. Testing machines in research and development laboratories should be equipped with a varied selection of clamps and fixtures so that many different sample configurations can be tested. In production or quality control laboratories, the predominance of a few standard sample configurations warrants constructing custom fixtures. A proper fixture should hold a specimen securely without damaging it and in a manner such that the test surface is normal to the direction of indentation. If specimens are improperly fixtured, microindentations will repeatedly appear asymmetrical. Loosely fixtured samples could move during load application or during translation of the mechanical stage. If clamping is too tight on certain kinds of thin

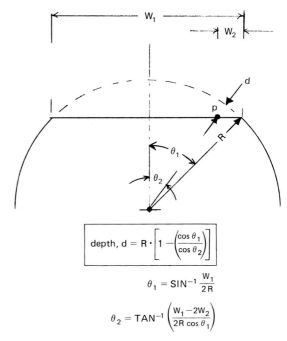

$$\text{depth, } d = R \cdot \left[1 - \left(\frac{\cos\theta_1}{\cos\theta_2}\right)\right]$$

$$\theta_1 = \sin^{-1}\frac{W_1}{2R}$$

$$\theta_2 = \tan^{-1}\left(\frac{W_1 - 2W_2}{2R\cos\theta_1}\right)$$

Fig. 9-14. Pseudo-taper section geometry giving an equation to calculate the radial distance d of a point P from the surface given the cylinder radius R and measured distances W_1 and W_2 on the plane of polish.

samples, buckling of the sample or undue stressing of coatings may result.

Appropriate fixturing can be as simple as a pair of leaf-spring clips or as complex as a multiaxis leveling vise. Before designing sophisticated fixtures, equipment manufacturers should be consulted. Many specialized accessories have already been developed. One common type of self-leveling fixture for standard metallographic mounts is represented schematically in Fig. 9-15. The ball and socket joint below the specimen and the flat-bottomed plate assure that even samples with nonparallel top and bottom surfaces are presented normal to the indenter. One drawback of this type of fixture is that the microscope objective (especially the higher magnifications) may sometimes bump into the edge of the hold-down plate and prevent examination of the edges of the sample. Clearance of both the microscope objectives and the indenter tip should be carefully checked when using this type of fixture.

When low loads are being used, it is possible to use various adhesives, including double-backed adhesive tape, to mount specimens. Before doing so, however, the operator should examine the mounting arrangement to determine that the back of the sample immediately below the indenter is securely supported. If specimens are very small or otherwise difficult to handle, it may be advisable to cast them in a quick-setting plastic mount. Once mounted, they are more easily handled and fixtured. Furthermore, they may be archivally stored in standardized containers for re-examination at a later date. Larger mounts are also easier to mark clearly.

Sometimes the mounting arrangement will be dictated by other uses to which the microindentation hardness samples are subjected. For example, samples may also be used for photomicrography or scanning electron microscopy. In the latter case, the mount may have to be electronically conductive as well as rigid, or its size and shape may have to conform to sample chamber requirements. Forethought in this area can do much to avoid remounting problems. Furthermore, when hardness testing is completed, the indenter can be used at higher loads to mark interesting areas for expedient targeting in the scanning electron microscope.

SUMMARY

Microindentation hardness tests, including both quasistatic and scratch tests, are sensitive to different combinations of basic surface and bulk properties. Interconversion of values from one hardness scale to another should not be attempted without enlightened attention to the composition, form, and microstructures of the materials being tested. Special techniques for sample preparation and fixturing have been presented, as well as citations of representative references to demonstrate how microindentation data have been used in a variety of specialized studies. Selection of a proper test method and testing machine has also been discussed.

REFERENCES

1. Bückle, H. *Mikrohartprüfung und Ihre Anwendung.* Stuttgart: Berliner Union, 1965.
2. Tabor, D. *The Hardness of Metals.* Oxford: Oxford at the Clarendon Press, 1951.
3. Mott, B. W. *Microindentation Hardness Testing.* London: Butterworths, 1956.
4. O'Neill, H. *Hardness of Metals and Its Measurement.* London: Chapman and Hall, 1967.
5. Knoop, F.; Peters, C. G.; and Emerson, W. B. A sensitive pyramid-diamond tool for indentation measurements. *National Bureau of Standards Journal of Research* 23:39–61 (1939).
6. Blau, P. J. Effect of heat treatment and electron-beam surface melting on the wear and friction behavior of a Cu–12 wt % Al Alloy. *Wear* 94:1–2 (1984).
7. Blau, P. J. Use of Knoop indentations for measuring microhardness near worn metal surfaces. *Scripta Metall.* 13:95–98 (1979).
8. Blau, P. J. Use of a two-diagonal method for reducing scatter in Knoop microhardness testing. *Scripta Metall.* 14:719–724 (1980).
9. Tate, D. A comparison of microhardness indentation tests. *Trans. A.S.M.* 35:374–384 (1945).
10. ASTM. Standard Test Method E-384. In *Annual Book of ASTM Standards,* vol. 03.01:496–517 (1983).
11. ASTM. Standard Test Method E-92. Ibid: 241–251.
12. Brookes, C. A. The indentation hardness of diamond and other crystals. *Industrial Diamond Rev.* 21:338–341 (1973).
13. ASTM. Standard Test Method D-1474. In *Annual Book of Standards,* Part 27 (1982).
14. Brookes, C., and Moxley, B. A. Pentagonal indenter for the measurement of hardness. *J. Appl. Phys.* E8:456 (1975).
15. Gasilin, V. V., and Kunchenko, V. V. Application of an indenter penetration recording device to the study of mechanical characteristics of materials. Zavodskaya Laboratoriya 46(4):353–356 (1980).
16. Pethica, J. B. Microhardness tests with penetration depths less than ion implanted layer thickness. Conference, "Modification of the Surface Properties of Metals by Ion Implantation." U.M.I.S.T., Manchester, England (1981).

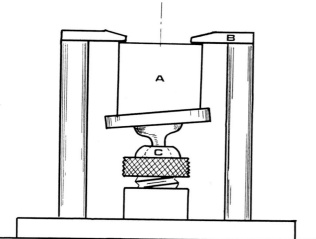

A = SAMPLE WITH NON-PARALLEL FACES
B = LEVEL HOLDDOWN PLATE
C = BALL SWIVEL

Fig. 9-15. General features of a self-leveling sample holder.

17. Lo, C. C. Microindentation and microscratch hardness tests for thin electroplates. *Plating* 60:247–250 (March 1973).
18. Tabor, D. The physical meaning of indentation and scratch hardness, *British J. of Appl. Phys.* 7:159–166 (May 1956).
19. Benjamin, P., and Weaver, C. Measurement of adhesion of thin films. *Proc. Roy. Soc.* 254A:163–176 (1960).
20. Hintermann, H. E. Exploitation of wear and corrosion-resistant CVD coatings. *Tribology International* 13:267–277 (1980).
21. Jonsson, B.; Hogmark, S.; and Vingsbo, O. University of Uppsala, Sweden. Private Communication (1982).
22. Fairbanks, C. J.; Polvani, R. S.; Wiederhorn, S. M.; Hockey, B. J.; and Lawn, B. R. Rate effects in hardness. *J. Mat. Sci. Letters* 1:391–393 (1982).
23. Torrance, A. A. A new approach to the mechanics of abrasion. *Wear* 67:233–257 (1981).
24. Lawn, B. R. Kinetics of shear-activated indentation crack initiation in soda-lime glass. *J. Materials Sci.* 18:2785–2797 (1983).
25. Ivanko, A. A. *Handbook of Hardness Data.* Kiev: Naukova Dumka (1968).
26. Taoka, T.; Nakajuna, F.; Hirano, Y.; Arita, K.; and Nishimura, K. Functional features of a hot micro-Vickers hardness tester with loading rate controllable device and its metallurgical applications. *Trans. Jap. Inst. of Metals* 21:773–780 (1980).
27. Vander Voort, G. *Metallography: Principles and Practice.* New York: McGraw-Hill Book Co., Inc. (1984).
28. Moore, A. W. J. Taper sectioning. *Metallurgia* 38:1–12 (1948).
29. Ruff, A. W., and Lashmore, D. S. Dry sliding wear studies of Ni–P and Cr coatings on 0–2 tool steel. *ASTM Spec. Tech. Pub.* 769:134–156 (1982).
30. Blau, P. J. Investigation of the nature of microindentation hardness gradients below sliding contracts in five copper alloys worn against 52100 steel. *J. Materials Sci.* 19:1957–1968 (1984).
31. Shives, T. R., and Smith, L. C. Microindentation hardness measurements on metal powder particles, in *ASTM Spec. Tech Publ.* 889:243–256 (1986).

10
THE SEM AS A METALLOGRAPHIC TOOL

George F. Vander Voort
Carpenter Technology Corp.

Since its commercial introduction about 20 years ago, the use of the scanning electron microscope (SEM) in materials science has grown to the point where it is a rather common tool. Indeed, its use rivals that of the light optical microscope (LOM). This popularity stems from the substantial progress made by instrument manufacturers, the wide range of instruments available, and the many advantages of the SEM—ease of specimen preparation (compared to other electron microscopy techniques), high resolution, extensive depth-of-field, and wide range of magnification. An additional advantage is that chemical analysis can also be performed by the use of suitable attachments to the SEM.

The SEM has largely replaced the transmission electron microscope (TEM) for fractographic investigations. Through the addition of energy-dispersive and wavelength-dispersive X-ray detectors and suitable signal processing units, the SEM can also perform phase analysis of bulk or extracted specimens, thus accommodating many of the chores traditionally limited to the TEM or the electron microprobe analyzer (EMPA).

Some SEM enthusiasts also view the SEM as a replacement for the light microscope. However, just as the SEM cannot totally replace either the TEM or the EMPA, it also cannot be regarded as a complete substitute for the light microscope. Indeed, errors in interpretation can be, and have been, made by examining specimens *only* with the SEM.

The contrast mechanisms that produce SEM images are different from those that produce images in the light microscope. Contrast produced from topographic features (surface roughness) results in similar images in both the SEM and the light microscope—a feature that makes interpretation of these images rather straight-forward. Other contrast mechanisms, however, are different. For example, LOM images may also be produced by light reflectivity differences, color differences, or polarization effects—none of which are detectable in the SEM. Likewise, SEM images may also be produced by contrast mechanisms not observable in the LOM, for example, atomic number contrast or electron channeling contrast. Consequently, microscopists must have a basic understanding of the composition of a specimen if they are to select the most suitable observational procedures for taking advantage of these contrast mechanisms.

As a general rule, it is always best to conduct the preliminary examination of specimens using the LOM first. The reason for this rule is one of practicality. First, if the specimen preparation is inadequate to observe the microstructure in the LOM, it will also be inadequate for SEM observation. Preparation artifacts that obscure the microstructure when viewed in the light microscope will similarly influence SEM examination. Second, the entire specimen surface can be examined far more quickly with the LOM than the SEM. Features of interest can be circled with an objective marker (an objective with a diamond-tipped stylus), thereby producing scribe marks of variable diameter that can be quickly located in the SEM. Many specimens can be thoroughly examined by light microscopy in the time required to examine a single specimen in the SEM.

This review will concentrate on the use of the SEM to examine microstructures in bulk specimens. There are many excellent publications that review the principles of SEM construction, operation, and performance.[1-8] Although the SEM applications literature is quite extensive, only a few publications have concentrated on the use of the SEM to study microstructures per se.[9-17] This paper will concentrate on this important aspect of SEM usage by comparing and contrasting results obtained by light microscopy and by scanning electron microscopy. Emphasis will be placed on the practical aspects of such work. This study shows that these two observational methods are complementary rather than competitive and, hopefully, will show the SEM user how to obtain more useful microstructural images.

SEM IMAGE CONTRAST FORMATION

Regardless of the tool used, LOM or SEM, the objective of the examination is to produce an image with the necessary magnification to permit resolution of the desired features. Resolution, however, is only one aspect of image visibility. Equally important is the contrast that is developed between different constituents. A contrast (brightness) difference of at least 3 to 7 percent must be present if the human eye is to observe the image details. Greater differences are obviously desired. Additionally, there should be an abrupt change in contrast at the edge of the features to produce "sharp" details. This is especially important if accurate measurements of the structure are to be made, particularly when automated systems are used. Irrespective of the resolution potential of the instrument, the sharpness of the image will establish the ultimate resolution obtained, i.e., resolution of features requires visible features with sharply defined interfaces. As with the LOM, the SEM depends heavily upon the nature of the specimen and the quality of specimen preparation. To obtain optimal LOM or SEM images, specimen preparation itself must be optimal.

The most common sources of SEM image contrast stem from topographic and compositional (atomic number) differences. Major emphasis, therefore, will be placed on these effects. The SEM has several operational modes for image production. Those modes most useful for microstructural examination of bulk specimens are (1) emissive, (2) reflective, (3) absorptive, or (4) X-ray. Newbury[18] has prepared an excellent review of contrast mechanisms that the interested reader may wish to consult for additional details.

The *emissive mode* utilizes low-energy secondary electrons emitted from the specimen surface to a depth of up to about 10 nm. Very high resolutions can be obtained with such images; about 4 nm is the typical limit for many modern SEMs.

The *reflective mode* employs the higher energy backscattered electrons which come from a much greater depth (up to several micrometers). Secondary electron (SE) images are primarily formed from topographic contrast, whereas backscattered electron (BSE) images arise from compositional (atomic number) contrast.

The *absorptive mode* is obtained by detecting the signal flowing through the specimen to ground, the procedure widely used in electron microprobe work. At a constant beam current, the absorbed current or sample current image is produced by variations in the emitted SE and BSE signals. Such images are essentially the inverse of BSE images.

The *X-ray mode* utilizes the characteristic X-rays generated by the incident beam to display or analyze the distribution of chemical elements at, and slightly below, the specimen surface. These signals may be detected using either an energy-dispersive or wavelength-dispersive detection system, the former being more common.

Secondary Electrons

When the electron beam strikes the specimen surface, it penetrates to a depth of about 1 micrometer. The penetration depth increases with increasing beam current. The incident electrons collide with atoms in the specimen, thereby displacing electrons from their orbits. These "secondary" electrons are produced throughout the excitation volume; because they possess very low energy, however, only those very close to the surface can escape to produce a detectable signal. Thus, this signal is generated from an area slightly larger than the beam diameter and to a depth of less than about 10 nm. Part of the signal, in the form of background noise, also comes from "remote" secondary electrons that are excited by escaping backscattered electrons.

Topographic Contrast

Topographical effects, i.e., surface roughness, alter the production of secondary electrons. A sharp edge in relief will promote the escape of secondary electrons, thereby enhancing image contrast. Likewise, tilting of the specimen relative to the beam will increase the generation of secondary electrons and enhance contrast. Because the path of these electrons to the detector does not require line-of-sight orientation to the detector, specimen tilting is widely utilized. However, although image visibility is enhanced, measurements are complicated because tilting increases the apparent feature spacing and the magnification varies across the image. Image distortion can also result, although modern SEMs correct this problem reasonably well.

Since surface topographic contrast produces light-dark shading contrast analogous to that observed by light microscopy, interpretation is easy. The SE image produces edge brightening, a feature not present in LOM images, but no confusion results. It is important to orient the photomicrographs properly (detector at the top) because elevated features will appear to be depressions if the picture is rotated 180° for viewing. This is less of a problem when the specimen is normal to the beam, but it is still present. Compositional contrast can also be detected in SE images because the SE detector invariably collects some primary backscattered electrons that contain atomic number information. This information is most noticeable when the specimen is tilted towards the Everhart–Thornley (E–T) detector, which is widely used for producing SE images.

The number of secondary electrons escaping from a surface is strongly influenced by the angle between the incident electron beam and the local normal to the specimen surface; this influence is particularly strong for angles of about 45°. As this angle is increased to 45°, the length of the primary electron path at the extreme surface (to a depth of 10 nm) increases, and the number of escaping secondary electrons also increases.

Backscattered Electrons

A portion of the high-energy electrons striking the specimen surface are subject to Rutherford scattering and re-

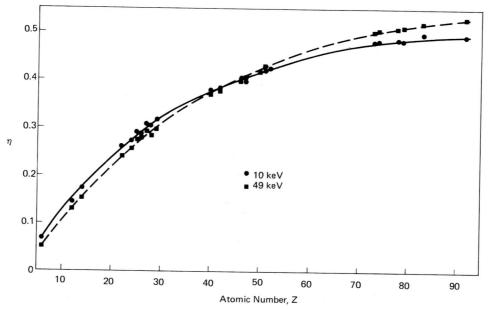

Fig. 10-1. Relationship between the backscatter coefficient, η, and atomic number for 10- and 49-keV beam potentials (data from Heinrich, Ref. 20).

emerge from the specimen surface with nearly as much energy as the incident electrons. These backscattered electrons come from a much greater depth than the secondary electrons. Consequently, the image resolution achievable with BSEs is not as good as with SEs but somewhat better than the best that can be achieved with the light microscope. Because they arise from a greater area and depth than the secondary electrons, BSEs usually provide little information about the specimen surface topography unless two or four opposed solid-state detectors are used for signal subtraction. In certain cases, depending on the nature of the specimen and the type of BSE detector, strong topographic contrast can be obtained.

Despite these negative aspects, backscattered electron images are highly informative because the efficiency of producing backscattered electrons depends on the average atomic number of the constituents and such images thus reflect the compositional differences present in the specimen. Because polished metallographic specimens are quite smooth unless heavily etched (not always desirable), topographic contrast will be quite weak, but atomic number contrast can provide very useful images.

Atomic Number (Compositional) Contrast

The generation of backscattered electrons has been shown to increase with increasing atomic number Z.[19-22] Figure 10-1 shows data developed by Heinrich[20] that demonstrates the relationship between the atomic number and the electron backscatter coefficient, η, a measure of the relative amount of backscattered electrons produced. The incident beam energy has little influence on the backscatter coefficient. Figure 10-2 shows a semilogarithmic plot of measurements of η as a function of Z for a constant accelerating potential of 30 keV using data from four studies.[19-22] For Z values above 13 (Al), the data is linear when plotted in this manner. Heinrich developed the following equation to permit calculation of the backscatter coefficient for a homogeneous alloy or compound:

$$\overline{\eta} = \sum_i C_i \eta_i \qquad (10\text{-}1)$$

where C_i represents the weight fractions of the components of the alloy or phase and η_i, the backscatter coefficients of the pure elements present in the alloy or compound.

To illustrate the use of secondary electrons and backscattered electrons to examine the microstructures of

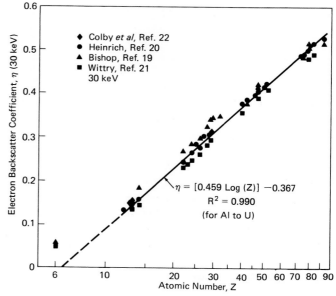

Fig. 10-2. Semilog plot of atomic number vs. backscatter coefficient, η, using 30-keV data from four sources.

142 APPLIED METALLOGRAPHY

specimens with high atomic-number differences, three examples will be given. Figure 10-3 shows an as-polished specimen of as-cast (monotectic) Al–27-wt.%Bi. The LOM micrograph reveals the microstructure by means of the light reflectance differences between the aluminum matrix and the bismuth particles. Polishing has produced minor relief between these two relatively pure phases, resulting in a dark line around the particles. Figure 10-3 shows the microstructure of this sample with the Everhart–Thornley detector and secondary electrons and with the solid-state detector (combined signal from both detectors, A + B) placed directly above the specimen. In both cases, the sample was normal to the beam (no tilt), and the potential of the incident beam was 20 keV. For aluminum ($Z = 13$) and bismuth ($Z = 83$), the difference in the backscattered coefficients is 0.369, or 71.9 percent. Because there is very little relief, and the specimen was not tilted, both images are virtually identical, i.e., image contrast was solely due to the compositional contrast between these two relatively pure phases.

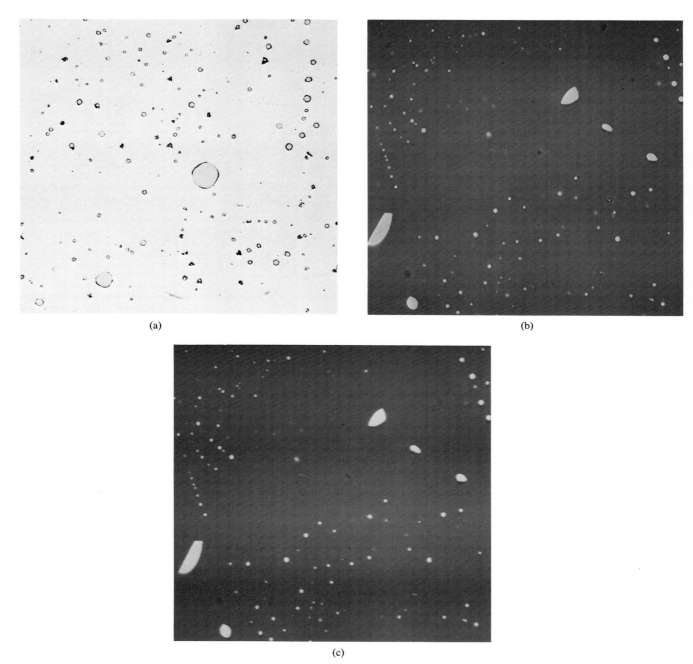

Fig. 10-3. LOM and SEM images of an as-cast Al–27-wt.% Bi alloy with high atomic-number contrast ($\Delta\eta = 0.369$, or 71.9%): (a) light microscope image, as-polished, 500×; (b) secondary electron image, as-polished, 20 keV, no tilt, 500×; and (c) backscattered electron image, as-polished, A+B, 20 keV, no tilt, 500×.

Note that the particle interfaces are not sharp but slightly blurred. One can also see some faint, small particles that probably arise from particles just beneath the surface.

Figure 10-4 shows the microstructure of as-cast (monotectic) Zn–10-wt.%Bi that was lightly etched after polishing. Light etching was helpful for producing good LOM images of these relatively pure phases (pure Bi in a pure Zn matrix) because the reflectivity difference was less than for the Al–Bi specimen. The SEM micrographs in Fig. 10-4 were taken with the specimen normal to the beam, a position that minimizes the etching topography. For this specimen, the differences in the backscatter coefficients for zinc ($Z = 30$) and bismuth ($Z = 83$) is 0.202, or 39.3 percent. The E–T detector was used to produce SE and BSE images of the microstructure using a beam potential of 20 keV. Again, contrast is due to the atomic number difference. The E–T BSE image is more contrasty than the SE image because the matrix is darker. A backscatter electron image was also produced using the two solid-state detectors (combined A + B image) at 30 keV,

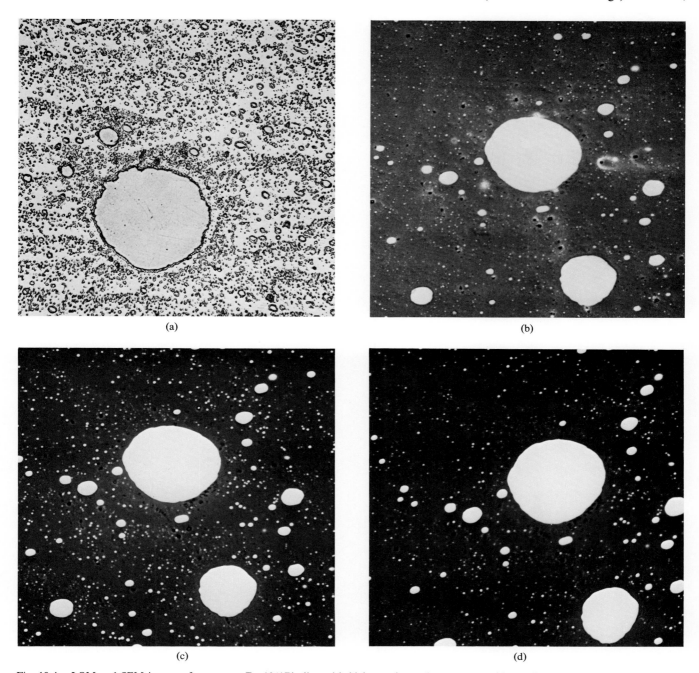

Fig. 10-4. LOM and SEM images of an as-cast Zn–10%Bi alloy with high atomic-number contrast, with no tilt used ($\Delta\eta = 0.202$, or 39.3%): (a) light microscope image, etched with 2% nital, 500×; (b) secondary electron image, 20 keV, 1000×; (c) backscattered electron image, 20 keV, 1000× (Everhart-Thornley detector); (d) backscattered electron image, 30 keV, 1000× (A+B solid-state detector).

and this resulted in a still greater contrast. Light etching helped to produce sharper interfaces in these micrographs than in the Al–Bi micrographs.

In these two examples, the minor phase had a much higher atomic number (high Z) than the matrix. Consequently, the high atomic-number phase appears bright, whereas the low atomic number phase (low Z) appears dark. The contrast will be reversed if a light atomic number phase is dispersed within a heavy atomic number matrix.

An example of such contrast is provided by graphitic cast irons where essentially pure carbon ($Z = 6$) is dis-

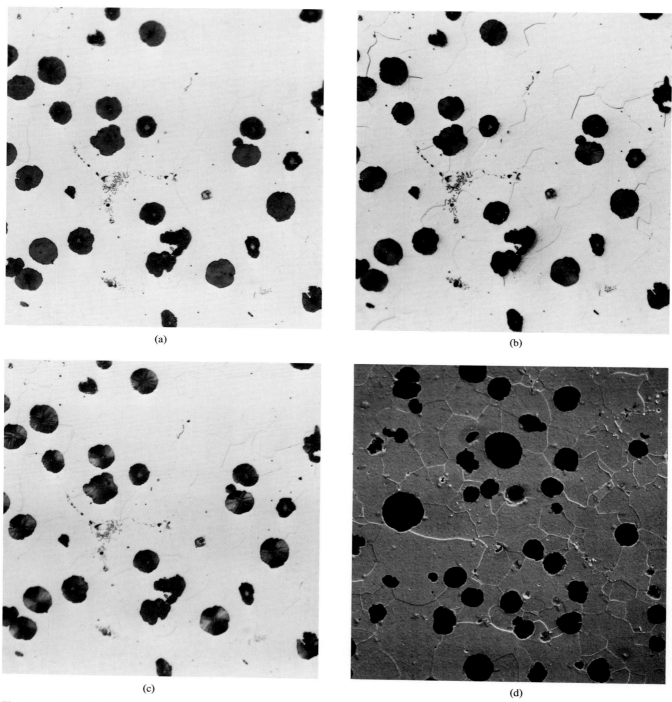

Fig. 10-5. LOM and SEM images of ductile cast iron containing graphite nodules ($\Delta\eta = 0.227$, or 79%): (a) light microscope image, 2% nital etch, bright field illumination, 400×; (b) light microscope image, differential interference contrast mode, 2% nital etch, 400×; (c) light microscope image, crossed polarized light, 2% nital etch, 400×; (d) secondary electron image, 2% nital etch, Everhart-Thornley detector, 500×; (e) backscattered electron image, Everhart-Thornley detector, 2% nital etch, 500×; (f) backscattered electron image, A+B solid-state detector, 500×; (g) backscattered electron image, A−B solid-state detector, 2% nital etch, 500×; and (h) specimen current image, 2% nital etch, 500×.

persed in iron ($Z = 26$, ignoring solid-solution elements). Figure 10-5 shows the microstructure of an annealed as-cast nodular iron where the difference in backscatter coefficients is about 0.227, or 79 percent. Three LOM images are shown using bright-field illumination, nearly crossed polarized light, and differential interference contrast (DIC). Polarized light reveals the substructure of the graphite nodules, which cannot be observed with the SEM unless special etching techniques are used to reveal this structure by producing topographic effects. The DIC image produces topographic contrast not present in the bright-field image.

The SEM SE and BSE images in Fig. 10-5 reveal the graphite nodules in strong contrast with the graphite particles, which appear to be black because of the very low backscatter coefficient for carbon. The sample was lightly etched, and topographic details are minimal, if observable, in the SE image but totally absent in the BSE image

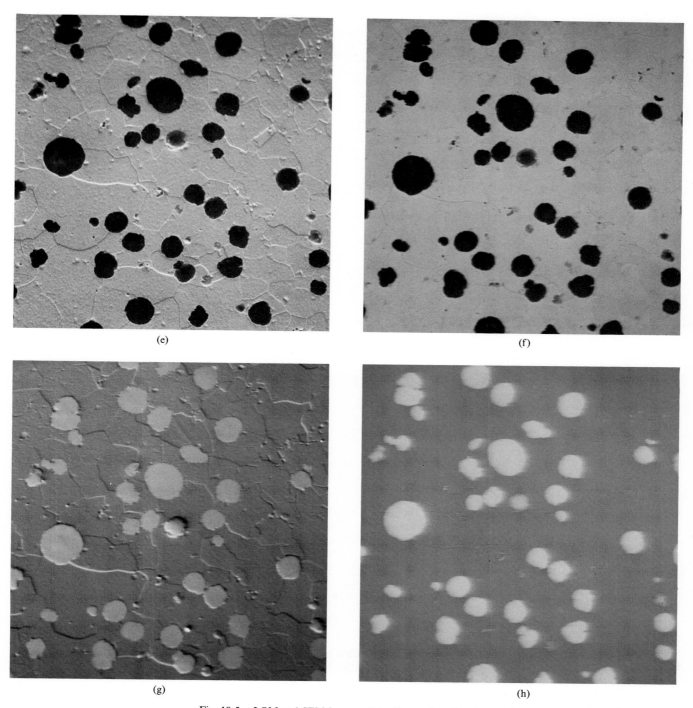

Fig. 10-5. LOM and SEM images of ductile cast iron (continued).

Fig. 10-6. Two examples of grain-orientation contrast with electropolished, unetched, single-phase 200 nickel using a quad-diode, solid-state backscatter detector at 30 keV (left) and the Taylor backscatter electron detector attachment at 10 keV (right), both at 100×.

produced with the solid-state (A + B) detector. By subtracting the two solid-state backscatter images (A − B), topographic detail is produced, and the image contrast, although minor, is reversed. In most such images, little atomic number contrast is observed. Thus, by comparing A + B and A − B solid-state BSE images, considerable structural detail can be obtained. The specimen current (SC) image, although not as contrasty as the BSE A + B image, produces acceptable contrast with the contrast reversed, as expected. The SC image is not quite as sharp as the other images. Better SC images can be achieved using high-quality SC signal amplifiers.[23]

Channeling Patterns

Backscattered electrons have also been utilized to produce crystallographic information through development of Kikuchilike electron-channeling patterns.[24-31] Such details are produced when the angle between the beam and the lattice planes in the specimen permit Bragg diffraction to occur. Although orientations of grains as small as 10-μm diameter can be determined,[28] areas of 50 to 100 μm in diameter are usually scanned, and this requires rather large grains or single crystals. Diffraction causes a greater portion of the incident beam to penetrate the specimen, thus reducing the generation of backscattered electrons and producing dark lines in the image in turn that can be related to the crystal structure and orientation.

Crystallographic Orientation Contrast

Backscattered electrons can also produce crystallographic orientation contrast as a result of diffraction effects in polycrystalline specimens. To produce such detail, a high-quality surface finish is needed; most authors recommend electropolishing. Figure 10-6 demonstrates this effect in polycrystalline nickel, an effect quite similar to that achieved by grain-orientation etchants. This sample, annealed 200 nickel, was electropolished (not etched) and examined using backscattered electrons detected by using a solid-state quad-diode detector and a Taylor high-resolution backscattered-electron phosphor-scintillator detector.* The latter is substituted for the E–T bias cage, scintillator, and retaining ring.

Figure 10-7 also demonstrated grain contrast brought about by electron channeling using a two-phase alpha-beta brass (Cu–40-wt.%Zn) specimen that was heated to 940°F (504°C) and water-quenched. This specimen was prepared by mechanical polishing with colloidal silica plus a small amount of 1-percent aqueous ferric nitrate as the final (attack) polishing step. Because there is virtually no atomic number contrast between the alpha and beta phase and little polishing relief, no image was obtained using the as-polished specimen. Hence, the specimen was tint-etched with Klemm's I reagent (saturated aqueous sodium thiosulfate plus 2-percent potassium metabisulfate), which colors the beta phase preferentially. Although several authors have stated that tint etchants are not useful for SEM examination and should be avoided, this is certainly not true. The SEM micrographs in Fig. 10-7 were taken with the specimen normal to the beam so that any topographic features present will not produce significant image contrast. The as-polished specimen exhibited no image because of the absence of significant atomic number contrast. However, the deposition of a sulfide stain on the beta phase produced a variation in backscattered electrons that shows up clearly in

---

* M. E. Taylor Engineering, Inc., 11506 Highview Ave., Wheaton, MD 20902.

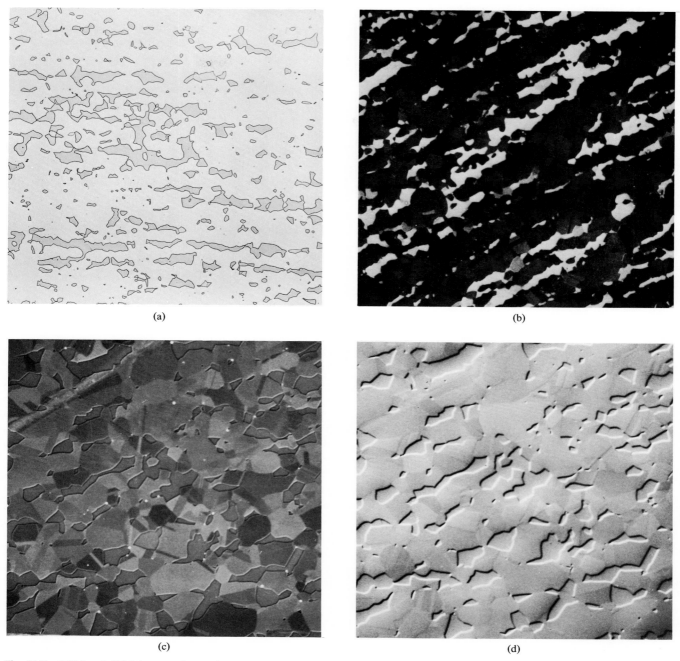

Fig. 10-7. LOM and SEM images of wrought alpha-beta brass (Cu–40%Zn) heat-treated at 940°F (504°C) for 1hr, water quench, and tint-etched with Klemm's I reagent after mechanical polishing ($\Delta\eta \simeq 0$): (a) light microscope image, 200×; (b) secondary electron image, 20 keV, 200×; (c) backscattered electron image, 20 keV, A+B solid-state detector, 500×; and (d) backscattered electron image, 20 keV, A−B solid-state detector, 500×.

the SE image. When the solid-state BSE detector was used, orientation contrast was observed, but there was little difference in contrast between the alpha and beta phases.

In Fig. 10-7, the LOM micrograph shows the preferential coloring of the beta phase with no detail present in the alpha matrix. The SE image reveals the beta phase as bright white with a slight amount of detail present in the dark matrix grains. In the A + B BSE image, all of the grains are observed because of orientation contrast. It is difficult to determine which grains are alpha and which beta. Examination of the A − B BSE image reveals the beta phase as depressions. By comparing the A + B and A − B BSE images of the same field, we note that the beta phase in the A − B image is a uniform medium gray and appears to stand above the matrix. In both the A + B and A − B BSE images, we observe annealing twins in some of the alpha grains.

Magnetic Contrast

Magnetic contrast can arise from two different mechanisms, referred to as Type I and II.[32-35] In both instances, the contrast is due to the influence of Lorentz forces on the electrons traversing a magnetic field. Type I contrast is produced by deflection effects caused by the passage of emerging electrons through the leakage field above the specimen surface. These effects produce a variation in the collection efficiency of secondary electrons, which in turn produces the image contrast. Because the deflection effects occur outside the sample, the specimen current image does not exhibit such contrast.

Type II magnetic contrast arises as a result of the scattering of the electrons within the specimen that is caused by their interactions with the magnetic field within the specimen. This leads to variations in the backscatter coefficient, thus producing image contrast. To detect such information, specimen tilting is desirable, and contrast increases with increasing accelerating potential.

The resolution of such images is rather poor compared to that for other types of SEM images. Nevertheless, the use of such a technique does permit collection of useful information. Yakowitz and Newbury[34] have described the operating conditions required to obtain magnetic contrast.

X-ray Contrast

Characteristic X-rays are produced from the elements in the specimen when they are excited by the high-energy incident electron beam. Along with the characteristic X-ray signals is the continuous background radiation. The characteristic X-rays occur as a series of signals of discrete energies that arise because of the energy-level transitions in the atoms. These characteristic X-rays are of great value because they permit determination of chemical analysis, either qualitatively or quantitatively, depending on the nature of the analysis procedure.

X-ray signals can be detected and analyzed in several ways. One very useful qualitative technique is to detect and display the X-ray signals, one element at a time, from the field of view. In this manner, a correlation is obtained between the microstructure and composition. These compositional maps for each phase will vary in brightness for each chemical element, depending on the composition of the specimen. Several factors, however, make it impossible to relate the brightness of the signal directly to the concentration of each element present. If the specimen is rough, shadowing effects will reduce the signal because X-ray signals travel in a straight line to the detector. Also, X-ray signals from different elements vary in strength for equal concentrations present because, in general, X-ray flux increases with increasing atomic number. The X-ray flux is also influenced by the relationship between the incident-electron energy and the absorption-edge energy. Despite these problems, elemental scans are very useful for identifying the number and location of various phases in the microstructure.

Figure 10-8 shows X-ray elemental scans for zinc and bismuth in the specimen illustrated in Fig. 10-4. The faint, very small white dots in the matrix for the bismuth scan are results of the background radiation. This technique can also be performed on rougher surfaces, as demonstrated by Fig. 10-9, which shows manganese-sulfide inclusions standing above the matrix of a deep-etched steel specimen.

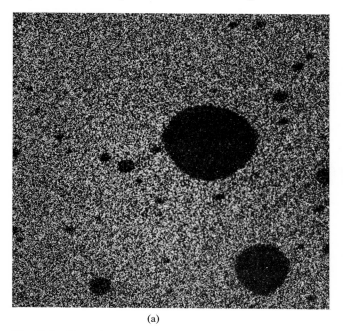

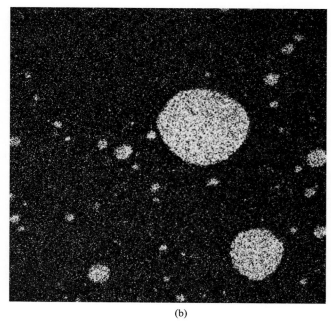

(a) (b)

Fig. 10-8. X-ray dot maps at 1000× for Zn and Bi in the as-cast Zn–10%Bi sample shown in Fig. 10-4: (a) X-ray image for zinc; and (b) X-ray image for bismuth.

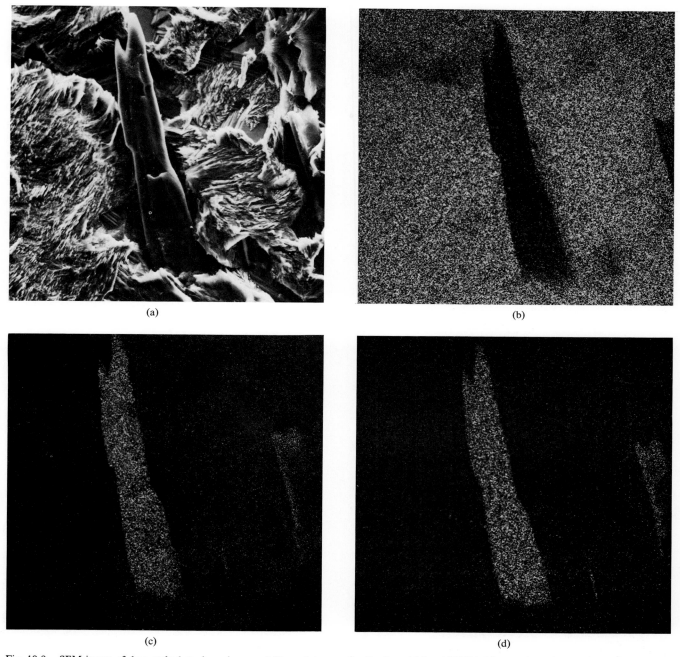

Fig. 10-9. SEM image of deep-etched steel specimen and X-ray dot maps for Fe, S, and Mn at 1000×: (a) secondary electron image, deep-etched with 5% bromine in methanol; (b) X-ray image for iron; (c) X-ray image for sulfur; and (d) X-ray image for manganese.

Another commonly used procedure, although not one that produces an image, is to detect the characteristic and continuous radiation and plot the data as a spectrum of intensity versus energy. A solid-state, energy-dispersive X-ray detector is used for such work. These detectors do not require focusing and can be used at low magnifications. Although wavelength-dispersive detectors are widely used on the electron microprobe, they require focusing, cannot operate at less than about 400× magnification, must be tuned to a specific angle to detect each element, and cannot work on rough surfaces. Nevertheless, they are more sensitive than energy-dispersive detectors and can be used to detect lightweight elements. Energy-dispersive detectors usually cannot detect elements with atomic weights below that of sodium unless special windowless detectors are used. Energy-dispersive analysis is also affected by a few interference problems caused by the overlapping of the peaks of several important elements, e.g., sulfur and molybdenum.

The X-ray dot maps mentioned above are produced by setting the analyzer for the desired element alone while the specimen surface is scanned. The white dots on the screen correspond to the location of the element of interest plus the background radiation at the same energy level as that of the element. A certain minimum concentration of each element is required to produce such information.

A somewhat more sensitive technique is to move the beam slowly across the sample along a horizontal line chosen to intercept desired microstructural contituents. During the scan, the intensity for a desired element is measured and displayed so that it can be superimposed over the SE or BSE image of the microstructure.

To obtain greater sensitivity and quantitative data, the electron beam must be fixed at a desired position so that longer counting times, up to a few minutes, can be used. Counting times must be longer to measure minor elements as compared to major elements. Minimum detectable limits are often in the range of 0.05 to 0.1 percent, depending on the element. To obtain quantitative results, the raw count data must be analyzed using appropriate corrections, aided by data from standards and a controlled, stable beam current. References 1 through 5 contain information required for such work.

RESOLUTION

Whereas the maximum obtainable resolution of the LOM is about 200 nm (0.2 μm), modern SEMs can achieve an ultimate resolution of 4 to 5 nm from ideal, high-contrast specimens. This greater resolution, which permits examination at very high magnifications, is one of the chief merits of the SEM. It must be emphasized, however, that such resolution can be achieved only in the emissive mode, i.e., with secondary electron images.

Resolution in the SEM is strongly influenced by beam size, the size of the excitation volume in the sample, and the available image contrast. Reducing beam size, however, also reduces beam intensity, and the latter must be adequate to obtain a useful image. As the probe size decreases, the number of separate picture elements in the surface increases and the rastered area decreases, factors that increase the useful magnification.

As with all microscopes, lens aberrations degrade the quality of the image. Chromatic aberration is controlled by using an electron source (the filament) with high spectral purity, i.e., one that produces electrons with an energy spread narrow enough to permit beam focusing to a single point. Increasing the accelerating potential is also beneficial. High beam currents are usually not a problem with bulk metallic specimens but will damage more delicate specimens, such as polymers. Higher accelerating potentials do produce a greater excitation volume, which will degrade resolution in the BSE made but not the SE mode. The nature of the filament material is important; lanthanum hexaboride (LaB_6), for example, produces better results than tungsten.

Because backscattered electrons are produced from a much larger specimen volume than secondary electrons, the resolution of BSE images is poorer than that of SE images. For example, fine polishing scratches are readily observed in SE images but more difficult to detect in BSE images. Also, the resolution of BSE images decreases with increasing accelerating potential, whereas the resolution of SE images is relatively unaffected. Signal-to-noise ratios also influence resolution.

The smallest feature spacing resolvable by the average human eye is about 0.3 mm. If the limit of resolution of the SEM using a SE image is 5 nm, a magnification of about 60,000× would therefore be required to enlarge parallel features (with adequate contrast) to produce spacings at this limit just resolvable by the average human eye. At this magnification, the image, or at least part of the image, should appear sharp to the eye. It is also necessary to focus the cathode ray tube (CRT) so that its spot size is less than 0.3 mm, a requirement that is easily achieved, and surpassed, by modern SEMs.

In practice, a great deal of SEM work is conducted at magnifications below 5000×. At such a magnification, the smallest feature spacing that can be enlarged to an apparent spacing of 0.3 mm would be 0.06 μm. Under these conditions, and at lower magnifications, a very small spot size of the incident electron beam is not required to obtain sharp images. Larger spot sizes produce increased beam currents and better picture quality, assuming that the specimen can tolerate higher currents and that charging artifacts are not a problem.

MICROSCOPE VARIABLES

To obtain high-quality images of microstructures, the SEM must be in proper working condition, the column and apertures must be clean and aligned, the most appropriate image mode must be chosen, and the operating conditions must be carefully selected and adjusted. Image quality will also depend on the quality of specimen preparation and the mounting method. Murphy[36] has prepared a very thorough review of specimen mounting procedures for all types of SEM work. The sample must be well grounded to the stub, and exposed polymeric mounting media (if used) should be fully covered with conductive paint. The paint must be fully dried before the specimen is placed in the SEM chamber. If oxide inclusions in the specimen are to be examined, it is a good procedure to vacuum-deposit a thin carbon layer on the surface to prevent charging artifacts.

Microscope variables that must be controlled include: aperture size, accelerating potential, beam current, spot size, working distance, scan rate, tilt angle, operating mode, and magnification. The magnification is established by excitation of the scan coils. When the image is focussed at a high magnification, the image will remain in focus when the magnification is lowered. For photographic work, it is best to focus at a higher magnification and then lower the magnification to that desired. Unlike the

TEM, the image does not rotate when the magnification is altered. Changing the working distance, however, will produce image rotation.

The picture point size on the CRT depends on magnification and the CRT spot size, which may be about 100 μm. The picture point size is calculated by dividing the CRT spot size (100 μm) by the magnification.[4] Table 10-1 lists picture point sizes for a variety of common SEM magnifications. An image will be in focus when the area excited by the incident beam is smaller than the picture element size. Beyond some magnification, images will appear blurred, and no more useful information will be obtained because of the overlapping of the picture elements. At low magnifications, the beam size can be enlarged; doing so increases the total signal available and information content without degrading image sharpness.

The depth of focus of the SEM is at least two orders of magnitude better than that of the light microscope at the same magnification—a very important advantage. When a rough surface is examined, features will be at a different distance from the final lens aperture, and the beam size at these locations will thus differ. To determine the depth of focus, we must determine the distance above and below the position of optimum focus where the beam is broadened to such an extent that picture point overlapping is sufficient to produce loss of focus. The depth of focus, D, is calculated from the equation,

$$D = \frac{0.2 \text{ mm}}{\alpha M} \quad (10\text{-}2)$$

where α is the beam divergence and M is the magnification.[4] The beam divergence angle, α, is a function of the radius R of the final lens aperture and the working distance WD, as follows:

$$\alpha = \frac{R}{WD} \quad (10\text{-}3)$$

Table 10-1 lists depth-of-focus values for a constant working distance (10 mm) for different magnifications and apertures of 100-, 200-, and 300-μm diameters.

Decreasing the final aperture size reduces the aperture angle, producing a smaller beam size and greater depth of focus. Increasing the working distance also increases the depth of focus, but this is a less satisfactory procedure since image rotation results and the resolution and magnification decrease at the same time that lens aberrations increase.

For examination of microstructures, it is best to work with the specimen normal to the beam. Although tilting improves topographic contrast, it distorts the image, and magnification varies across the image. These problems make measurement of microstructural features more difficult. Rough surfaces, even when the specimen is normal to the beam, exhibit effective magnification variations between peaks and valleys.

Both the accelerating potential and beam current influence image detail. For example, fine surface detail is masked in SE images at high accelerating potentials. At higher accelerating potentials, more backscattered electrons are generated, thus increasing the excitation of secondary electrons. Charging effects are also enhanced at high accelerating potentials. Hence, to examine the fine surface features with secondary electrons, a low accelerating potential, such as 10 keV, is preferred, along with low tilt angles and short working distances.

For best results with BSE images, the working distance should be 10 to 14 mm; accelerating potentials of 20 to 30 keV are preferred; a large spot size must be chosen; and tilting should be avoided. Solid-state detectors with large surface areas that are mounted concentrically around the beam provide better results than the Everhart–Thornley detector. Both 200- and 300-μm final aperture sizes are commonly used. Because the resolution of BSE images is poorer than that of SE images, a small incident beam size is not necessary.

To illustrate the influence of spot size on resolution and contrast, Figure 10-10 shows the microstructure of an Al–33-wt.%Cu (eutectic) as-cast alloy at 5000× magnification using secondary electrons at 20 keV potential and 0° tilt (sample lightly etched with Keller's reagent). Image contrast arises from both topographic and compo-

Table 10-1. Picture Point Size and Depth of Field at Various Magnifications.

		DEPTH OF FIELD*		
MAGNIFICATION	PICTURE POINT SIZE	100-μM APERTURE ($\alpha = 5 \times 10^{-3}$ RAD.)	200-μM APERTURE ($\alpha = 10^{-2}$ RAD.)	300-μM APERTURE ($\alpha = 1.5 \times 10^{-2}$ RAD.)
10×	10 um	4 mm	2 mm	1.33 mm
100×	1 um	0.4 mm	0.2 mm	0.133 mm
500×	0.5 um	80 um	40 um	26.7 um
1000×	0.1 um	40 um	20 um	13.3 um
5000×	50 nm	8 um	4 um	2.7 um
10,000×	10 nm	4 um	2 um	1.33 um
50,000×	5 nm	0.8 um	0.4 um	0.27 um
100,000×	1 nm	0.4 um	0.2 um	0.13 um

* 10-mm working distance.

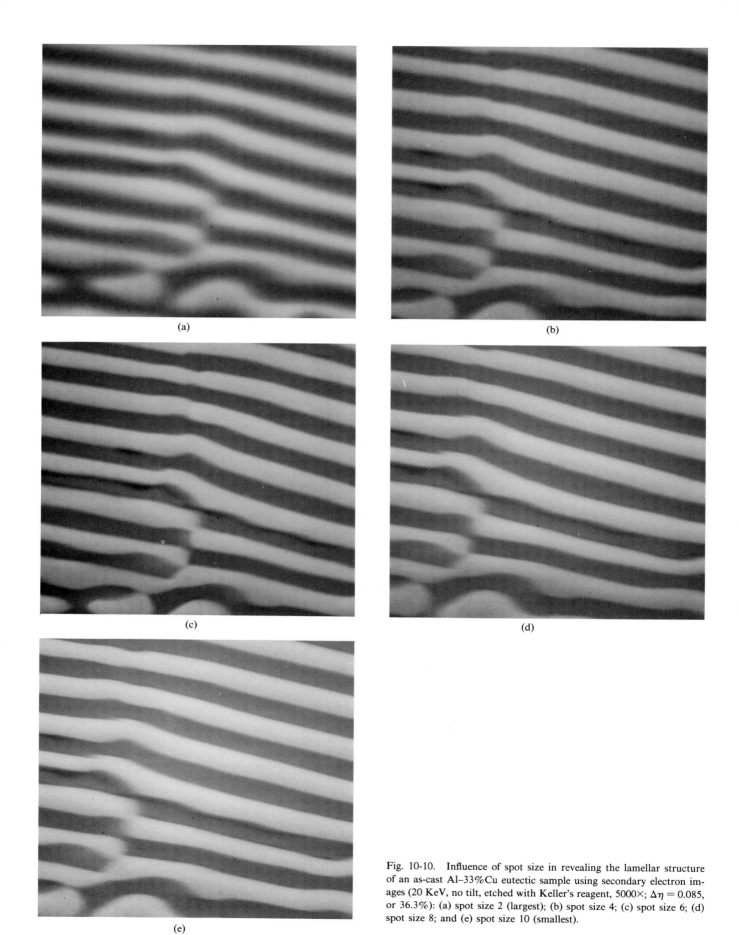

Fig. 10-10. Influence of spot size in revealing the lamellar structure of an as-cast Al–33%Cu eutectic sample using secondary electron images (20 KeV, no tilt, etched with Keller's reagent, 5000×; $\Delta\eta = 0.085$, or 36.3%): (a) spot size 2 (largest); (b) spot size 4; (c) spot size 6; (d) spot size 8; and (e) spot size 10 (smallest).

sitional effects. The CuAl₂ lamellae appear brighter than the aluminum-rich matrix because of the greater average atomic number of this phase. The difference in backscatter coefficients between the aluminum-rich matrix and CuAl₂ is 0.085, or 36.3 percent. As the spot size decreases, resolution improves.

Figure 10-11 shows the same Al–33-wt.%Cu eutectic sample, again etched lightly with Keller's reagent, but viewed with a minor amount of tilt at 1000×. Secondary electron images are shown at 20 keV and a large spot size and at 10 keV with a smaller spot size. Both images are quite good, but the latter exhibits minor edge brightness around the CuAl₂. The third image is a BSE image produced with the E–T detector and reveals minor relief (sample tilted slightly) and strong atomic number contrast.

Figure 10-12 shows the same sample photographed using the solid-state BSE detector. Both images were taken at 20 keV and no tilt. The image on the left shows some topographic information because only one of the two detectors was used. The image on the right, which used the signal from both detectors, shows stronger atomic-

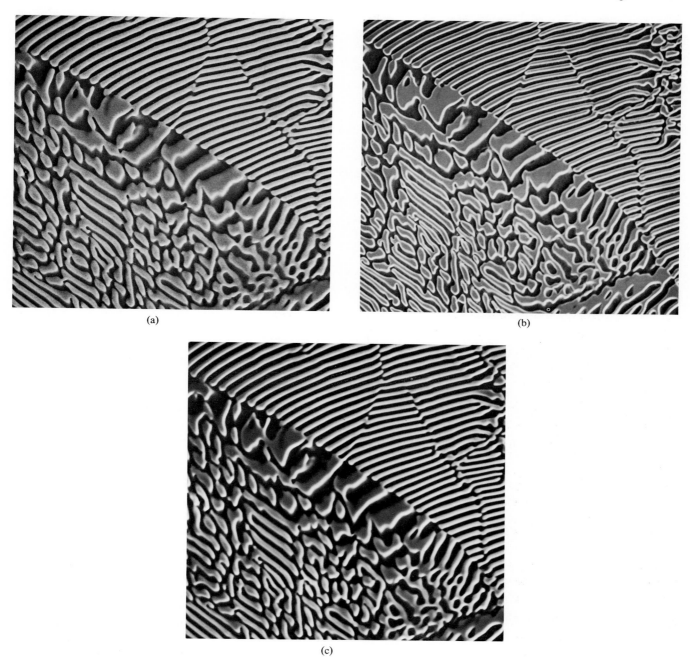

Fig. 10-11. SEM images of an as-cast Al–33%Cu eutectic specimen etched with Keller's reagent, 1000×: (a) secondary electron image, 20 keV, large spot size; (b) secondary electron image, 10 keV, smaller spot size; and (c) backscattered electron image, 20 keV, Everhart-Thornley detector.

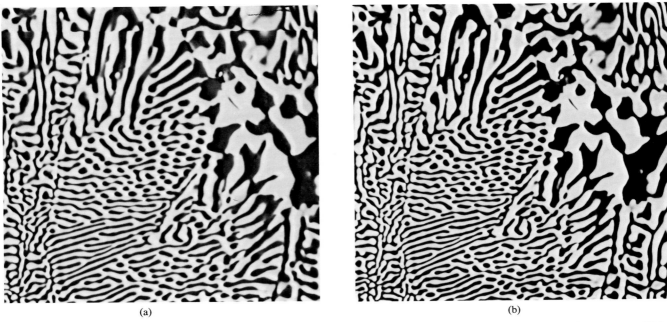

Fig. 10-12. Backscattered electron images (solid-state detector) of the as-cast Al–33%Cu eutectic specimen etched with Keller's reagent, 1000×: (a) backscattered electron image, 20 keV, A only; and (b) backscattered electron image, 20 keV, A+B.

number contrast but no topographic information. Note that the image is quite sharp at 1000×. This specimen was mechanically polished, colloidal silica being used as the final abrasive.

Figure 10-13 shows the same specimen after being polished but not etched, as photographed with a 300-μm aperture, a large spot size, 10-mm working distance, and no tilt. Two A + B BSE images are presented, one at 30 keV and the other at 20 keV. Note that the 20-keV image is superior as a result of the slight blurring of the 30-keV image caused by the greater production of backscattered elections. The topographic relief effected by polishing is revealed in the A − B image. The topographic features are stronger in this image than in the SE image of the lightly etched view of this specimen shown in Fig. 10-11 because of this deliberate relief polishing.

Figure 10-14 shows E–T SE and BSE images of this sample in the relief polished condition (as in Fig. 10-13) taken with a 200-μm aperture and a 20-keV potential. The sample was tilted towards the E–T detector, at 20° for the SE image and 36° for the BSE image. Neither image is quite as good as that for the etched sample shown in Fig. 10-11. Increasing the potential to 30 keV produced a poorer BSE E–T image. It is not quite as crisp as the solid-state A + B BSE images in Fig. 10-13.

SPECIMEN PREPARATION

Metallographic specimen preparation for SEM examination should be of the same high quality as for LOM examination despite some comments in the SEM literature that suggest that specimen preparation for SEM work can be less rigorous.[9,10] The aim of the preparation should be to produce a scratch- and artifact-free surface. One exception to this rule would be situations where quite deep etching to permit observation of the morphology of second-phase particles by SE images is required. For such work, polishing up to a 1-μm diamond-paste finish is usually adequate.

Electrolytic polishing is often recommended for preparing specimens, particularly if electron channeling contrast is desired. Radavich,[37] for example, uses electrolytic polishing extensively in his examination of superalloys. Electropolishing is highly useful for studies of pure metals or single-phase alloys, particularly those that are soft or difficult to polish mechanically without forming mechanical twins. For the preparation of alloys containing several phases, mechanical polishing is often preferred to electropolishing because the latter often produces preferential attack at one of the phases rather than uniform polishing.

For most of the specimens shown in this chapter, samples were ground and polished to a 1-μm diamond finish using the Struers Abramatic system. Grinding employed water-cooled SiC papers (120, 240, 320, 400, and 600 grits) followed by two diamond-polishing steps. Final polishing was conducted using colloidal silica abrasive with either the Leco Fini-Pol device or the Syntron vibratory polisher. In a few cases, an attack-polishing agent was added to the colloidal silica. Samples were carefully cleaned to remove any surface contaminants from polishing or etching. This practice produced high-quality specimen surfaces for both LOM and SEM examination. The

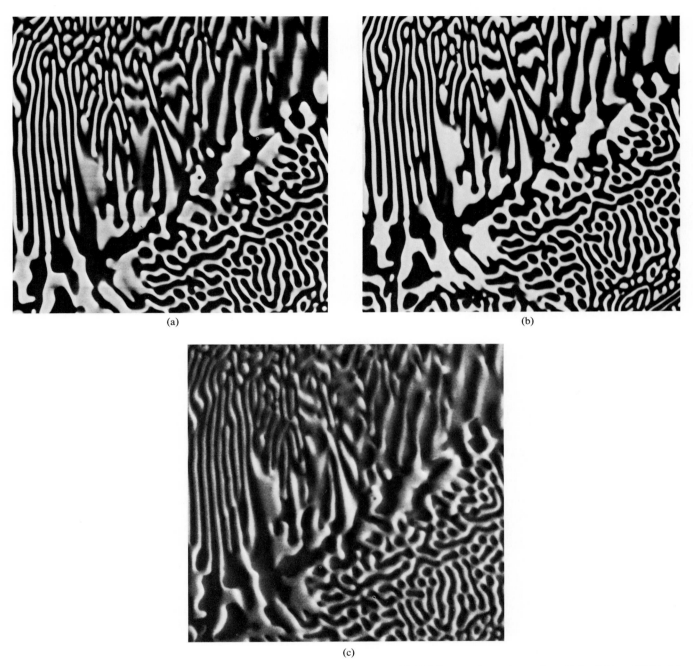

Fig. 10-13. Backscattered electron (solid-state detector) SEM images of the as-cast Al–33%Cu eutectic specimen after relief polishing, 1000×: (a) A+B, 30 keV; (b) A+B, 20 keV; (c) A−B, 30 keV.

only deviation from the practice was the use of MgO as the final abrasive for the Zn–10-wt.%Bi specimen.

For specimens that exhibit good atomic-number contrast using BSE images, etching was unnecessary. Initial focusing of as-polished, relief-free specimens using the SE mode can be rather difficult unless such samples either contain a small amount of inclusions or a scribe mark or hardness impression is present. If atomic-number contrast is too limited to be useful, etching should be employed or, perhaps, election channeling contrast. Several types of etchants have been used in this study. Careful cleaning and drying after etching are necessary to prevent staining and eliminate water marks or etching residue.

In some instances, etching can produce poorer SEM images than those obtained by atomic-number contrast, as shown in Fig. 10-15, which reveals the eutectic microstructure of as-cast Sn–38.1-wt.%Pb. The LOM micrograph shows its microstructure after a light etch with 2-percent nital. The black particles are lead and the white phase, tin. The SE images of the etched specimen is blurred, mostly because of the roughness of the recessed lead particles. Tilting improved the image quality slightly.

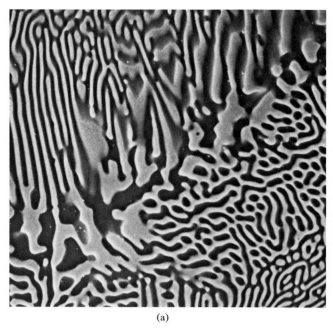 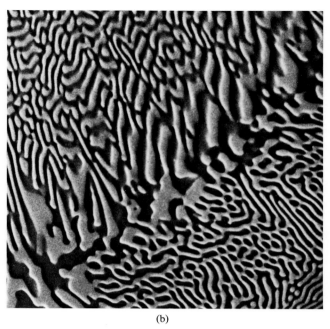

Fig. 10-14. Everhart-Thornley detector SE and BSE images of the as-cast Al–33%Cu eutectic sample after relief polishing, 1000×: (a) secondary electron image, 20 keV; and (b) backscattered electron image, 20 keV.

Much better results, however, were obtained by examining this specimen as-polished. Because the as-polished specimen is quite free of relief, and tilting was avoided, the contrast comes solely from the atomic-number difference between the tin ($Z = 50$) and lead ($Z = 82$), where the difference in backscatter coefficients is about 0.117, or 22.9 percent. As a result, the SE and BSE images are virtually identical.

When atomic-number differences in multiphase alloys are substantial, atomic-number contrast is an invaluable aid to the SEM microscopist. When such contrast is absent, however, as shown in Fig. 10-7 for alpha-beta brass, the microscopist can resort to etching. To illustrate this problem, we will show LOM and SEM images of several alloys with virtually no atomic number difference: Al($Z = 12$)–Si($Z = 13$); Ni($Z = 28$)–Cu($Z = 29$); and Pb($Z = 82$)–Bi($Z = 83$). The difference in BSE coefficients for these specimens is very small and decreases with increasing atomic number (see Fig. 10-1). Hence, without special decoration techniques,[38] atomic-number contrast cannot be utilized although electron-channeling contrast may sometimes be obtained.

Figure 10-16 shows the microstructure of as-cast Al–12-wt.%Si, which consists of aluminum-rich dendrites and an Al–Si eutectic. Since silicon has very little solubility in aluminum, and vice versa, nearly all of the silicon is precipitated as particles of nearly pure silicon. Again referring to Fig. 10-1, we can see that although the atomic numbers of Al and Si differ only by 1, there is a decided difference (about 0.01, or 6.7 percent) between the backscatter coefficients for pure Al vs. those for pure Si. Development of atomic-number contrast in this specimen is a very good test of a BSE detector.

In the as-polished condition, there is adequate color contrast between Al and Si to obtain reasonably good LOM micrographs, as shown in Fig. 10-16. Because there is very little relief, however, there was inadequate topographic contrast to obtain a good SE image. Moreover, because of the very low atomic-number contrast, we were able to obtain only a faint BSE image with a quad-diode solid-state detector (not shown) and none at all with a dual-diode detector. With the Taylor BSE detector attachment, a faint image was present (bright spots indicate Fe–Al particles).

A number of etching procedures can be used to color silicon preferentially in Al–Si alloys.[39] Because we were unable to get good etching results with the reagent suggested by Paul and Bauer,[38] we etched the sample by immersion in a solution containing 90-ml water, 4-ml HF, 4-ml H_2SO_4, and 2-g CrO_3, which colors the silicon blue.[39] As shown in Fig. 10-16, an excellent image was obtained with the Taylor BSE attachment. Although the BSE image at 5000× was not quite so sharp as the SE image at 5000×, the results were quite satisfactory and clearly demonstrates the value of tint etchants.

Figure 10-17 shows the microstructure of as-cast Monel (Ni–30-wt.%Cu). Although nickel and copper are mutually soluble in each other at all concentrations, the as-cast condition is not homogeneous but exhibits "coring," thus revealing the dendritic freezing pattern. The coring is clearly revealed by tint etching with Beraha's reagent,[39] which consists of 66-ml water, 33-ml HCl and 1-g of

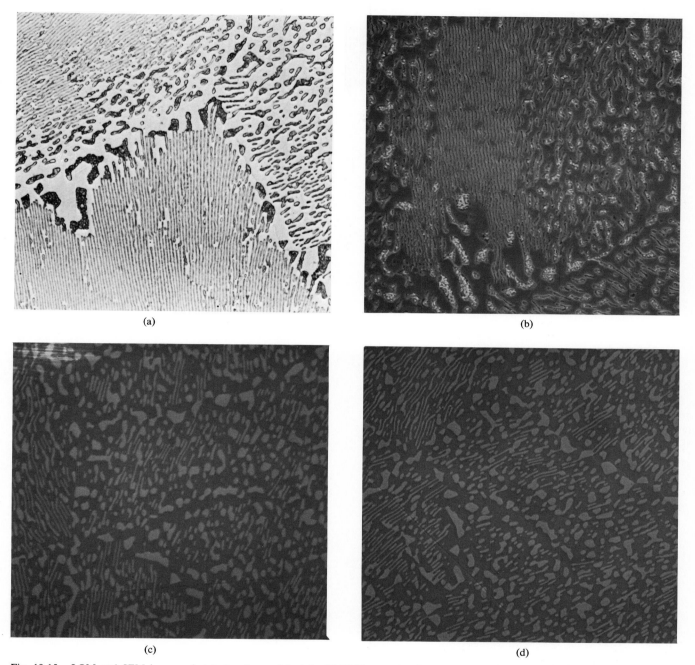

Fig. 10-15. LOM and SEM images of etched and as-polished Sn–38.1%Pb eutectic alloy (particles are Pb); note that the best SEM image is obtained in the as-polished condition ($\Delta\eta \simeq 0.117$, or 22.9%) at 1000×: (a) light optical microscope image, 2%-nital etch; (b) secondary electron image, 2% nital etch, 20 keV, no tilt; (c) backscattered electron image, A+B, 20 keV, as-polished; and (d) secondary electron image, 20 keV, no tilt, as-polished.

potassium metabisulfite. Light microscopy produced excellent contrast, as did the Taylor BSE attachment. The atomic number contrast of this sample is even less than that for the Al–12-wt.%Si sample.

Figure 10-18 shows the microstructure of a Pb–55.5-wt.%Bi sample that exhibits dendritic particles of nearly pure beta phase ($BiPb_3$), a eutectic of nearly pure Bi particles and beta, and larger (white in LOM view) Bi particles. There is virtually no atomic-number contrast because the difference in backscatter coefficient between Bi and $BiPb_3$ is about 0.002, or 0.4 percent. No useful images were obtained from the as-polished sample using the SEM because of the lack of topographic or atomic-number contrast, although a good LOM image was obtained. Consequently, the sample was etched in 2-percent nital. The LOM micrograph at 100× displays excellent contrast, but the eutectic is too fine to observe at 100×. As shown in Fig. 10-18, an SE image at 100× produced very poor contrast compared to the LOM image. At 1000×, however, a useful SE image was obtained, better than that achieved with the LOM; the solid-state BSE image is quite poor.

Figure 10-19 shows the microstructure of an annealed Fe–1-wt.%C binary alloy that contains grain-boundary cementite and pearlite. The atomic-number contrast between cementite and ferrite in the pearlite is low; the difference in BSE coefficients is about 0.015, or 5.5 percent. Only a very faint BSE image could be obtained in the as-polished condition with the solid-state detector. Consequently, the specimen shown in Fig. 10-19 was etched for 30 s with 4-percent picral, which uniformly attacks the ferrite and leaves the carbide in relief. Low-magnification SEM examination produced poorer image contrast than the LOM. At high magnifications, however, particularly those beyond the capability of the LOM, the SEM images were extremely useful, e.g., to permit measurement of the interlamellar spacing. The BSE image (A + B) using the solid-state detector produced good

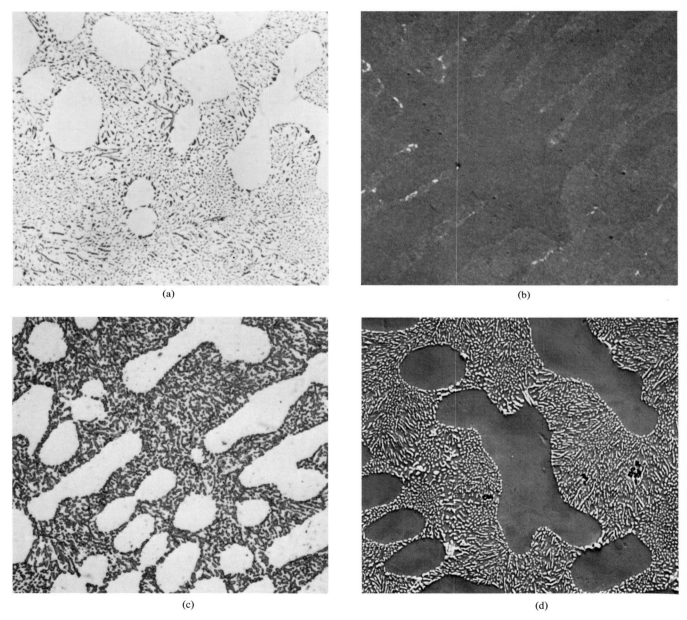

Fig. 10-16. LOM and SEM micrographs of an as-cast Al–12% Si alloy with one atomic number difference and very low atomic-number contrast ($\Delta\eta = 0.01$, or 6.7%). In the as-polished condition, a reasonable LOM image is obtained by reflectivity differences. With the SEM, however, no image could be obtained with the dual-diode BSE detector, and only a faint image was obtained with the Taylor BSE attachment [see (b)]. Stain etching produced good SE and BSE SEM images. All at 1000×: (a) light optical micrograph, as-polished, 1000×; (b) backscattered electron image, as-polished, 1000× (bright spots are from iron impurities); (c) light optical micrograph, stain etched, 1000×; (d) backscattered electron image, stain etched, 1000×; (e) backscattered electron image, stain etched, 5000×; and (f) secondary electron image, stain etched, 5000×.

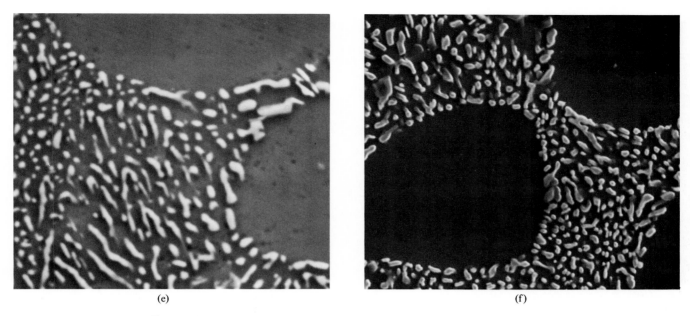

Fig. 10-16. LOM and SEM micrographs of an as-cast AL–12% Si alloy (continued).

contrast and sharp images even at 5000×. The SE image clearly reveals the carbides in relief. The E–T BSE image exhibits more contrast than the E–T SE image but less than the solid-state BSE image.

Figure 10-20 shows the same Fe–1-wt.%C alloy tint-etched with Beraha's sodium molybdate reagent that colors the cementite (grain boundary cementite is reddish brown; cementite in pearlite is bluish).[39] Although many authors have stated that tint etchants should not be used for SEM work, this example, as well as others referred to previously, clearly indicates that this restriction is unnecessary provided the sample is properly polished and tint-etched. Figure 10-20 shows two SE images, one at 15 keV and the other at 20 keV. The higher potential produces charging of the cementite, thereby degrading the image. Two E–T BSE images are also shown, one with a small spot size and the other with a larger spot size. The larger spot size also reduced the image quality as a result of the increased signal strength from the cementite that reduced resolution slightly.

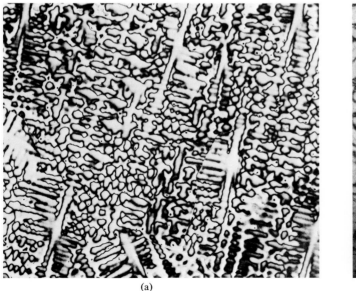
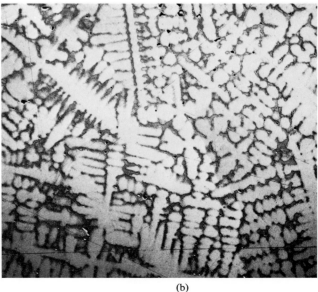

Fig. 10-17. LOM and SEM micrographs of as-cast Monel (Ni–30%Cu) with dendritic segregation ("coring"). This sample has virtually no atomic number contrast ($\Delta\eta \simeq 0$) but can be viewed with the SEM after tint etching with Beraha's reagent: (a) light optical micrograph, tint etched; and (b) Taylor backscatter electron-detector attachment, 10 keV, stain-etched. Both at 50×.

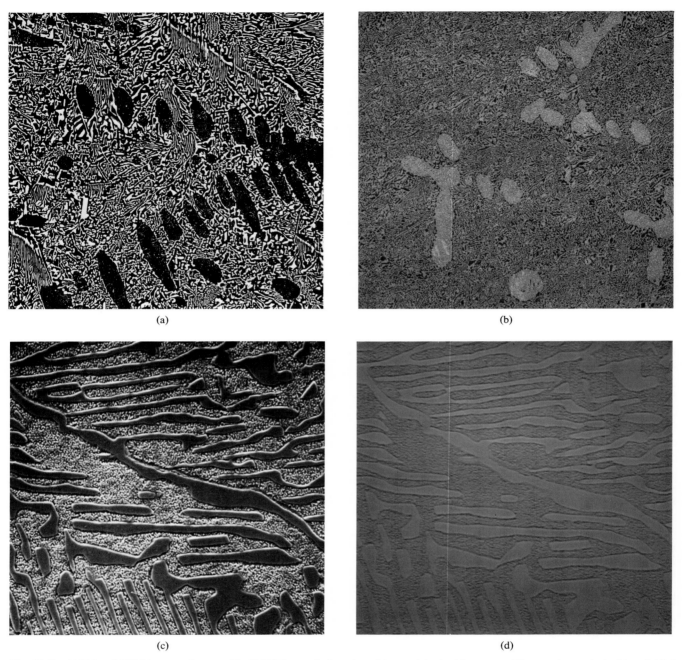

Fig. 10-18. LOM and SEM images of as-cast Pb–55.5%Bi sample [massive white particles in (a) are pure Bi; eutectic matrix in (c) and (d) is BiPb$_3$ and Bi] with virtually no atomic-number contrast ($\Delta\eta \simeq 0.002$, or 0.4%): (a) light optical micrograph, 2% nital etch, 100×; (b) secondary electron image, 10 keV, 15° tilt, 2%-nital etch, 100×; (c) secondary electron image, 20 keV, no tilt, 2%-nital etch, 1000×; and (d) backscattered electron image, A+B, 20 keV, no tilt, 2%-nital etch, 1000×.

Two-Surface Analysis

Investigators have also used the SEM to examine two prepared specimen surfaces simultaneously. The greater depth of field of the SEM permits such work, whereas the LOM, except sequentially, does not. One technique combined examination of the fracture surface and the underlying microstructure by polishing a plane cut at an angle to the fracture,[40] whereas another used two intersecting polished planes.[41] The former procedure was used to relate fracture characteristics to the microstructure, whereas the latter was used to obtain three-dimensional structural information.

Deep Etching

Many investigators have utilized deep etching to reveal the morphology of second-phase constituents. References

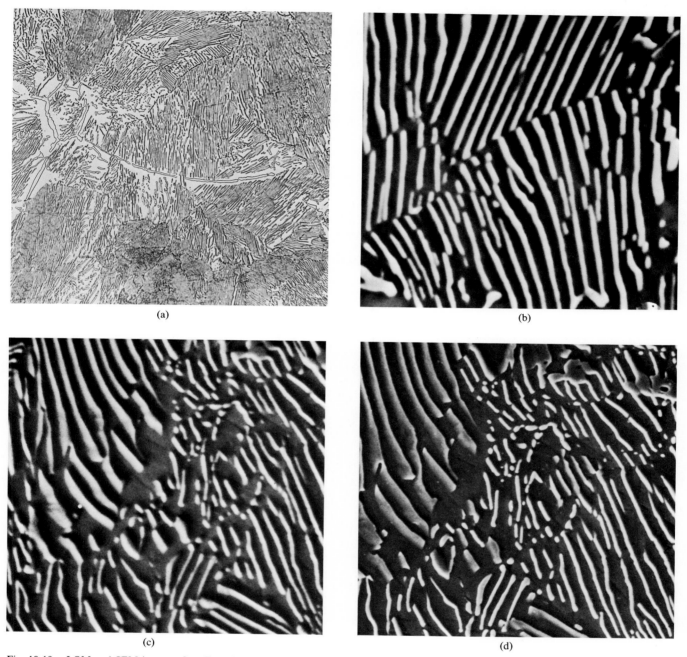

Fig. 10-19. LOM and SEM images of an Fe–1%C high-purity alloy etched with 4% picral ($\Delta\eta \simeq 0.015$, or 5.5%): (a) light optical micrograph, 500×; (b) backscattered electron image, A+B, 30 keV, 5000×; (c) secondary electron image, 15 keV, 5000×; and (d) backscattered electron image, Everhart-Thornley detector, 15 keV, 5000×.

42 to 49 present a sample of some of these studies. Deeply etched samples cannot be examined in the LOM but are easily examined in the SEM and many metals and alloys have been examined in this way. Table 10-2 lists some of the deep etching reagents that have been utilized.

Figure 10-9 demonstrated the use of deep etching to study sulfide inclusions in a steel. The SEM has been used to study inclusion morphology by the examination of deeply etched surfaces,[50] bulk extractions,[51] or ductile fractures.[52,53]

Two additional examples of deep etching are presented in Figs. 10-21 and 10-22 to demonstrate the correlation between LOM images and SEM deep-etched SE images. Figure 10-21 shows the microstructure of AISI 8121 alloy steel, which consists of equiaxed ferrite and bainite. Deep etching reveals the fineness of the carbide in the bainite but suppresses the equiaxed ferrite. Figure 10-22 shows the microstructure of Nb-treated AISI 1137, which consists of grain-boundary proeutectoid ferrite and pearlite. Deep etching reveals the blocky nature of the pearlite,

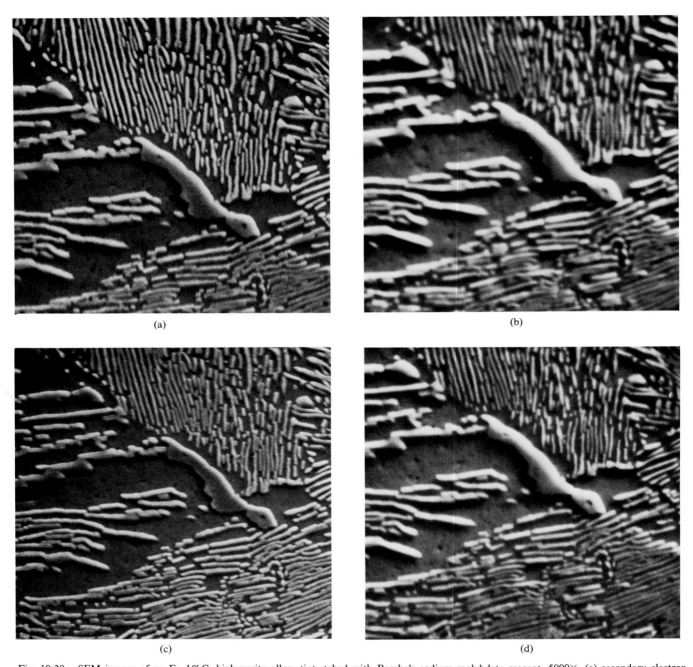

Fig. 10-20. SEM images of an Fe–1%C, high-purity alloy, tint-etched with Beraha's sodium molybdate reagent, 5000×: (a) secondary electron image, 15 keV; (b) secondary electron image, 20 keV; (c) backscattered electron image, Everhart-Thornley detector, small spot size, 20 keV; and (d) backscattered electron image, Everhart-Thornley detector, large spot size, 20 keV.

but, again, the ferrite phase is less noticeable. At 500×, the pearlite lamellae are too fine to fully resolve.

Etching of Fractures

Use has also been made of SEM examination of etched fracture surfaces both to determine how the fracture features correlate with the microstructure and to permit analysis of microstructural constituents intersected by the fracture surface.[54-57] In these studies, standard microstructural reagents were employed but usually with somewhat longer etching times.

SEM Replicas

Replicas are occasionally examined in the SEM, and a number of replication techniques have been published.[58-63] SEM replication is generally reserved for fracture examination when the specimen either cannot be cut to fit the

THE SEM AS A METALLOGRAPHIC TOOL

Table 10-2. Examples of Deep Etching Techniques for SEM Examination.

MATERIAL	ETCH COMPOSITION	COMMENTS
1. Aluminum Alloys Al–Si eutectics	(a) 10%-aq. HCl (b) 2.5-ml HCl 2.0-ml HNO₃ 0.75-ml HF 40-ml water*	Etched 24 hr in (a), then 1 hr in (b) to dissolve Al matrix and reveal Si [Bell and Winegard, *Nature* 208:177 (Oct. 9, 1965)]
	Dilute-aq. HCl	Reveals Si [Day and Hellawell, *J. Inst. Metals* 95:377 (1967)]
	25%-aq. HCl	Wash repeatedly in warm alcohol, then acetone, and dry; dissolves Al, revealing Si [Day, *J. Metals* 21:31–34 (April 1969)]
	(a) 15-ml HCl 10-ml HF 90-ml water (b) 25-ml HNO₃ 75-ml water	Etch with (a) for 30–60 min, then dip in (b) for 2–4 min; dissolves Al, revealing Si [Holmanova, *Pract. Metallogr.* 11:155–159 (Mar., 1974)]
	10-ml HCl 90-ml water	Immerse about 60 min; dissolves Al, revealing Si [Paul and Mürrle, *Pract. Metallogr.* 18:413–423 (Sept. 1981)]
Al alloys	2-g KI 100-ml methanol	Electrolytic etch at 2–5 V (0.1–0.3A for 5 × 5 × 40-mm sample) for 5–20 min; wash in methanol; dissolves Al matrix [Ryvola and Morris, *Microstructural Science* 5:203–208 (1977)]
	2.5-g KI 100-ml methanol	Modification of above method; electrolytic etch, Al-sheet cathode, 3 × 3-mm area etched; raise voltage rapidly from 0 to 30 V, etch 30 s, stir gently, about 0.3A/mm²; clean ultrasonically in methanol for 30 s; dry [Howe, *Metallography* 16:275–286 (1983)]
Al–Ge eutectic	25-ml HCl 75-ml water	Dissolves Al; Ge not attacked [Hellawell, *Trans. AIME* 239:1049–1055 (July, 1967)]
Al–Al₃Ni eutectic	3-ml HCl 50-ml water 50-ml alcohol	Electrolytic etch (no other details) [El-Mahallawy and Farag, *Met. Trans.* 10A:707–714 (June 1979)]
2. Copper alloys Cu-Cu₂S eutectic	20% acidic FeCl₃	Etched for 6 min; dissolves Cu to a depth of 2–3 μm [Marich and Jaffrey, *Met. Trans.* 2:2681–2689 (Sept. 1971)]
3. Iron-based alloys Carbon steel	5-ml Bromine 95-ml methanol	Immerse 6 min to dissolve iron matrix, revealing carbides and inclusions (very popular for steels) [Rege, et al. *Met. Trans.* 1:2652–2653 (Sept. 1970)]
Steels	10-ml Bromine 90-ml alcohol	Dissolves matrix for study of carbides and inclusions [Fremunt, et al., *Pract. Metallogr.* 17:497–508 (Oct. 1980)]
Cast iron	10-ml HCl 90-ml alcohol	Dissolves matrix, revealing graphite [Day, *J. Metals* 21:31–34 (Apr., 1969)]
Fe–C–Si	40-ml HCl 40-ml alcohol 20-ml water	Used to reveal graphite [Lux et al., *Pract. Metallogr.* 5:587–603 (Nov. 1968)]
Al–Si–Cu alloyed cast iron	As above	Used to reveal graphite [Janakiev, *Pract. Metallogr.* 8:543–546 (Sept., 1971)]
Gray cast iron	20–35 ml HCl 80–65 ml water	Used to reveal graphite; conc. depends on the size of the sample and type of graphite; after

Table 10-2 (continued)

MATERIAL	ETCH COMPOSITION	COMMENTS
		etching, wash in petroleum ether; dry [Ruff and Wallace, *Trans. AFS* 85:167–170 (1977)]
Compacted graphitic cast iron	5-ml HNO_3 95-ml alcohol	Used to reveal graphite [Cooper and Loper, *Trans. AFS* 86:267–272 (1978)]
Cast iron	10–20-ml HCl 90–80-ml water	Used to reveal graphite; conc. and time depend on alloy and graphite type; after etching, rinse with 5%-aq. HF, then repeatedly rinse with alcohol, then acetone; dry [Liu and Loper, *SEM/1980/I*, pp. 407–418 (1980)]
Stainless steels†	10-ml HF 15-ml HCl 30-ml HNO_3 45-ml water	Used to reveal carbides when precipitation is extensive; several-minute immersion; Br methanol used when minor amount of carbides are present [Wilson, *JISI* 209:126–130 (Feb., 1971)]
AISI 316	10-ml bromine 90-ml methanol	Samples lightly etched to reveal second phases [Weiss, et al., *Pract. Metallogr.* 8:523–542 (Sept., 1971)]
AISI 316	10-ml HCl 3-g tartaric acid 90-ml methanol	Used electrolytically at 20 mA/cm² for 30 min to 5 hr to reveal $M_{23}C_6$ [Kny, et al., *Pract. Metallogr.* 14:512–520 (1977)]
Fe–26%Cr	5-ml acetic acid 5-ml HNO_3 15-ml HCl	Swab etched for 45 s to reveal second phases [Grubb and Wright, *Met. Trans.* 10A:1247–1255 (Sept. 1979)]
4. Lead-tin alloys	1-ml HCl 99-ml water	Dissolves tin-rich phase [Hillmer, *Pract. Metallogr.* 16:465–479 (Oct. 1979)]
5. Nickel-base alloys		
Ni–C eutectic	40-ml HCl 50-ml alcohol 20-ml water	Etch 5 hr; dissolves nickel, revealing graphite [Lux, et al., *Pract. Metallogr.* 5:567–571 (Oct., 1968)]
Ni–0.5-to-4% Al	10-ml bromine 90-ml methanol	Used at 30°C to reveal oxidation layer ($Al_2O_3 \cdot NiAl_2O_4$-NiO) at surface or interior structure [Hindam and Whittle, *J. Materials Science* 18:1389–1404 (1983)]
NiAl–Cr eutectic	100-ml water 40-ml HCl 5-g CrO_3	Dissolves NiAl, revealing Cr [Walter and Cline, *Met. Trans.* 1:1221–1229 (May 1970)]
IN–738	10-ml HCl 3-g tartaric acid 90-ml methanol	Used electrolytically at 20 mA/cm² for 30 min to 5 hr to reveal second phases [Kny, et al., *Pract. Metallogr.* 14:512–520 (1977)]
Inconel 617	5-g $CuCl_2$ 100-ml HCl 100-ml ethanol	Waterless Kallings reagent used for 30 min to reveal second phases [Kihara et al., *Met. Trans.* 11A:1019–1031 (June 1980)]
Superalloys with high Mo and W	5-ml HCl 95-ml methanol	Used electrolytically (no details given) [Long et al., *Pract. Metallogr.* 19:561–572 (1982)]
Superalloys	(a) 50-ml HCl 100-ml glycerin 1050-ml methanol	Use (a) at 50 mA/cm² and −5 to −10°C for 30–60 min; dissolves γ and γ', revealing carbides.
	(b) 10-g $(NH_4)_2SO_4$ 10-g citric acid 1200-ml water	Use (b) at 10–30 mA/cm² and 5°C or potentiostatically at 960–1060 mV vs SCE; dissolves γ, revealing γ'.
	(c) 10-ml H_3PO_4 90-ml water	Use (c) at 10 mA/cm² or potentiostatically at 1000–1050 mV vs SCE; dissolves γ, revealing γ', carbide, and sigma.
	(d) 5-g $(NH_4)_2SO_4$ 15-ml HNO_3	Use (d) at 30 mA/cm² and 15–20°C or potentiostatically at 1020–1060 mV vs SCE; dissolves

Table 10-2 (concluded)

MATERIAL	ETCH COMPOSITION	COMMENTS
	35-g citric acid 1000-ml water	γ, revealing γ'. [Hum-De and Jing-Yun, *Pract. Metallogr.* 17:608–618 (1980)]
6. Sintered carbides WC–Co and others	20-ml HCl 80-ml water	Suspend sample in boiling solution for several hr; dissolves Co binder [Warren, *J. Inst. Metals* 100:176–181 (1972)]
WC–Co WC–TiC–Co	50-ml HCl 50-ml water	Electrolytic etch at 50 mA for 80–120 s; dissolves Co binder [Schreiner, et al. 17:547–553 (1980)]
7. Zinc-based alloys		
Zn–In Zn–Bi	50-ml HCl 50-ml water	Dissolves matrix phase (Passerone, et al., *Metal Science* 13:359–365 (June 1979)]
8. Zirconium-based alloys		
Zr–8.6% Al	45-ml HNO_3 5-ml HF 45-ml water	Dissolves matrix [Schulson and Trottier, *Met. Trans.* 11A:1459–1464 (Aug. 1980)]

* Whenever water is specified, use distilled water.
† Potentiostatic deep-etching methods for stainless steels have been reported by: Rughunathan, et al., *Met. Trans.* 10A:1683–1689 (Nov., 1979) and Herbsleb and Schwaab, *Pract. Metallogr.* 20:53–63 (Feb., 1983).

chamber or cannot be removed from the failure site. Another common application is the chemical identification of second-phase constituents in extraction replicas. This procedure is quite useful because matrix chemistry influences can be avoided, thus making the analysis of precipitates smaller than a few micrometers in diameter more reliable. The third, and least common SEM replica application, is the examination of the microstructure of specimens.[64]

SEM QUANTITATIVE METALLOGRAPHY

The development of LOM image analysis equipment in the early 1960's spurred similar developments in electron metallography, beginning with procedures for the electron microprobe analyzer.[65-69] Because of the many advantages of the SEM relative to the microprobe, however, such work has been transferred to the SEM with considerable enthusiasm.[70-87]

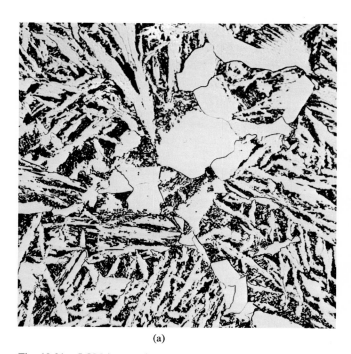
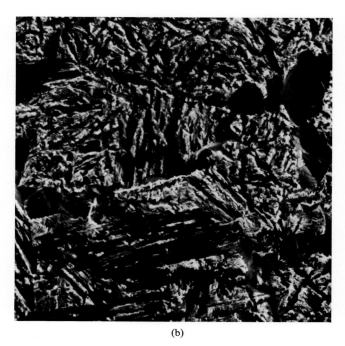

Fig. 10-21. LOM image of AISI 8121 alloy steel and SEM image after deep etching, 500×: (a) light microscope image, etched with 4% picral and 2% nital; and (b) secondary electron image, deep-etched with 5%Br–methanol.

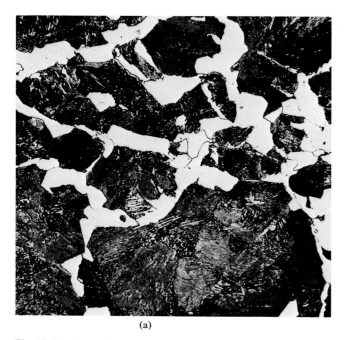 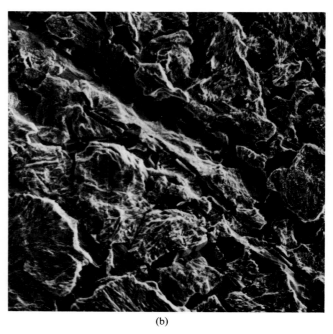

(a)   (b)

Fig. 10-22. LOM image of Nb-treated AISI 1137 alloy steel and SEM image after deep etching, 500×: (a) light microscope image, etched with 4% picral and 2% nital; and (b) secondary electron image, deep-etched with 5% Br–methanol.

The application of quantitative metallography to the SEM has chiefly centered on particle sizing and chemical analysis as well as volume fraction analysis, whereas LOM image-analysis work has concentrated on broader, more stereological-based applications. The reason for this difference lies in the strengths and weaknesses of the two instruments. The high-magnification and chemical-analysis ability of the SEM is well suited for identifying and measuring fine particles. On the other hand, because stereological measurements are highly influenced by the homogeneity of the material being tested, best results are obtained by distributing many measurements over a large surface area. This type of measurement is well suited for the LOM image analyzers, in which stage movement and focusing can be automated to permit rapid measurement of parameters on many fields. LOM-based instruments lack chemical-analysis capability altogether and detect features based on gray-level differences instead, these being produced by natural reflectivity differences between phases or by selective etching. The LOM image analyzer is not effective for individual particle measurement when the average particle size is less than a few micrometers in diameter. Hence, the two instruments are complementary.

In SEM image-analysis work, the image is sent to either the energy-dispersive X-ray analysis (EDXA) system with appropriate measurement software or to a specially designed image-analysis system. Most EDXA systems do have optional software for particle sizing and other measurements. The image is usually provided by either SE, BSE, or X-ray detectors although other signals have been employed.

Samples for image-analysis work may be examined in the etched or unetched condition, the latter being preferred for inclusion studies. The desired features are usually detected by gray-level differences (thresholding) of the image on the CRT, but the composition may also be used as the selection criterion.

When etching is required, light etching is preferred because the deep etching commonly employed in SEM work will increase the apparent amount and size of second-phase particles and decrease the spacing between them. Indeed, whenever atomic-number contrast alone can be used to define the feature of interest, as-polished specimens can be employed with a BSE image. Because the BSE detector requires line-of-sight orientation, the most suitable orientation of dual or quad solid-state SE detectors is around the electron-beam entrance. With this orientation, the sample is normal to both the beam and the detector, thereby eliminating distortion and variable magnification effects when samples are tilted.

To illustrate the infuence of etching degree on the apparent amount and size of second phases, Fig. 10-23 presents the microstructure of AISI D2 tool steel containing M_7C_3 carbides. In the as-polished condition, there is adequate atomic number contrast to permit discrimination of the carbides. A light etch with 2-percent nital also produces a good SE image with no particle enlargement. Heavier etching with 4-percent picral plus HCl reveals a greater amount of carbide, with the larger particles in relief and with a bright edge around them—an undesirable feature. Deep etching with 5-percent bromine in methanol reveals a very high amount of carbide, with the contrast between the carbide and the matrix very poor. The first

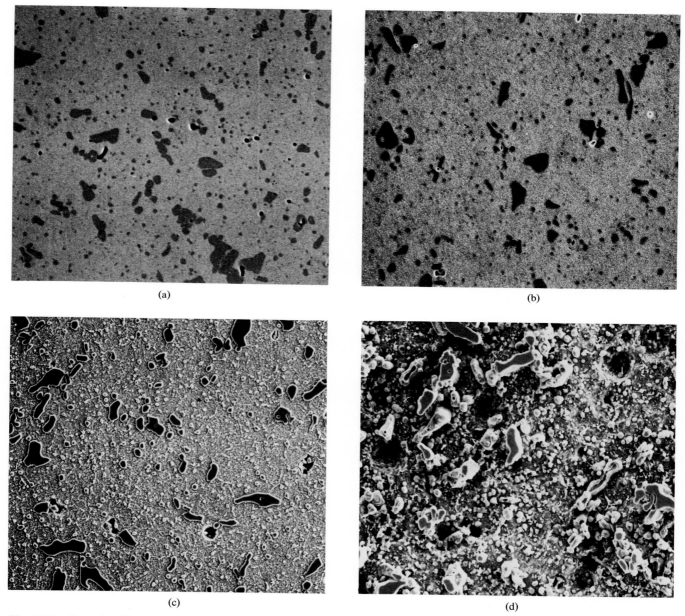

Fig. 10-23. Example of influence of etchant type and degree of etching on apparent number, size, spacing, and volume fraction of alloy carbides in AISI D2 tool steel, 500×: (a) backscattered electron image, as-polished (unetched); (b) secondary electron image, 2% nital (light etch); (c) secondary electron image, 4% picral + HCl (heavy etch); and (d) secondary electron image, 5% bromine in methanol (deep etch).

two images are highly desirable for image analysis; the latter two, undesirable. In these micrographs, the sample was perpendicular to the beam.

SUMMARY

In this review, we have shown how the SEM and the LOM are complementary tools, each with specific but different advantages and disadvantages. It should also be obvious from the illustrations given that high-quality metallographic sample preparation is just as valuable for SEM studies as it is for light microscopy. Nothing is to be gained from sloppy or inadequate sample preparation, regardless of the examination method.

LOM and SEM examination rely upon specific image-contrast formation mechanisms, which are different. The SEM user should understand the ways in which image contrast can be formed and determine which mechanisms are most appropriate for each sample. As with LOM work, etching is not always required for SEM work. Indeed, as shown, a better rendition of the structure can sometimes be obtained with as-polished samples than after etching. If etching is required, one should give consideration to the use of tint etchants despite numerous comments in the literature claiming that they should be avoided. Tint etchants do provide very precise delineation of the structure, which is generally highly feature-specific, and usually do not attack the matrix. Thus, the true size of the features on the plane-of-polish is preserved. Deep etching should be employed only when the shape of the second-phase constituents is of interest.

Although the vast majority of SEM examination is conducted using secondary-electron images, the value of backscattered electron images for microstructural examination should not be ignored. And although the degree of resolution obtainable with BSE images is not as good as with SE images, it is adequate for most microstructural examinations.

Acknowledgment. The writer expresses his thanks to L. L. Hahn and J. W. Bowman, Jr., for taking the SEM micrographs used in this chapter.

REFERENCES

1. Hearle, J. W. S.; Sparrow, J. T.; and Cross, P. M. *The Use of the Scanning Electron Microscope*, Oxford: Pergamon Press, 1972.
2. Wells, O. C. *Scanning Electron Microscopy*. New York: McGraw-Hill Book Co., 1974.
3. Goldstein, J. I., and Yakowitz, H., eds. *Practical Scanning Electron Microscopy*. New York: Plenum Press, 1975.
4. Goldstein, J. I., et al. *Scanning Electron Microscopy and X-Ray Microanalysis*. New York: Plenum Press, 1981.
5. Heinrich, K. F. J. *Electron Beam X-Ray Microanalysis*. New York: Van Nostrand Reinhold Co., 1981.
6. Pease, R. F. W. Fundamentals of scanning electron microscopy. *Scanning Electron Microscopy/1971/Part I*. Chicago, IL; IIT Res. Inst., pp. 9–16 (1971).
7. Joy, D. The scanning electron microscope—principle and applications. *Scanning Electron Microscopy/1973*. Chicago, IL: IIT Res. Inst., pp. 743–750 (1973).
8. Russ, J. L. Scanning electron microscopy. In *Systematic Materials Analysis* 2:159–181. New York: Academic Press (1974).
9. Johari, O., et al. Sample preparation for scanning electron microscope metallography. *Proc. First Annual Technical Meeting of the Intl. Metallographic Society*, Nov. 11–13, 1968, Denver, Co. pp. 127–131 (1969).
10. Johari, O., et al. Microstructures of common metals and alloys as observed by the SEM. *Scanning Electron Microscopy/1969*. Chicago, IL: IIT Res. Inst., pp. 277–284 (1969).
11. Johari, O. Microstructure with the SEM—a new approach through SEM fractography. In *Electron Microscopy and Structure of Materials*, Berkeley, CA: University of California Press, 1972, pp. 313–332.
12. McCall, J. L. Scanning electron microscopy for microstructural analysis. In *Microstructural Analysis: Tools and Techniques*. New York: Plenum Press, 1973, pp. 93–124.
13. Yeoman Walker, D. E. Some comments on conventional metallography with the Stereoscan. *Scanning Electron Microscopy: Systems and Applications 1973*. The Inst. of Physics, Conf. Ser. No. 18, London, pp. 202–207 (1973).
14. Österlund, R., and Vingsbo, O. Scanning electron microscope and optical microscope studies of the topography of some etched steel structures. *Ultramicroscopy* 4:155–162 (1979).
15. Meiley, S. L. Optical, SEM, and TEM study of transformed steel structures. *Microscope* 27:41–52 (1979).
16. Hall, M. G., and Hutchinson, W. B. Smooth surface metallography using the scanning electron microscope. *The Metallurgist and Materials Technologist* 12:371–375 (July, 1980).
17. Hall, M. G. Metallography in the SEM. *Scanning Electron Microscopy/1981/I*. AMF O-Hare, IL: SEM Inc., pp. 409–422 (1981).
18. Newbury, D. E. Fundamentals of scanning electron microscopy for physicist: Contrast mechanisms. *Scanning Electron Microscopy/1977, I*. Chicago, IL: IIT Res. Inst., pp. 553–568 (1977).
19. Bishop, H. E. Some electron backscattering measurements for solid targets. *4th Intl. Conf. X-Ray Optics and Microanalysis,* Hermann, Paris, pp. 153–158 (1966).
20. Heinrich, K. F. J. Electron probe microanalysis by specimen current measurement. *4th Intl. Conf. X-Ray Optics and Microanalysis,* Hermann, Paris, pp. 159–167 (1966).
21. Wittry, D. B. Secondary electron emission in the electron probe. *4th Intl. Conf. X-Ray Optics and Microanalysis,* Hermann, Paris, 168–180 (1966).
22. Colby, J. W.; Wise, W. N.; and Conley, D. K. Quantitative microprobe analysis by means of target current measurements. *Advances in X-Ray Analysis* 10. New York: Plenum Press, pp. 447–461 (1967).
23. Yakowitz, H., et al. Implications of specimen current and time differentiated imaging in scanning electron microscopy. *Scanning Electron Microscopy/1973, I*. Chicago, IL: IIT Res. Inst., pp. 173–180 (1973).
24. Coates, D. G. Kikuchi-like reflection patterns obtained with the scanning electron microscope. *Philo. Mag.* 16, No. 144:1179–1184 (1967).
25. Booker, G. R., et al. Some comments on the interpretation of the 'Kikuchi-like reflection patterns' observed by scanning electron microscopy. *Philo. Mag.* 16, No 144:1185–1191 (1967).
26. Van Essen, C. G., and Schulson, E. M. Selected area channelling patterns in the scanning electron microscope. *J. Materials Science* 4:336–339 (1969).
27. Weiss, B., et al. SEM-techniques for the microcharacterization of metals and alloys, I. *Practical Metallography* 8:477–491 (1971).
28. Klaffke, D. The determination of the orientation of cubic crystals relative to any desired direction using electron channelling patterns in the scanning electron microscope. *Practical Metallography* 10:615–627 (1973).
29. Durlu, T. N. The use of selected area channelling patterns to determine orientation relationships in martensitic Fe–Ni–C alloys. *Scanning Electron Microscopy: Systems and Applications, 1973,* Inst. of Physics, London, pp. 320–323 (1973).
30. Davidson, D. L. The quantification of deformation using electron channelling. *Scanning Electron Microscopy/1983/III,* Chicago, IL: SEM Inc., pp. 1043–1050 (1983).
31. Davidson, D. L. Uses of electron channelling in studying material deformation. *Intl. Metals Review* 29:75–95 (1984).
32. Fathers, D. J., et al. Magnetic domain contrast from cubic materials in the scanning electron microscope. *Philo. Mag.* 27:765–768 (1973).
33. Zwilling, G. Observation of magnetic domains in the scanning electron microscope. *Practical Metallography* 11:716–728 (1974).
34. Yakowitz, H., and Newbury, D. E. Magnetic domain structures in Fe-3.2 Si revealed by scanning electron microscopy—a photo essay. *J. Testing and Evaluation* 3:75–78 (Jan. 1975).
35. Jones, G. A. Magnetic contrast in the scanning electron microscope: An appraisal of techniques and their applications. *J. Magnetism and Magnetic Materials* 8:263–285 (1978).
36. Murphy, J. A. Considerations, materials, and procedures for specimen mounting prior to scanning electron microscopic examination. *Scanning Electron Microscopy/1982/II*. Chicago, IL: SEM Inc., pp. 657–696 (1982).
37. Radavitch, J. F. Private communication.
38. Paul, J., and Bauer, B. Contrast techniques for phase separation in the scanning electron microscope. *Practical Metallography* 20:213–221 (May, 1983).

39. Vander Voort, G. F. *Metallography: Principles and Practice.* New York: McGraw-Hill Book Co., 1984.
40. Almond, E. A. et al. Interpretation of SEM fracture surface detail using a sectioning technique. *Metallography* 3:379–382 (1970).
41. Hillnhagen, E., and Schauf, E. Two-surface analysis of polished sections in the scanning electron microscope. *Practical Metallography* 17:29–34 (1980).
42. Kny, E., et al. Electrolytic deep etching—a valuable complementary method to standard metallographic procedures. *Practical Metallography* 14:512–520 (1977).
43. Fřemunt, P., et al. The study of sulphides by the method of gradual etching. *Practical Metallography* 17:497–508 (1980).
44. Paul, J., and Mürrle, U. Assessing the changes in shape of structural components in deep etched specimens. *Practical Metallography* 18:413–423 (1981).
45. Datta, M., et al. Selective dissolution of dendritic or interdendritic phase in Sn–Al alloys. *Practical Metallography* 20:394–405 (1983).
46. Ross, L. R. Selective etching techniques for SEM examination of superalloys. *Microstructural Science* 3A:351–356 (1975).
47. Wilson, F. G. The morphology of grain-and twin-boundary carbides in austenitic steels. JISI 209:126–130 (Feb., 1971).
48. Gill, T. P. S., and Gnanamoorthy, J. B. A method for quantitative analysis of delta-ferrite, sigma and $M_{23}C_6$ carbide phases in heat treated type 316 stainless steel weldments. *J. Materials Science* 17:1513–1518 (1982).
49. Herbsleb, G., and Schwaab, P., Delineation of the microstructure of high alloy CrNi, CrNiW and CrNiCoW steels by deep etching. *Practical Metallography* 20:53–63 (1983).
50. Rege, R. A., et al. Three-dimensional view of alumina clusters in aluminum-killed low-carbon steel. *Met. Trans.* 1:2652–2653 (Sept., 1970).
51. Okohira, K. et al. Observation of three-dimensional shapes of inclusions in low-carbon aluminum-killed steel by scanning electron microscope. *Trans. ISIJ* 14:102–109 (1974).
52. Baker, T. J. Use of scanning electron microscopy in studying sulphide morphology on fracture surfaces. *Sulfide Inclusions In Steel.* Metals Park, OH: American Society for Metals, 1975, pp. 135–158.
53. Wilson, A. D. Application of the SEM to the investigation of inclusion behavior in steels. *Scanning Electron Microscopy/1977/I.* Chicago, IL: IIT Research Inst., pp. 121–128 (1977).
54. Inckle, A. Etching of fracture surfaces. *J. Materials Science* 5:86–88 (Jan., 1970).
55. Tvrdik, Z., and Vinckier, A. The use of the scanning electron microscope for the investigation of the aging of carbon steel. *Practical Metallography* 10:87–93 (1973).
56. Klimesch, B., et al. The use of the scanning electron microscope (SEM) in the assessment of suspect case hardening. *Practical Metallography* 12:417–427 (1975).
57. Dudek, H. J., and Ziegler, G. Identification of phases in metallographically prepared specimens and fracture surfaces using the energy dispersive X-ray microanalysis. *Practical Metallography* 13:521–533 (1976).
58. Marlow, P., et al. Some replica techniques for the scanning electron microscope. *Micron* 2:139–147 (1970).
59. Wu, W., et al. A metallic replica technique for scanning electron microscopy. *J. Materials Science* 8:1670–1672 (Nov., 1973).
60. Stirland, D. J. A replica method for the examination of large specimens in the scanning electron microscope. *J. Microscopy* 92:31–36 (Aug., 1970).
61. Barnes, I. E. Replica models for the scanning electron microscope. A new impression technique. *British Dental J.* 133:337–342 (Oct. 17, 1972).
62. Neville, G. Replica technique for the scanning electron microscope. *J. Physics E. (Scientific Instruments)* 5:743–744 (Aug., 1972).
63. Eckert, J. D., and Caveney, R. J. A replica technique for conventional and scanning electron microscopy. *J. Scientific Instruments* 3:413–414 (May 1970).
64. Wendler, B., and Neubauer, B. Increased information from microstructural replicas through the application of a scanning electron microscope. *Practical Metallography* 16:3–10 (1979).
65. Melford, D. A., and Whittington, K. R. Application of the scanning microanalyser to inclusion counting and identification. *4th Intl. Cong. on X-ray Optics and Microanalysis,* Hermann Press, Paris, pp. 497–505 (1966).
66. Dörfler, G. Quantitative evaluation methods for alloy microstructures by microprobe analysis. *Quantitative Electron Probe Microanalysis.* NBS Special Publ. 298, pp. 215–267 (Oct., 1968).
67. Waldman, J. et al. Electronprobe determination of phase volume fractions. *Trans. ASM* 62:818–819 (1969).
68. Jones, M. P. Quantitative determination of phase and stereological parameters by electron microprobe. *Micron* 2:125–138 (1970).
69. Jones, M. P. Automatic stereological analysis by electron probe X-ray microanalyser. In *Quantitative Scanning Electron Microscopy.* London: Academic Press, pp. 531–549 (1974).
70. White, E. W., et al. Particle size distributions of particulate aluminas from computer-processed SEM images. *Scanning Electron Microscopy/1970.* Chicago: IIT Research Institute, pp. 57–64 (1970).
71. Braggins, D. W., et al. The applications of image analysis techniques to scanning electron microscopy. *Scanning Electron Microscopy/1971 (Part I).* Chicago: IIT Research Institute, pp. 393–400 (1971).
72. Kupcis, O. A.; Woo, O. T.; and Ramaswami, B. The determination of size and spacing of second-phase particles by scanning electron microscopy. *Materials Science and Engineering* 9:47–49 (1972).
73. Johari, O., and Samudra, A. V. Measurement of retained austenite using scanning electron metallography. *Micron* 3:238–246 (1972).
74. Gibbard, D. W. The application of image analysis techniques to scanning electron microscopy and microanalysis. In *Quantitative Scanning Electron Microscopy.* London: Academic Press, pp. 75–92 (1974).
75. Ekelund, S., and Werlefors, T. A system for the quantitative characterization of microstructures by combined image analysis and X-ray discrimination in the scanning electron microscope. *Scanning Electron Microscopy/1976 (Part III).* Chicago: IIT Research Institute, pp. 417–424 (1976).
76. Baumgartl, S., and Bühler, H. E. Quantitative microstructural analysis by means of focussed electron beams. *Practical Metallography.* 13:263–288 (1976).
77. Werlefors, T., and Ekelund, S. Automatic multiparameter characterization of non-metallic inclusions-an evaluation of PASEM. *Scandinavian J. of Metallurgy* 7:60–70 (1978).
78. Werlefors, T., and Eskilsson, C. Automatic multiparameter characterization of non-metallic inclusions—practical applications of PASEM in the study of inclusions in steel. *Scandinavian J. of Metallurgy* 7:215–222 (1978).
79. Stott, W. R., and Chatfield, E. J. A precision SEM image analysis system with full-feature EDXA characterization. *Scanning Electron Microscopy/1979/II.* AMF O'Hare, IL: SEM Inc., pp. 53–59 (1979).
80. Werlefors, T., Eskilsson, C., and Ekelund, S. A method for the automatic assessment of carbides in high speed steels with a computer controlled scanning electron microscope. *Scandinavian J. of Metallurgy* 8:221–231 (1979).
81. Lee, R. J., and Kelly, J. F. Overview of SEM-based automated image analysis. *Scanning Electron Microscopy/1980, Part I.* Chicago, IL: SEM Inc., pp. 303–310 (1980).
82. Lee, R. J.; Spitzig, W. A.; Kelly, J. F.; and Fisher, R. M. Quantitative metallography by computer-controlled scanning electron microscopy. *J. of Metals* 33:20–25 (Mar., 1981).
83. Werlefors, T. PASEM (the particle analysing scanning electron mi-

croscope) applied to the study of non-metallic inclusions in steel. *Swedish Symposium on Non-Metallic Inclusions in Steel.* Hagfors, Sweden: Uddeholms AB, pp. 234–241 (1981).
84. Kim, C. et al. A new procedure for determining volume fraction of primary carbides in high-speed and related tool steels. *Met. Trans.* 13A:185–191 (Feb., 1982).
85. Isaacs, A. M. Limits to quantitation in particle analysis: Some empirical determinations. *Microbeam Analysis—1983.* San Francisco, CA: San Francisco Press, Inc., pp. 202–208 (1983).
86. Jeulin, D. Morphological SEM picture processing. *J. Microsc. Spectrosc. Electron.* 8:1–18 (1983).
87. Lee, R. J., et al. Quantitative metallography by computer-controlled scanning electron microscopy. *Practical Metallography* 21:27–41 (1984).

11
METALLOGRAPHY IN THE SCANNING TRANSMISSION ELECTRON MICROSCOPE

D. B. Williams

Department of Materials Science and Engineering
Lehigh University

The scanning transmission electron microscope (STEM) is a combination of the conventional transmission electron microscope (TEM) and scanning electron microscope (SEM) that has revolutionized the amount of information that can be obtained from a thin-foil TEM specimen. Figure 11-1(a) shows the range of signals generated when a high-energy (usually ~ 100 kV) electron beam interacts with a thin-foil specimen. Although all of these signals are theoretically accessible in the STEM, visible light and absorbed and Auger electrons are rarely sought. In practice, detectors are used to pick up the remaining signals [see Fig. 11-1(b)], as described in appropriate sections of this chapter. It will be assumed that the reader is familiar with the conventional TEM and SEM and the information that is available about them. Several standard texts describing TEM and SEM techniques are available, for example, see References 1, 2, and 3.

The development of the STEM and associated microanalytical techniques [which has given rise to the term "analytical electron microscopy" (AEM)] has also been the subject of two recent texts, in which far more detail, both theoretical[4] and practical,[5] is presented than can be condensed into a single chapter. What we will concentrate on here is how to make best use of the STEM to perform classical metallographic functions such as identification of second phases and other heterogeneities in crystalline materials. To this end, it is important to understand the principles of the STEM instrument, the precautions necessary to prepare representative thin samples, and finally the practical use and limitations of the various imaging, diffraction, and microanalytical techniques available on the instrument. It should be noted at the start that a typical STEM is usually a converted TEM operating in the range of 100 to 200 kV and is capable of all the routine high-resolution imaging and selected-area-diffraction (SAD) operations that permit identification of defects in crystalline materials. The STEM also suffers the same limitation as the TEM, namely, that all the information obtained is averaged through the thickness of the thin specimen, and therefore the techniques to be described here are all insensitive to any depth variations in structure and chemistry. The STEM can also generate high-resolution surface images and carry out X-ray microanalysis in the same manner as a combined SEM/electron-probe microanalyzer (EPMA). Given the similarity between a STEM and the combined properties of a TEM and SEM, we will only emphasize aspects of the former that offer an improvement over the conventional TEM and SEM.

THE STEM INSTRUMENT

The conventional TEM illuminates a few square micrometers of a thin-film specimen with a parallel beam of high-kV electrons, as shown in Fig. 11-2(a). The objective lens recombines the transmitted (undeviated) and most of the scattered (mainly diffracted) electrons to form a diffraction pattern in the back focal plane, (AB) in Fig. 11-2(a), and an image in the image plane, (CD). The imaging system transfers the electron intensity distribution in AB or CD to the viewing screen at the desired magnification, from which it can also be recorded photographically. Use of an aperture in the back focal plane (AB) permits the choice of which electrons (transmitted or diffracted) contribute to the image (bright or dark field, respectively). Similarly, an aperture in the image plane (CD) permits selection of which area of the specimen contributes to the diffraction pattern on the viewing screen (SAD patterns).

The STEM operates on principles similar to those shown in Fig. 11-2(b). Instead of a parallel beam, however, a convergent probe of electrons (down to ~ 2 nm, or less, in diameter) is used. When a small probe is formed, conventional TEM image formation is impossible; thus, imaging is performed by scanning the probe over the re-

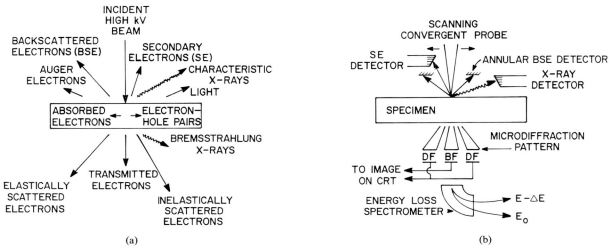

Fig. 11-1. (a) Possible signals generated when a high-energy electron beam interacts with a thin foil in the STEM (the diagram only approximately relates to the direction of the various signals). (b) The usual range of detectors placed in a STEM, indicating the major signals available for study.

gion of interest on the specimen, just as in a conventional SEM. The transmitted and diffracted electrons are still brought separately to the back focal plane and form a diffraction pattern. The information content in the back focal plane is time-dependent, however, since it varies as the probe samples different regions of the area of interest on the specimen. In order to reconstruct a useful image, the time-dependent signal in a specific area of the back focal plane is picked up by an electron detector (positioned in a plane conjugate with the back focal plane) and converted to charge pulses that modulate a cathode ray tube (CRT). The image can be recorded directly from the CRT using Polaroid film. A range of detectors and suitable adjustment of the imaging lenses permits a choice of which electrons impinge on the detector and form the image, thus acting like the objective aperture in TEM. Suitable adjustment of the scan coils permits choice of which area contributes to the image and diffraction pattern in a manner analogous to that of the SA aperture in the TEM.

As we shall see, the images formed in a STEM and TEM can be made equivalent,[5,6] although in most cases the STEM image offers little or no advantage over the TEM. The main advantage of STEM over TEM comes when the scanning probe is stopped and positioned on an area of interest (e.g., precipitate, grain boundary, or other heterogeneity) in the thin foil, as shown in Fig. 11-1(b). Obviously the scanning image information is lost, but what remains is the ability to form a diffraction pattern from a very small region (microdiffraction), and what is gained is the chance to obtain chemical information (microanalysis) using energy-dispersive spectrometry (EDS) of X-rays or electron-energy-loss spectrometry (EELS) of inelastically scattered electrons. The combination of a fine probe and a thin TEM specimen means that microdiffraction and microanalysis can routinely be obtained from regions <50 nm in diameter and often <10 nm diameter. These dimensions contrast with conventional TEM diffraction resolution and microanalysis resolution in the SEM/EPMA, which are both ~1 μm. Therefore, high spatial-resolution microdiffraction and microanalysis are the main modes of operation of the STEM. Since most STEM operations basically involve signal detection and/or processing, and the useful nature of the signal is limited by the signal-to-noise ratio, it is important always to maximize the signal of interest. The various signals are ultimately controlled by the signal put into the specimen, i.e., the electron current density in the probe. The electron density from a conventional tungsten (W) filament gun is often sufficient to generate suitable signals. Lanthanum hexaboride (LaB_6) sources are an order of magnitude more intense, however, and are strongly recommended, despite their increased expense. Field-emission guns (FEGs) are available on a very select number of STEMs but are exceedingly expensive and require an ultra-high vacuum (UHV) system. The FEG is essential only when the ultimate available sensitivity and spatial resolution are sought and is not required for most routine STEM operations.

Before we consider routine STEM operations, it is appropriate to address the question of correct specimen preparation to ensure that the information gained in the STEM is characteristic of the bulk specimen. Also, some idea of the important parameter, spatial resolution, must be given.

SPECIMEN PREPARATION

Thin-foil specimen preparation by standard methods, such as electropolishing or ion-beam thinning, is known to introduce artifacts into the TEM specimen that are not characteristic of the bulk. With due care and experi-

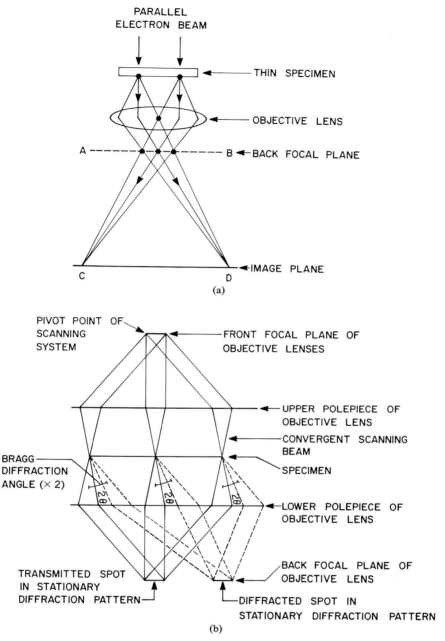

Fig. 11-2. (a) Ray diagram showing the basic optics of a TEM: Electrons scattered by the specimen are focused into a diffraction pattern in the back focal plane and then recombined into an image. Either the diffraction pattern or the image can be viewed by suitable adjustment of subsequent lenses in the imaging system of the microscope. (b) Ray diagram showing the basic optics of a TEM/STEM: The scanned beam generates a time-dependent stationary diffraction pattern in the back focal plane, which can be imaged on the TEM screen if necessary. Images are formed by allowing the transmitted or diffracted rays to fall on a detector in a conjugate plane. The varying signal falling on the detector is translated into varying intensity (contrast) on a CRT.

ence, however, the effects can be recognized and/or minimized[7] in a way that permits correct interpretation of results to be straightforward. Similarly, any STEM imaging applications involving the internal defect structure of the specimen can proceed without difficulty. What is more important in the STEM is the state of the surface of the thin specimen, particularly in terms of changes in the solute concentration compared with the bulk and the presence of contaminants that decompose to carbonaceous deposits when struck by the electron probe. Surface-solute enrichments are well documented[8,9,10] and in very thin specimens can give rise to erroneous microanalytical data. There are various ways to remove or minimize the effects, however, including chemical cleaning,[8] ion beam "dusting," or sputtering in a UHV system.[10] Alternatively, if the defect structure is irrelevant, use of ultramicrotomy has been successful.[11] Different alloys are variably susceptible to surface solute enrichment effects and require various treatments to minimize them.

The operator should be aware of the possibility of surface-solute enrichment and check to see if the composition of a nominally homogeneous specimen varies with thickness, as shown schematically in Fig. 11-3. If so, any or all of the various methods previously described to remove surface films should be attempted prior to proceeding with quantitative microanalysis.

Carbon contamination is often present on specimens prepared by conventional means.[12] It is undesirable in microdiffraction, where it masks fine detail. In X-ray microanalysis, contamination degrades spatial resolution and absorbs low atomic-number (Z) X-rays, and, in energy-loss spectrometry, it raises the background intensity and masks any carbon content in the specimen. Carbon contamination can be minimized in a variety of ways. In particular, the specimen and specimen holder should be clean and dry and never be touched by hand (particularly inside any O-ring seals). The specimen can be "flooded" by the electron beam prior to observation to "pin" any contaminant, or it can be cooled to ~ −45°C to prevent contamination from migrating to the probe position or heated to +60°C to remove any build-up on the surface. Appropriate low-background, double-tilt, low-temperature specimen holders are available for most STEMs. These are strongly recommended for the serious analytical microscopist, since they reduce contamination to negligible levels, thus ensuring the highest spatial resolution, the most sensitive use of microdiffraction, and the most accurate low-Z elemental microanalysis by X-ray EDS or EELS.

Extraction Replication

Because of the need to analyze quantitatively very small regions such as precipitates, conventional specimen preparation is often not suitable. As we have noted, all the signals detected in STEM microanalysis and microdiffraction are averaged through the foil thickness. Therefore, it is not possible to analyze an embedded precipitate in a quantitative manner. Often the best solution is to extract the precipitate and support it on a suitable thin film (e.g., evaporated C or Al films). Extraction replication is an old specimen preparation technique[7] now undergoing a resurgence of interest because of its ability to present small particles for analysis free of their surrounding matrix. If such a technique can extract very small (<10-nm) particles, then this, in effect, lowers the spatial resolution of analysis (see next section). Resolution is no longer limited by the probe size and specimen thickness but by the dimensions of the extracted region of interest, as shown in Fig. 11-4.

SPATIAL RESOLUTION IN THE STEM

One of the major advantages of microanalysis and microdiffraction in the STEM is the significant improvement in spatial resolution realized as compared with conventional microanalysis in the SEM/EPMA and SAD in the TEM. This improvement in resolution arises as a result of a combination of small probe sizes and thin specimens. There is still a lot of active research on the question of spatial resolution,[5,13] but it is generally accepted to be the diameter of the base of the electron beam-specimen interaction volume (approximating a truncated cone in shape) that contains 90 percent of the electrons leaving the foil (Fig. 11-5). There are a variety of models to determine this diameter,[13] and much experimental work has been reported.[14,15] In practice, however, the simplest approach proves as reasonable as most and, to a first approximation, gives a good figure for the spatial resolution. This approach, attributed to Reed in the classical paper by Goldstein et al.,[16] is known as the "single scattering equation" and predicts a spatial resolution b (in cm) due to an infinitely small probe and given by the following:

$$b = 625\left(\frac{Z}{E_0}\right)\left(\frac{\rho}{A}\right)^{1/2} t^{3/2} \quad (11\text{-}1)$$

where Z is the atomic number, E_0 is the electron beam energy (in kV), ρ is the specimen density (g cm^{-3}), t is the specimen thickness (cm), and A is the atomic weight.

If a finite probe size of d is assumed, Reed[17] recommends that the relation, $b_{\text{actual}} = \sqrt{b^2 + d^2}$, should be used to estimate the spatial resolution. Determination of d is best accomplished by direct imaging of the STEM probe on the TEM screen. Conventional manufacturers'

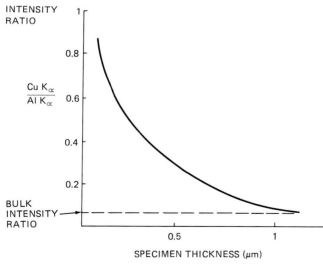

Fig. 11-3. Plot showing the variation of CuK$_\alpha$/AlK$_\alpha$ X-ray intensity ratio as a function of increasing thickness in a nominally homogeneous thin foil and exhibiting the effects of a Cu-rich film on the specimen surface.

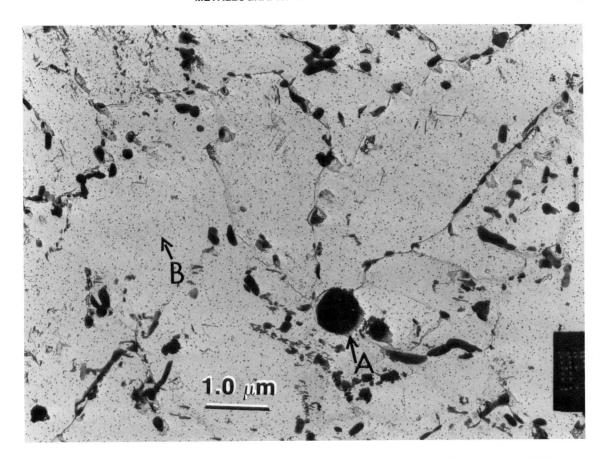

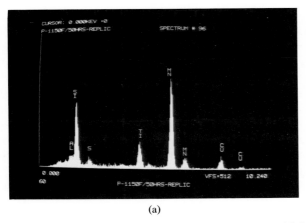

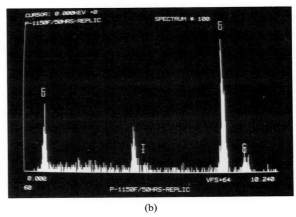

(a)   (b)

Fig. 11-4. Carbon extraction replica (on a titanium support grid) from a pressure-vessel steel weldment aged 50 hr at 1150°F showing a range of particle sizes. Each particle acts as its own spatial resolution limit, and the X-ray data in the two spectra came from the arrowed particles of ∼ 1 to 3 μm and 15 nm in diameter, (a) and (b), respectively. The spectra show, qualitatively, that the large inclusion is probably a manganese silicate from the flux used during welding, whereas the small particles are copper precipitates that formed during the post-weld heat treatment. (Courtesy, R. Dias.)

estimates are the full-width half maximum (FWHM) of an (assumed) Gaussian probe. Since the value for b considers 90 percent of the scattered intensity, d values should similarly consider 90 percent of the Gaussian probe, i.e., ∼1.8 × FWHM, or nearly twice the manufacturers' figures. For a known specimen, the spatial resolution should be determined beforehand in order to ensure that the detected microanalytical signal or the microdiffraction pattern comes just from the feature being investigated and that no extraneous signals are involved (unless they are being sought, for example, in trying to determine a particle-matrix orientation from a single diffraction pattern).

The spatial resolution limited by beam spreading is most important in X-ray analysis and microdiffraction since both these techniques detect the signals emanating

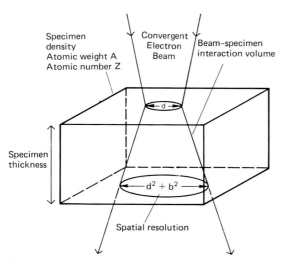

Fig. 11-5. The generally accepted definition of spatial resolution in the STEM. A convergent electron beam of diameter d spreads through the foil of thickness t, density ρ, atomic weight A, and atomic number Z to give a beam broadening parameter b, the base diameter of a cone containing 90 percent of the scattered electrons. The actual spatial resolution is given by $\sqrt{b^2+d^2}$. (Courtesy, Philips Electronic Instruments.)

from the entire interaction volume. EELS, on the other hand, has an inherently higher spatial resolution since the spectrometer collects and analyzes only those electrons with a very small angular deviation (<10 mrad) from the optic axis. If spatial resolution caused by specimen thickness is a problem, it can be avoided only by suitable preparation of very thin specimens, use of higher voltages or, if feasible, extraction of the region of interest as already described in the section on specimen preparation. If a small precipitate is extracted onto a thin carbon film, the particle size itself is the effective spatial resolution, provided that it is well separated from other particles (see Fig. 11-4). When a very small region is analyzed, however, it should be borne in mind that this invariably results in a reduction in the intensity of signal and a concomitant increase in the error of any quantification. This is because a much smaller volume is contributing to the detected signal. A similar effect happens when a small probe is used to minimize spatial resolution because the probe current (i_p) varies as $d^{8/3}$, where d is the probe diameter. Therefore, quantitative high-resolution microdiffraction and X-ray microanalysis in the STEM is always a compromise between desired spatial resolution and desired accuracy. Use of LaB_6 or FEGs as well as improved X-ray detectors (with higher take-off and collection angles and greater surface area positioned closer to the specimen) will help this problem. Similarly, a move to instruments of intermediate (200 to 500 kV) or higher voltage will generate less beam spreading. Generally speaking, however, there are a large range of metallographic problems in which the dimensions of interest are between 50 and 2000 nm; under these circumstances, current 100- to 200-kV STEMs using W guns and routinely thinned specimens are ideally suited to the generation of high-quality, quantitative information with no spatial-resolution limitations.

IMAGING IN THE STEM

There are many ways to obtain an image in the STEM. Basically, any of the signals generated when a high-kV electron beam hits a specimen (Fig. 11-1) can be detected and used to modulate the STEM CRT. In practice, however, transmitted-electron bright field (BF), scattered-electron dark field (DF), secondary-electron (SE), and backscattered-electron (BSE) images are the major operating modes in STEM. Since all of these images are also available on a conventional TEM or SEM, it is important to know the advantages to be gained by forming the image in a STEM, and we will emphasize these. Also, any of the scanning images can be used as a basis for selecting a position for the fine probe to carry out microanalysis or microdiffraction (see appropriate sections below), although the standard BF image is usually used for this purpose.

We will also discuss briefly the possibility of forming chemical images (maps) using X-rays and the information obtained from electron energy-loss (EEL) images.

Bright Field (BF) Imaging

Forming a scanning image with the transmitted electrons offers no advantages over TEM when the specimen is crystalline since diffraction contrast effects are generally lower in STEM.[5] Furthermore, image resolution and the general image quality is usually poorer than in TEM,[18] as shown in Fig. 11-6. For noncrystalline specimens, however, mass-thickness (amplitude) contrast can be enhanced by collecting the maximum signal available on the BF detector and using the electronic contrast-enhancement techniques that are available for all scanning images. For beam-sensitive specimens (such as polymers), STEM imaging usually offers a significantly improved lifetime in the beam compared with that achievable with conventional static imaging in the TEM.[19]

It is worth noting in this context that often, in practice, there are circumstances where it is desirable to remove dynamical diffraction-contrast effects in STEM images of crystalline specimens and to enhance amplitude contrast. As shown in Fig. 11-7, which compares TEM and STEM images from a thin section of a steel, the grain-boundary precipitates are much clearer in the STEM image since they are not masked by strong diffraction contrast from the matrix as they are in the TEM image. Since these precipitates are the areas of interest for microdiffraction and microanalysis using a stationary fine probe, it is easiest to operate in STEM BF image mode

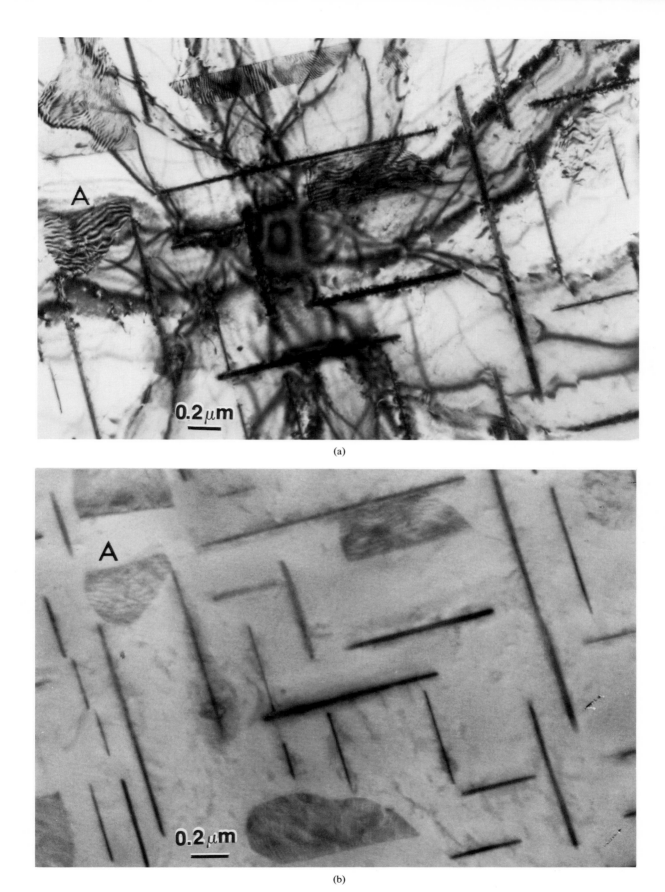

Fig. 11-6. Comparison of (a) TEM and (b) STEM BF images of Al–4-wt.%Cu aged to give θ' particles on {100} planes. The TEM image shows strong dynamical diffraction contrast (the bend center) that is absent in the STEM image. The STEM image shows good mass-thickness contrast between the θ' and the matrix but is generally of poorer quality than the TEM image. Point A is the same in both images.

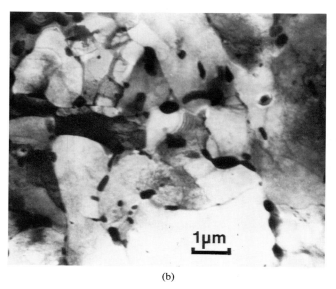

Fig. 11-7. Comparison of (a) TEM and (b) STEM BF images from an alloy steel aged to produce coarse grain-boundary precipitates. The precipitate contrast is partially masked in the TEM image by the strong dynamical diffraction contrast in the grain interiors. In the STEM image, the dynamical contrast is significantly reduced, making the presence of the precipitates far more obvious.

and simply position the probe on the precipitates for further study.

Dark Field (DF) Imaging

As in the BF image, diffraction contrast from crystalline specimens is generally poor, and the STEM image quality does not compare with the equivalent TEM image. Annular DF imaging of noncrystalline specimens, however, offers significant improvement in intensity and contrast over standard TEM DF images since the annular detector can collect a much larger signal than the objective aperture in a TEM.

Thus, in summary, both BF and DF STEM images are most useful for noncrystalline and/or beam-sensitive specimens. For crystalline specimens, diffraction-contrast information is usually poorer than in TEM, and the main role of the STEM image from crystalline specimens is to locate regions of interest for microdiffraction and/or microanalysis. Since the quality of STEM images recorded on Polaroid from the CRT is invariably poorer than that of images recorded on a photographic plate in TEM, it is usual practice to record a TEM image of the area of interest and indicate where microanalysis, etc., was performed.

Secondary Electron (SE) Imaging

SE images give topographic information about the surface of the specimen. In STEM, because of the high kV and very small probe sizes, SE imaging offers improved resolution over conventional SEM images. Furthermore, the objective lens prefield acts as an efficient collector of SEs, and, because of the position of the SE detector (within the upper objective polepiece), back-scattered electrons (BSE) cannot enter it. Therefore, the signal level of SE images is higher and the noise level lower in STEM than in SEM. The net result is that SE images in STEM offer a resolution of $\sim$ 2 to 5 nm (see Fig. 11-8) compared with 6 to 10 nm in a typical SEM. The drawback to SE imaging in STEM compared with SEM is that large specimens cannot be observed. The confines of the TEM/STEM stage limit specimen dimensions to $\sim 5 \times 5 \times 1$ mm compared with equivalent dimensions of several centimeters in a SEM.

Backscattered Electron (BSE) Imaging

BSE images offer atomic number and topographic information. In the conventional SEM, image resolution is $\sim 1 \mu m$ with a low-efficiency detector but may be much smaller with a high-efficiency detector. Again, the resolution limit is a signal/noise problem, and the high-kV, small probes in STEM therefore improve on the resolution of the SEM. Furthermore, the use of thin specimens in STEM can enhance resolution by cutting down on the amount of electron backscatter from regions far away from the probe. To maximize the probe current with thin specimens and generate a sufficient signal from the region close to the probe for good quality images with $\sim$ 10-nm resolution (see Fig. 11-9), a LaB_6 gun is desirable. The space limitations of the STEM stage are again a major reason why few high resolution BSE studies have been performed in STEM.

Multiple Imaging

A final advantage of STEM is its ability to combine conventional TEM and SEM information into a single

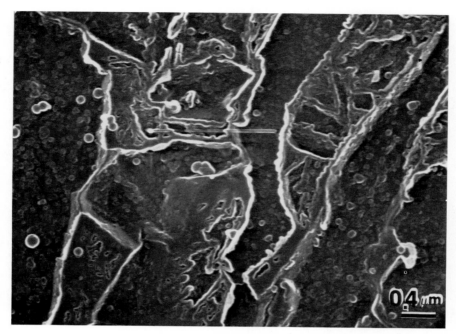

Fig. 11-8. SE image of a carbon-extraction replica from a high-strength low-alloy steel showing the topography of the sample caused by the presence of small extracted particles. Fine detail on a scale of <5 nm can be resolved. (Courtesy, Philips Electronic Instruments.)

micrograph.[20] An example is shown in Fig. 11-10, which is a combined BF and SE image of a fracture surface thinned from one side to TEM transparency. The combination image permits a close correlation between the internal defect structure visible in the BF part of the image and the surface topography visible from the SE component. Similar image combinations, or multiple signal display, are possible using any of the variety of signals detected in the STEM.[5]

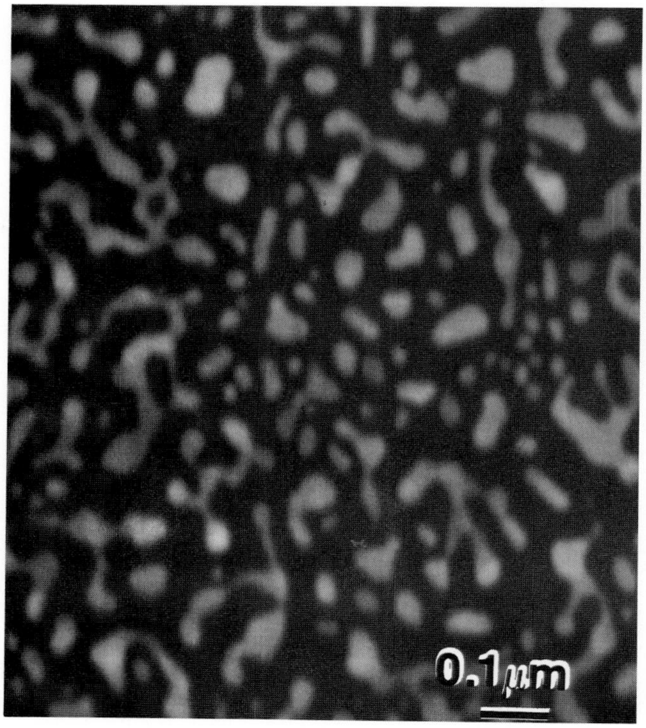

X-ray Imaging (maps)

It is possible to use the characteristic X-ray signal from a specific element to modulate the CRT, thus producing an image whose intensity depends on the amount of that element present. Such X-ray maps are a common feature of SEM and EPMA instruments. With a thin specimen in STEM, it is possible, in theory, to obtain a direct (quantitative) relationship between intensity and composition if the thin-foil criterion is obeyed (see the section on X-ray microanalysis). In practice, however, because of the low count rates from thin foils, even with LaB_6 guns, the quality of X-ray maps obtained is usually poor. Between 2×10^5 and 5×10^5 X-ray photons are needed for a good image,[3] and this may take many minutes to generate. As a result, X-ray maps in STEM are not of great use to the metallographer. Higher quality images, albeit with a much lower spatial resolution, can be obtained on a SEM/EPMA.

Electron Energy Loss (EEL) Images

In principle, images can be formed from electrons that have suffered a specific energy loss because of the ionization of a particular atom. These images should therefore be chemical images, just like X-ray maps. In practice,

Fig. 11-9. BSE image of gold islands on a carbon film showing a spatial resolution of ~ 10 nm. The source of contrast is the large atomic-number difference between gold and carbon. (Courtesy, JEOL USA.)

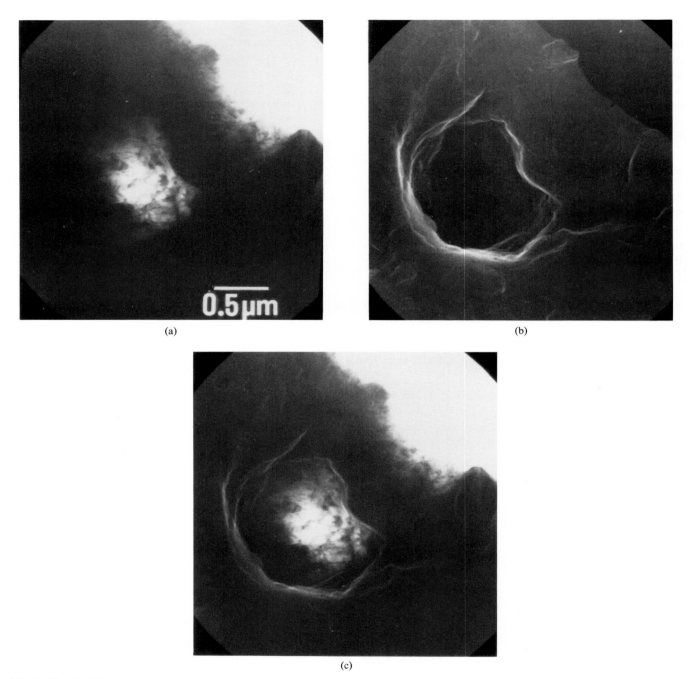

Fig. 11-10. (a) STEM, (b) SEM, and (c) combined STEM/SEM image showing both the internal defect structure and surface topography of a manganese-modified 316 stainless steel, back-thinned to local transparency around a dimple in the fracture surface. (Courtesy, R. M. Allen, reproduced by permission of SEM, Inc.)

however, the large background contribution to the EEL spectrum means that thickness effects usually override the chemical contrast entirely.[5] True chemical images require obtaining a spectrum and then subtracting the spectral background at each picture point (pixel). For a good 256 × 256 pixel image, this procedure requires either impractical lengths of time or enormous on-line computing capability.[21] EEL imaging is thus not at present a practical mode of STEM operation for the average user.

There is, however, the possibility of using the EELS to filter out unwanted (energy-loss) electrons, leaving a STEM image of "pure" transmitted (BF) or elastically scattered (DF) electrons. Such images show improved contrast, but again this is best observed in annular DF images of noncrystalline samples for which the signal is still large after this energy-filtering process.[5]

Similarly, the EEL signal can be used to produce a composite image that is a ratio of the elastic signal (annular DF detector) to the inelastic signal (BF detector with transmitted electrons—no loss—removed). The resultant signal is directly proportional to the atomic number, Z. Such Z contrast is very useful for observing very tiny particles and is the method by which single atoms, using FEG STEM instruments, can be imaged.[22] This technique is not within the capability of a typical TEM/STEM, however.

MICRODIFFRACTION

Several forms of microdiffraction are available to the STEM user,[5] but the most convenient and the one with the most information is that of convergent beam diffraction (CBD). This is the form of the diffraction pattern that is invariably present in the back focal plane of a TEM/STEM whether the probe is scanning an area or stationed on a point of interest, as shown in Fig. 11-2(a). (The latter, stationary mode is, in practice, the only sensible way to obtain the diffraction pattern.) The term "microdiffraction" is appropriate because the pattern comes only from the volume of specimen that interacts with the fine probe. The spatial resolution is much less than that of the conventional SAD, therefore, and, in the extreme case, using a field emission gun (so-called nanodiffraction[23]) can give information from single-unit cells. The most obvious use of microdiffraction is in a manner similar to one in SAD, namely, identification of phases in terms of their crystal structure (or lack of it). From that point of view, analysis of the patterns often involves nothing more than conventional indexing of the zero-order diffraction maxima, which are typical of any electron diffraction pattern. A typical CBD pattern is shown in Fig. 11-11. This pattern is a $\langle \bar{1}12 \rangle$ zone-axis

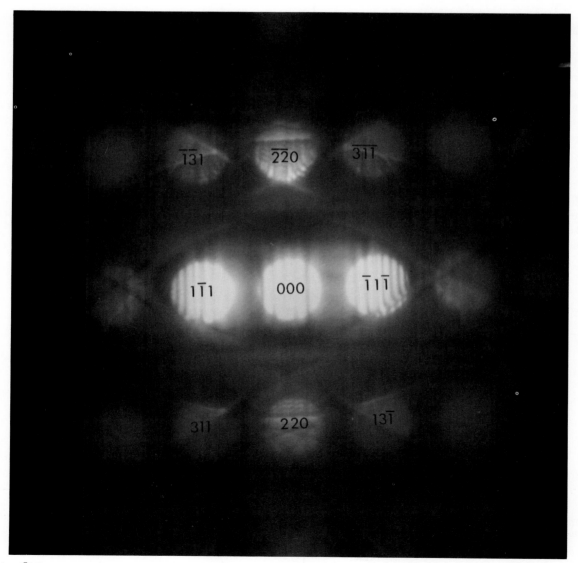

Fig. 11-11. $\langle \bar{1}12 \rangle$ convergent-beam zone-axis pattern from pure aluminum: Each disc corresponds to diffraction from the indicated {hkl} plane. Faint Kikuchi lines are visible outside the discs as is dynamical contrast information within them.

pattern from an fcc structure, the indexing having been performed using the conventional ratio method.[1] When used in this manner, CBD offers only a spatial resolution advantage over SAD in the TEM. As can also be seen in Fig. 11-11, however, there is often more intensity information in the diffraction maxima. The maxima are discs, not points as in SAD, because of the convergence of the beam (see Fig. 11-2). Within, and associated with these discs, there is dynamical electron-scattering contrast (in specimens > ~50 nm thick) that can give important metallographic information such as (a) specimen thickness; (b) accurate, relative lattice parameter measurements; (c) crystal lattice spacings in the third dimension; and (d) the full crystal symmetry of the specimen (point and space group). Clearly, a CBD pattern can be much more than just a high spatial-resolution SAD pattern for identifying the specimen orientation.

To obtain thickness information (which, as discussed at the end of this chapter, is an important specimen parameter in analytical electron microscopy), it is simply necessary to tilt the specimen until only one beam is diffracting strongly (i.e., two-beam condition). Parallel (Kossel-Möllenstedt) fringes appear when the probe is focussed at the plane of the specimen, as shown in Fig. 11-12. (This contrast effect is very sensitive to the correct focus of the objective lens.) The specimen thickness is then easily determined from measurement of the fringe spacing.[24,25]

Under certain very well-defined conditions,[26] it is possible to observe fine dark lines within the central 000 disc of the CBD as shown in Fig. 11-13. These so-called higher order Laue zone (HOLZ) lines are due to elastic scatter from higher order planes, but the lines behave like conventional inelastic Kikuchi lines, i.e., they move as if attached to the sample and are therefore a very sensitive indicator of the exact specimen orientation. By observing changes in the symmetry of these lines, lattice parameters can be determined within an accuracy of $\sim 2 \times 10^{-4}$ nm.[26,27] In practice, it is usual to enhance the low intensity of scatter from higher order zones by cooling the specimen to $\sim -196°C$.

If the CBD pattern is observed at very low magnifications (which are only routinely attainable in modern TEMs), then high-angle diffraction maxima from higher order atomic planes are often visible. The result is a ring of (HOLZ) intensity maxima around the normal zero-order zone (see Fig. 11-14). The radius of this ring can be measured and related to the spacing of the lattice planes parallel to the electron beam.[5,28] This spacing, in combination with measurements made in the zero-order pattern, gives 3D crystal spacing information in a single 2D pattern.[29]

Direct Phase Identification Using CBD

Certain phases may be identified by CBD in a "fingerprint" manner, that is, their CBD pattern is so distinctive in certain orientations that it can be recognized immediately upon inspection. There is some limitation to this technique insofar as increases in thickness lead to increased intensity variations within the diffraction maxima. As can be seen in Figs. 11-15(a) and (b), however, the [001] pattern from sigma phase in steel[30,31] shows a clear similarity when obtained from thin (a) as well as from thick (b) portions of the specimen.

If a phase does not retain a distinctive fingerprint over a range of thicknesses, it is often still possible to identify the phase by using symmetry and intensity variations

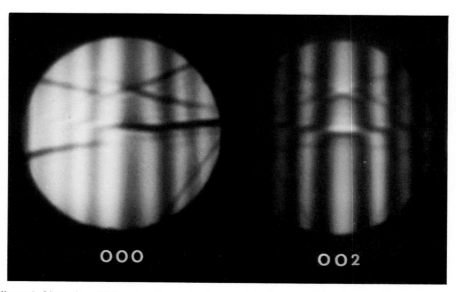

Fig. 11-12. Kossel-Möllenstedt fringes in a CBD pattern from aluminum under two-beam diffracting conditions. Note the symmetry of the fringes in the diffracted 002 spot and the asymmetry in the transmitted 000 spot. The nonparallel lines crossing the discs are due to diffraction effects from higher order Laue zones in the crystal.

Fig. 11-13. HOLZ lines crossing the 000 disc of a CBD pattern from nickel under ⟨114⟩ zone-axis conditions: The pattern symmetry shows a single mirror plane that is characteristic of the whole (3D) crystal symmetry in the [114] direction. (Courtesy, J. R. Michael.)

within the CBD discs. Figures 11-16(a) and (b) are from a Laves-type phase in austenitic stainless steel,[30] and, when examined at both low (a) and high (b) magnifications, it is evident that the CBD pattern symmetry is 6 mm in both cases. It is quite possible, therefore, that the point group may be 6 mm or 6/mmm.[32] Examination of other orientations shows conclusively that 6/mmm is the only possibility.[30] Further investigations permit direct determination of the space group,[32] and this often suffices to identify the phase unambiguously without having to

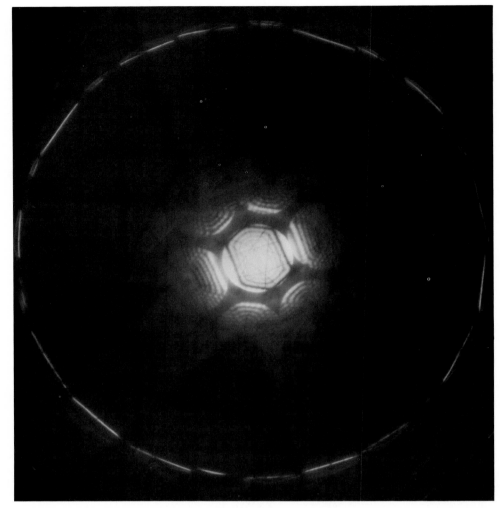

Fig. 11-14. Low-camera-length diffraction pattern showing a ring of HOLZ intensity around a ⟨114⟩ zone-axis pattern in nickel: Each bright line in the ring is parallel to a dark line in the central 000 disc, shown magnified in Fig. 11-13. (Courtesy, J. R. Michael.)

resort to chemical analysis and/or lattice parameter determination. In this approach to indirect chemical analysis, there is no atomic-number limitation; both high and low Z phases can be examined. The thickness limitation is the reverse of that for EDS and EELS insofar as specimens must not be too thin but rather thick enough for dynamical scattering to occur ($> \sim 50$ nm).

In summary, CBD can act as a straightforward microdiffraction technique offering the same capabilities as SAD in the TEM but with a higher resolution. More generally, however, it can be utilized to provide a wealth of detailed information on the thickness and crystallographic characteristics of the specimen.

X-RAY MICROANALYSIS

X-ray analysis in the STEM is invariably performed using a solid-state energy-dispersive spectrometer (EDS), which in its conventional form can detect elements of atomic number $Z \geq 10$ (Ne). If specimens containing elements of lower Z are being studied, alternative techniques such as "windowless" or "ultra-thin window" EDS, EELS (see next section), or indirect diffraction procedures must be used. Quantitative EDS analysis in the STEM can be achieved with an accuracy of $< \pm 10$ percent relative, and a minimum mass fraction of ~ 0.1 wt% is detectable. As a consequence, trace element and high-accuracy microanalysis cannot be performed using a STEM. The main advantages of STEM X-ray analysis over other more accurate and sensitive microanalytical techniques (microprobe, secondary-ion mass spectrometry, etc.) are its high spatial resolution and the relative ease of quantification of data it affords.

Qualitative X-ray Microanalysis

Often qualitative or semiquantitative chemical information is all that is required for conclusive identification of a particular phase in a microstructure. For example, there may be only one or two possible intermediate com-

(a)

(b)

Fig. 11-15. CBD pattern from the ⟨001⟩ axis of a sigma-phase particle in stainless steel showing the similarity in the pattern even when the specimen is (a) thin or (b) thick. The symmetry of these patterns is distinctive enough to be a conclusive "fingerprint" of the sigma phase. (Courtesy, J. Mansfield and N. S. Evans. Reproduced by permission of Adam Hilger, Ltd., Bristol.)

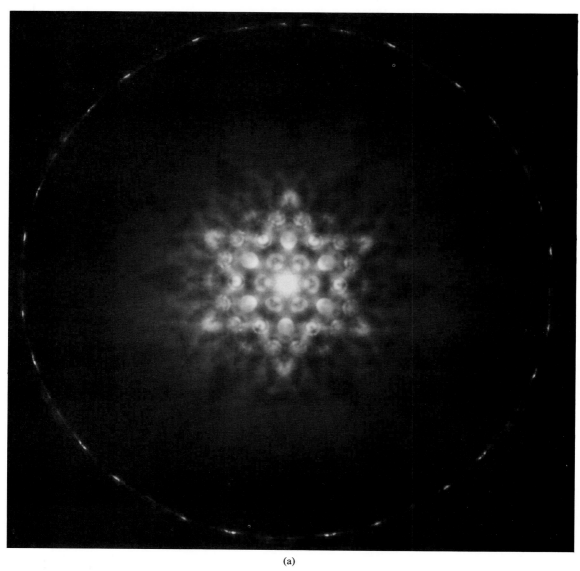

(a)

Fig. 11-16. The ⟨0001⟩ pattern from a Laves-phase intermetallic in 316 stainless steel. The whole pattern in (a) and the HOLZ lines in (b) both show 6-mm symmetry. Making use of this fact and studies of ⟨10$\bar{1}$0⟩ patterns, it becomes possible to deduce the point group as 6/mmm. Further study of the detailed contrast within the CBD discs and the disc spacing permits the space group to be determined as P6$_3$/mmc. (Courtesy, J. Mansfield and N. S. Evans. Reproduced by permission of Adam Hilger, Ltd., Bristol.)

pounds in a particular alloy system, and these may differ sufficiently in composition for a single EDS spectrum to be a conclusive fingerprint. This approach is still feasible, particularly in conjunction with diffraction, if the phases contain low Z elements not routinely detectable by EDS (e.g., C, N, O).

Qualitative microanalysis is performed by placing the probe on the region of interest in the STEM image and adjusting both the probe and the detector parameters so that a reasonable X-ray count rate is ensured (several thousand counts in any major characteristic peak in ~ 100 s). The spectrum can then be examined visually and an approximate analysis performed by estimating relative peak heights. As shown in Fig. 11-17, one of the greatest advantages of X-ray microanalysis in the STEM is the ability to obtain detailed, semiquantitative chemical information from exceedingly complex microstructures in a matter of minutes. In Fig. 11-17, the relative amounts of Mo, Mn, and Cu in the microstructure can be estimated very quickly. This approach to microstructural characterization is probably the most widely used technique in the STEM. Often such a qualitative study of the microstructure is all that is required to find out why a material is behaving in a specific manner. Quantitative microanalysis is not always essential. If the specimen is thin enough for the absorption or fluorescence of X-rays to be assumed insignificant (< ~ 5–10%), than quantification can be achieved in a relatively simple manner. The assumption

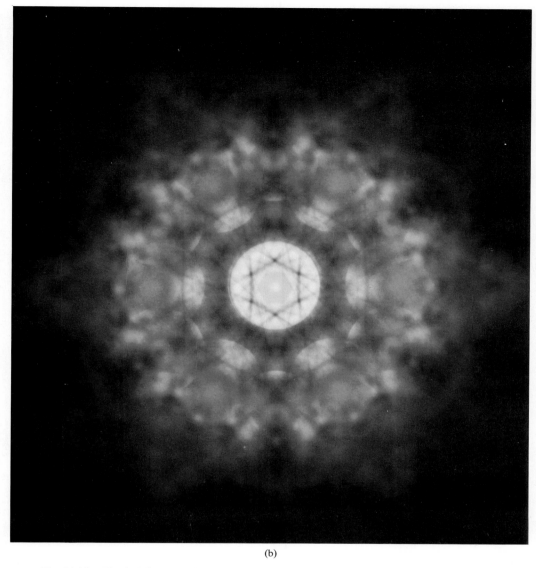

(b)

Fig. 11-16. The ⟨0001⟩ pattern from a Laves-phase intermetallic in 316 stainless steel (continued).

that absorption and fluorescence are negligible is often known as the "thin foil" criterion.[33]

Quantitative X-ray Microanalysis

For a binary system A-B obeying the thin foil criterion, the characteristic X-ray intensity above background, $I_{A/B}$, can be related to the composition, $C_{A/B}$, through the following equation:

$$\frac{C_A}{C_B} = k_{AB} \frac{I_A}{I_B} \qquad (11\text{-}2)$$

where k_{AB} is a sensitivity factor often termed the "Cliff-Lorimer factor."[32] There are many sources for values of k_{AB}, both experimental[5,34] and theoretical.[34] Usually, however, it is worth generating a suitable known standard for the system A-B, and then from this standard, to determine C_A and C_B in heterogeneous regions of the microstructure. In practice, the specimen can often act as its own standard. If, for example, the specimen is known to be chemically homogeneous in the bulk with variation in composition present only around, say, grain boundaries, then the central region of known composition can be used to generate k_{AB}, and this, in turn, can be used to determine C_A/C_B in regions of unknown composition.

Given a suitable value of k_{AB} for a particular binary system, it is a simple matter to obtain measurements of I_A and I_B using the computerized, multichannel analyzer system invariably attached to the EDS detector. Quantification in terms of C_A/C_B values is then trivial, and, if $C_A + C_B = 1$, absolute values of C_A and C_B can be determined. Care must be taken, however, to ensure that the microanalysis is carried out using the same conditions

188 APPLIED METALLOGRAPHY

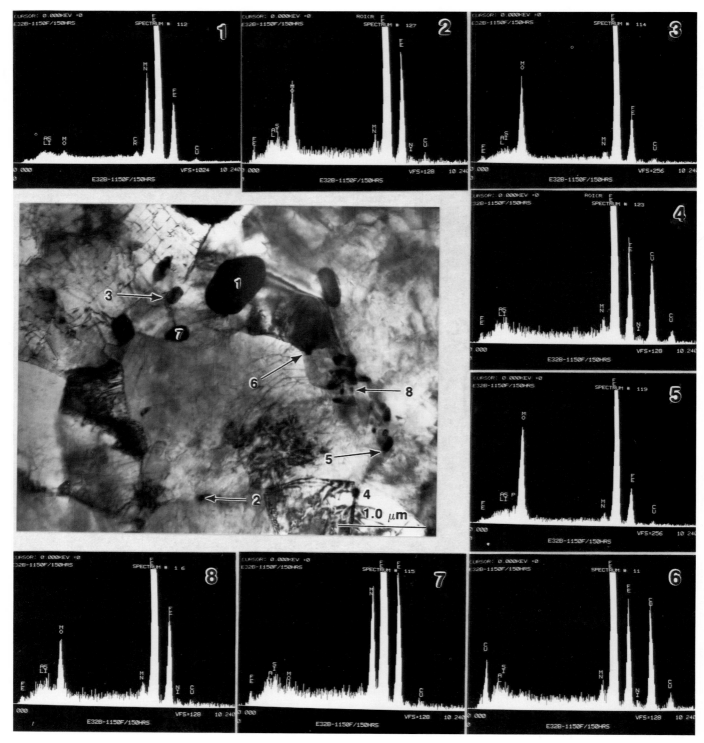

Fig. 11-17. Qualitative chemical analyses of eight particles in a thin-foil weld specimen aged for 150 hr at 1150°F: A quick study of the various X-ray spectra reveals the variety of precipitates and the predominant chemistry. (Courtesy, R. Dias.)

under which the k_{AB} factor was obtained. Since differences in operating voltage, X-ray detector, and X-ray data handling (e.g., different MCA systems) between different users or laboratories will result in differences in quantification, care should be taken when using data from standards obtained by others.

For ternary and higher order systems, Eq. 11-2 can be extended as follows:

$$C_A/C_B = k_{AB}\, I_A/I_B \qquad (11\text{-}3)$$
$$C_B/C_C = k_{BC}\, I_B/I_C \qquad (11\text{-}4)$$
$$C_A + C_B + C_C = 1 \qquad (11\text{-}5)$$

If the thin foil criterion is not obeyed, it will then be necessary to correct the measured values of I_A and I_B for the effects of absorption or fluorescence. Equations exist for both cases[35,36] that modify either the k_{AB} factor or the intensity ratio. Most MCA computer systems also possess these equations and can automatically make the corrections for absorption and fluorescence, with the former being the most usual requirement.

An example of the power of X-ray microanalysis in the AEM for metallographic purposes is shown in Fig. 11-18. Figure 11-18(a) is a micrograph showing three phases in a Ni–Cr–Mo alloy aged at 850°C for 100 hr. This aging treatment corresponds to a three-phase region of the ternary Ni–Cr–Mo phase diagram. Quantification of the X-ray spectra in Figs. 11-18(b), (c), and (d) permits definition of the three corners of the three-phase triangle in a single experiment, as shown in Fig. 11-18(e).[37] In addition, microdiffraction of the three phases can be performed *simultaneously* [Figs. 11-18(f), (g), and (h)], identifying the phases unambiguously as the parent, γ, and the intermetallic phases, P and μ. Such an experiment exemplifies the power of AEM as a metallographic tool, particularly in terms of the *combination* of techniques available.

Other more straightforward applications of STEM X-ray microanalysis involve the generation of an EDS spectrum from a single region of the microstructure or, more often, from a series of points across a region of interest. This permits generation of a composition profile, as shown in Fig. 11-19. From such profiles, which can be generated from specimens aged at very low temperatures because of the high spatial resolution of the microanalysis, important information such as diffusion coefficients (D_v is shown in Fig. 11-19), the state of equilibrium, and the phase diagram can be determined.[38]

It is still worth sounding a note of caution about possible spurious contributions to the X-ray spectrum obtained in the STEM. Because of the electron scatter around the specimen and X-ray generation within the specimen, the possibility exists of X-ray excitation or fluorescence remote from the area of interest. These spurious X-rays may enter the detector, reducing the precision and sensitivity of microanalysis. Well-defined methods exist for minimizing these effects, and these methods should be known and practiced routinely.[5] At the current state of STEM instrumentation, however, it is still very difficult to identify the presence of small (~1%) amounts of a particular element conclusively if it is also present in large amounts elsewhere in the specimen or the microscope stage. Copper seems to present a particularly difficult case since post-specimen electron scatter and interaction with the specimen surrounds often gives rise to a very small Cu peak in the EDS spectrum.

Light Element Analysis by EDS

Recent technological developments in EDS and STEM instruments have permitted the detectability limits to be extended to $Z \geq 5$(B). These developments involve removing the Be window (i.e., "windowless") or replacing it with an ultra-thin window (UTW) of a gold-coated polymeric film. Either advance requires significant modification to the microscope. Detection efficiency is low and X-ray generation is very limited. Quantification of the data is more difficult, less accurate, and almost invariably requires an absorption correction. Since low count rates require the use of larger probes ($\rightarrow$ 100 nm), spatial resolution is poor. Similarly, very thin specimens (< 20 nm) or very small particles do not give enough X-rays for quantification in sensible counting times. Under such circumstances, EELS is an essential tool for light element analysis (see the next section). Nevertheless, for the positive identification of reasonably sized carbides, nitrides, or oxides (see Fig. 11-20), the UTW detector is an invaluable addition to the AEM.

ELECTRON ENERGY-LOSS SPECTROMETRY (EELS)

Inelastic interactions between high-kV electrons and atoms in the specimen result in a spectrum of energy-loss electrons. This spectrum contains a wealth of information[5,39] such as chemical bonding, electron momentum transfer, surface interactions, and electronic structural detail. Since, from a metallographic standpoint, EELS permits elemental identification particularly of those low Z-elements undetectable by EDS, we will consider this aspect only. The technique is not as accurate as EDS, with $\pm$ 20 percent relative being a reasonable error, nor is it as sensitive as EDS for detecting small amounts of an element in a matrix of another, with ~3 wt% being the current limitation. In theory, however, it offers a higher spatial resolution, is not limited by beam spreading, and can easily detect the presence of elements, such as Li and Be, beyond the reach of windowless X-ray analysis.

EELS, however, is not yet reliably quantitative,[40] and much work remains to be done. Its primary function, therefore, is to act in conjunction with EDS and microdiffraction as a low-Z qualitative microanalysis technique. As is true of EDS, judicious use of qualitative data can lead to an unambiguous phase identification.

The EELS spectrum usually consists of a zero-energy loss peak (caused by transmitted electrons) followed by a plasmon-energy loss peak of about 15 to 25 eV (caused by interactions with conduction-band electrons). Follow-

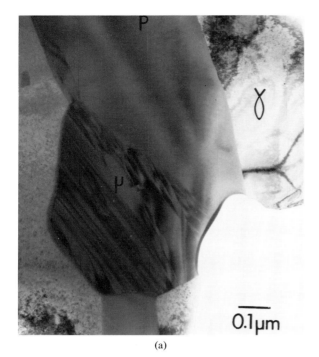

(a)

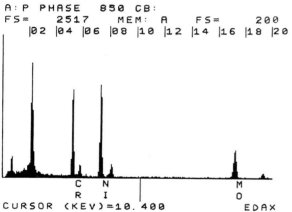

(d)

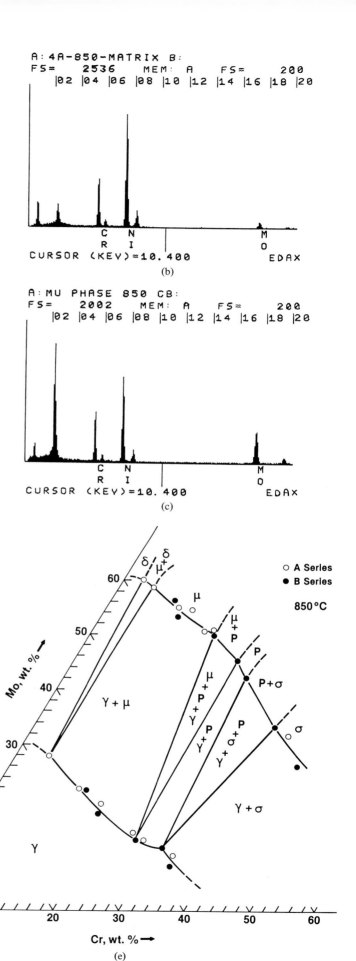

Fig. 11-18. (a) TEM image of the three phases [matrix (γ), mu phase (μ), and p-phase (P)] present in a Ni–Cr–Mo alloy aged at 850°C for 100 hr. The X-ray spectra from the three phases are shown in (b), (c), and (d), respectively. The γ matrix is clearly the most Ni-rich phase, whereas the μ and P phases have almost identical Ni compositions but differ slightly in their Cr and Mo content. Quantitative analysis of the X-ray spectra yields compositions that define the three corners of the $\gamma + P + \mu$ three-phase triangle in the 1523K section of the ternary phase diagram shown in Fig. 11-18(e). In Figs. 11-18(f), (g), and (h), microdiffraction patterns are shown. Although there is little difference between the chemistry of the phases, the diffraction patterns are clearly distinguishable and act as a "fingerprint" to identify the phases. In particular, the rhombohedral symmetry of the μ phase is obvious from Fig. 11-18(g). (Courtesy, M. Raghavan. Figures 11-18(e), (g), and (h) reproduced by permission of the American Society for Metals.)

(f)

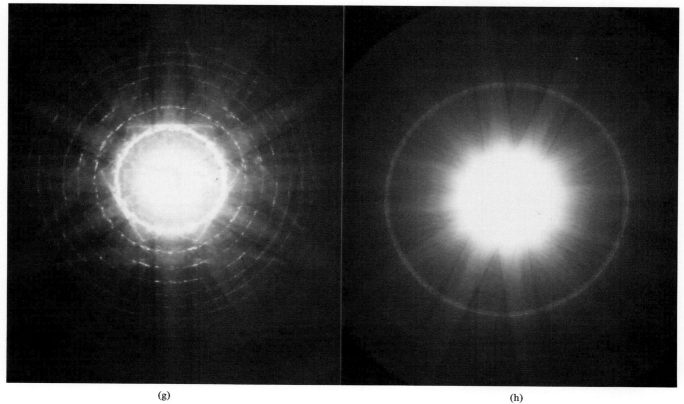

(g)　(h)

Fig. 11-18. Microanalysis of Ni–Cr–Mo alloy (continued).

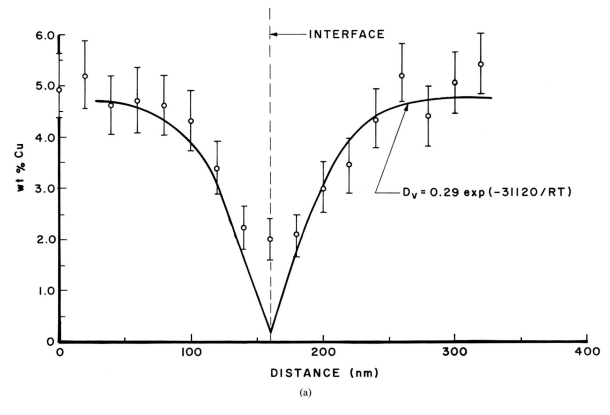

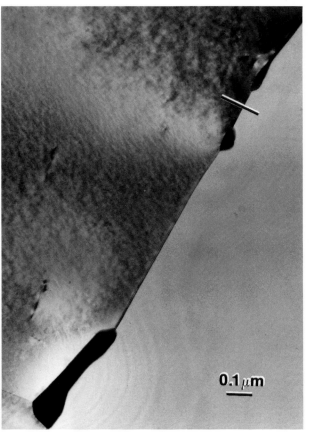

Fig. 11-19. Composition profile (a) across a grain boundary and (b) in Al–4%Cu aged for 90 s at 250°C to generate the equilibrium θ phase on the grain boundaries. The Cu composition profile, which has been mathematically modeled using the Grube equation, reflects the solute depletion accompanying θ precipitation. (Courtesy, J. R. Michael.)

ing the plasmon peak there is a rapid drop off in intensity of electrons suffering losses greater than 25 eV. An EELS spectrum is shown in Fig. 11-21(a). On this rapidly falling intensity are superimposed ionization edges; these represent small increases in intensity that correspond to the presence of electrons that have lost more than a critical amount of energy in ionizing a particular atom. The presence of these ionization edges is the important feature of the spectrum. As shown in Fig. 11-21(b), however, the edges are small, and electronic display enhancement is often needed to observe them. Observation of the edges is sufficient for qualitative microanalysis. Quantitative microanalysis can be attempted, either in an absolute manner or by a ratio technique[41] similar to that used in EDS. Either way, errors of $\sim \pm 20$ percent and often $> \pm 100$ percent are common, and EELS, as a result, is not yet a routine microanalysis technique. Nevertheless, as shown in Fig. 11-21(b), it is capable of identifying very light elements, and, a major use, of determining the presence of carbon or nitrogen in small percipitates in steels. It is often not too difficult to infer the correct type of carbide or nitride present.

Although the ionization edges are very small, the peak-

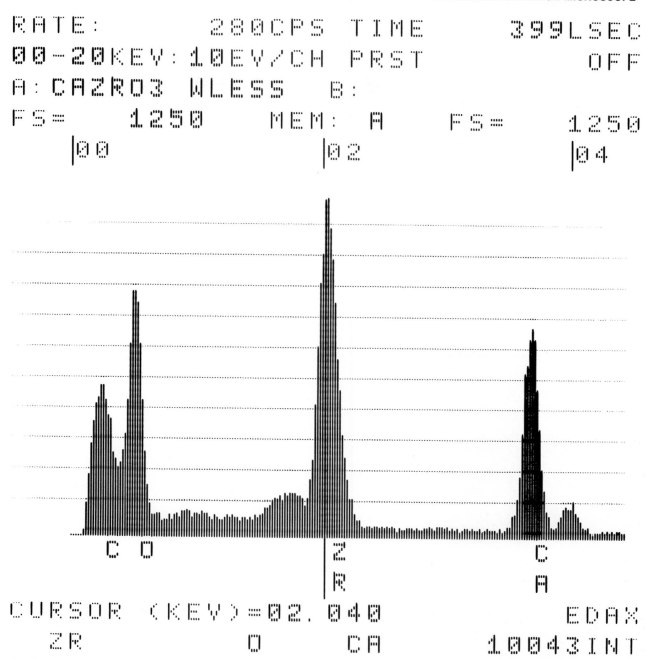

Fig. 11-20. Windowless EDS spectrum from a thin foil of CaZrO$_3$ coated with a carbon film. The calcium K$_\alpha$ and the zirconium L$_\alpha$ would normally be the only visible X-ray peaks in the conventional spectrum. Removal of the Be window in an ion-pumped STEM permits detection of the oxygen and carbon K$_\alpha$ peak also.

to-background ratio can be improved by using very thin specimens (< 20 nm) and operating at the highest available voltage. In the latter regard, EELS is not the best technique at 100 kV, but the advent of instruments of intermediate voltage (300 to 400 kV) should offer a significant improvement.

In summary, EELS is a semiquantitative microanalytical technique at best but essential for those seeking to determine the presence of low-Z elements.

SPECIMEN THICKNESS

It should be apparent from the preceding sections that the thickness of the thin-foil specimen is perhaps the most important metallographic parameter in the STEM. Determination of the foil thickness is essential when the object is to minimize and quantify spatial resolution and X-ray absorption. Knowledge of foil thickness is also essential for observing HOLZ lines, etc., in CBD patterns, and

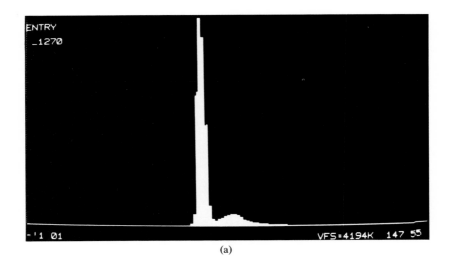

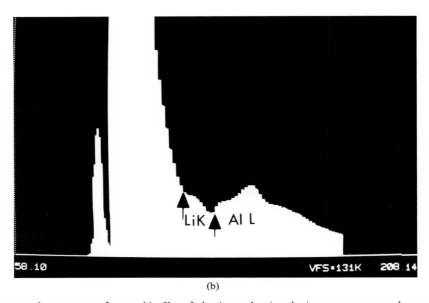

Fig. 11-21. (a) A typical energy loss spectrum from a thin film of aluminum showing the intense zero-energy loss peak followed by a plasmon loss peak at ~ 15 eV: The intensity falls off rapidly after the plasmon peak; in order to discern any information after the plasmon loss peak, the spectrum display has to be electronically increased in intensity. (b) Energy loss spectrum from an aluminum-lithium alloy: The increased display range shows a rapidly decreasing intensity with increasing energy loss. Small peaks (or edges) that perturb the decreasing intensity can be identified in terms of the elements present in the spectrum. In this case, the LiK edge at ~ 52 eV, corresponding to energy losses accompanying the ionization of Li atoms, and the AlL_{23} edge, corresponding to Al ionization, can be seen. (Courtesy, H. M. Chan.)

also for considering whether or not to pursue EELS or windowless EDS. Since reviews of methods to determine the thickness are available,[5,35] their details will not be repeated here. The most common involve use either of contamination deposits on the upper and lower surfaces of the specimen[42] or of Kossel-Möllenstedt fringes in CBD patterns as already described in the microdiffraction section. All methods have their limitations, and none are both straightforward and precise. The lack of a method for determining thickness *in situ* that is simple, universally applicable, and precise remains a fundamental limitation in the pursuit of accurate quantitative microanalysis and microdiffraction in the STEM.

SUMMARY

The STEM offers imaging, diffraction, and microanalytical information about thin specimens. It combines the role of a TEM and SEM with increased resolution for chemical and crystallographic information. It also performs the major metallographic operations of phase and defect identification along with characterization of other structural and chemical inhomogeneities. STEM is most useful when combinations of the many available techniques are taken advantage of, but it should always be used in conjunction with other techniques that overcome its limitations such as small samples, limited analysis regions (i.e., poor sampling statistics), and lack of depth resolution.

Acknowledgments. The author wishes to thank his colleagues and students, who contributed many of the examples given in this chapter. He also wishes to acknowledge the financial support for STEM/AEM research at Lehigh provided through the NSF(DMR 84–00427, DOE(DE–AC02–83ER45016), and NASA(NAG9–45).

REFERENCES

1. Edington, J. W. *Practical Electron Microscopy in Materials Science,* vols. 1–5. London: MacMillan Press Ltd. (Philips Technical Library), 1974.
2. Hirsch, P. B.; Howie, A.; Nicholson, R. B.; Pashley, D. W.; and Whelan, M. J. *Electron Microscopy of Thin Crystals.* New York: Krieger, 1977.
3. Goldstein, J. I.; Newbury, D. E.; Echlin, P.; Joy, D. C.; Fiori, C. E.; and Lifshin, E. *Scanning Electron Microscopy and X-ray Microanalysis.* New York: Plenum Press, 1981.
4. Joy, D. C.; Romig, A. D., Jr.; and Goldstein, J. I. *Principles of Analytical Electron Microscopy.* New York: Plenum Press, 1986.
5. Williams, D. B. *Practical Analytical Electron Microscopy in Materials Science.* Mahwah. NJ: Philips Electron Optics Publishing Group, 1984.
6. Maher, D. M., and Joy, D. C. The formation and interpretation of defect images from crystalline materials in the scanning transmission electron microscope. *Ultramicroscopy* 1:239–253 (1976).
7. Thompson-Russell, K. C., and Edington, J. W. *Electron Microscope Specimen Preparation Techniques in Material Science.* Eindhoven: N. V. Philips Gloelampenfabrieken, 1977.
8. Thompson, M. N.; Doig, P.; Edington, J. W.; and Flewitt, P. E. J. The influence of specimen thickness on X-ray count rates in STEM microanalysis. *Phil. Mag.* 35:1537–1542 (1977).
9. Pountney, J. M., and Loretto, M. H. The influence of surface layers in STEM X-ray microanalysis. *Electron Microscopy* 3:180–181. Leiden, The Netherlands, 7th European Congress on Electron Microscopy Foundation, 1980.
10. Fraser, H. L., and McCarthy, J. P. Specimen preparation limitations in quantitative thin foil microanalysis. In *Microbeam Analysis.* San Francisco, CA: San Francisco Press, 1982, pp. 93–96.
11. Ball, M. D., and Furneaux, R. C. Ultramicrotomy as a specimen preparation technique for thin foil microanalysis. In *Developments in Electron Microscopy and Analysis 1981.* Bristol and London: The Institute of Physics, 1982, pp. 179–180.
12. Hren, J. J. Barriers to AEM: Contamination and etching. In *Introduction to Analytical Electron Microscopy.* New York: Plenum Press, 1979, pp. 481–505.
13. Newbury, D. E. Beam broadening in the analytical electron microscope. In *Microbeam Analysis.* San Francisco, CA: San Francisco Press, 1982, pp. 79–83.
14. Romig, A. D., Jr., and Goldstein, J. I. Detectability limit and spatial resolution in STEM X-ray microanalysis: Application to Fe-Ni alloys. In *Microbeam Analysis.* San Francisco, CA: San Francisco Press, 1979, pp. 124–128.
15. Jones, I. P., and Loretto, M. H. Some aspects of quantitative STEM X-ray microanalysis. *J. Microsc.* 124:3–13 (1981).
16. Goldstein, J. I.; Costley, J. L.; Lorimer, G. W.; and Reed, S. J. B. Quantitative X-ray analysis in the electron microscope. *Scanning Electron Microscopy* 1:315–324. Chicago IITRI (1977).
17. Reed, S. J. B. The single scattering model and spatial resolution in X-ray analysis of thin foils. *Ultramicroscopy* 7:405–410 (1982).
18. Brown, L. M. Progress and prospects for STEM in materials science. In *Developments in Electron Microscopy and Analysis.* Bristol and London: The Institute of Physics, 1977, pp. 141–148.
19. Vesely, D. The electron beam damage of synthetic polymers. In *Developments in Electron Microscopy and Analysis.* Bristol and London: The Institute of Physics, 1977, pp. 389–394.
20. Allen, R. M. Secondary electron imaging in the STEM. *Scanning Electron Microscopy* 3:905–918. AMF O'Hare, IL: SEM Inc. (1985).
21. Gorlen, K. E.; Barden, L. K.; Del Priore, J. S.; Fiori, C. E.; Gibson, C. G.; and Leapman, R. D. Computerized analytical electron microscope for elemental imaging. *Rev. Sci. Instrum.* 55:912–921 (1986).
22. Isaacson, M.; Ohtsuki, M.; and Utlaut, M. Electron microscopy of individual atoms. In *Introduction to Analytical Electron Microscopy.* New York: Plenum Press, 1979, pp. 343–367.
23. Cowley, J. M. Coherent interference effects in SIEM and CBED. *Ultramicroscopy* 7:19–26 (1981).
24. Kelly, P. M.; Jostsons, A.; Blake, R. G.; and Napier, J. G. The determination of foil thickness by scanning transmission electron microscopy. *Phys. Stat. Sol.* A31:771–780 (1975).
25. Allen, S. M. Foil thickness measurements from convergent beam diffraction patterns. Phil. Mag. A43:325–335 (1981).
26. Steeds, J. W. Microanalysis by convergent beam electron diffraction. In *Quantitative Microanalysis with High Spatial Resolution.* London: The Metals Society, 1981, pp. 210–216.
27. Ecob, R. C.; Shaw, M. P.; Porter, A. J.; and Ralph, B. The application of convergent beam electron diffraction to the detection of small symmetry changes accompanying phase transformations. *Phil. Mag.* 44A:1117–1133 (1981).
28. Steeds, J. W. Convergent beam electron diffraction. In *Introduction to Analytical Electron Microscopy.* New York: Plenum Press, 1979, pp. 387–427.
29. Raghavan, M.; Koo, J. Y.; and Petkovic-Luton, R. Some applications of convergent beam electron diffraction in metallurgical research. *Journal of Metals* 36(6):44–53 (1983).
30. Mansfield, J. Convergent beam electron diffraction of alloy phases. Bristol (U.K.): Adam Hilger, 1984.
31. Steeds, J. W., and Evans, N. S. Practical examples of point and space group determination in convergent beam diffraction. In *Proc. 38th EMSA Meeting.* Baton Rouge, LA: Claitors Publishing Division, 1980, pp. 188–191.
32. Buxton, B. F.; Eades, J. A.; Steeds, J. W.; and Rackham, G. M. The symmetry of electron diffraction zone axis patterns. *Phil. Trans. Roy. Soc.* 281:171–194 (1976).
33. Cliff, G., and Lorimer, G. W. The quantitative analysis of thin specimens. J. Microsc. 103:203–207 (1975).
34. Wood, J. E.; Williams, D. B.; and Goldstein, J. I. Experimental and theoretical determination of k_{AFe} factors for quantitative X-ray microanalysis in the analytical electron microscope. *J. Microsc.* 133:255–274 (1984).
35. Williams, D. B., and Goldstein, J. I. Absorption effects in quantitative thin film X-ray microanalysis. In *Analytical Electron Microscopy.* San Francisco, CA: San Francisco Press, 1981, pp. 39–46.
36. Nockolds, C.; Nasir, M. J.; Cliff, G.; and Lorimer, G. W. X-ray fluorescence correction in thin foil analysis and direct methods for foil thickness determination. In *Electron Microscopy and Analysis 1979.* Bristol and London: The Institute of Physics, 1980, pp. 417–420.
37. Raghavan, M.; Mueller, R. R.; Vaughn, G. A.; and Floreen, S. Determination of isothermal sections of nickel rich portion of Ni–

Cr–Mo system by analytical electron microscopy. *Met. Trans.* 15A:783–792 (1984).
38. Romig, A. D., Jr., and Goldstein, J. I. Determination of the Fe–Ni and Fe–Ni–P phase diagrams at low temperature (700° to 300°C). *Met. Trans.* 11A:1151–1159 (1980).
39. Joy, D. C. The basic principles of electron energy loss spectroscopy. In *Introduction to Analytical Electron Microscopy.* New York: Plenum Press, 1979, pp. 223–244.
40. Joy, D. C., and Newbury, D. E. A "round robin" test on ELS quantitation. In *Analytical Electron Microscopy 1981.* San Francisco, CA: San Francisco Press, 1981, pp. 178–180.
41. Egerton, R. F. Formulae for light element analysis by electron energy loss spectrometry. *Ultramicroscopy* 3:243–251 (1978).
42. Lorimer, G. W.; Cliff, G.; and Clark, J. N. Determination of the thickness and spatial resolution for the quantitative analysis of thin foils. In *Developments in Electron Microscopy and Analysis.* London: Academic Press, 1976, pp. 153–156.

12
METALLOGRAPHY AND WELDING PROCESS CONTROL

C. E. Cross, O. Grong, S. Liu, and J. F. Capes
Center for Welding Research
Colorado School of Mines

Although welding metallurgy is concerned with the application of well-known metallurgical principles, the conditions existing during welding will be highly different from those prevailing during refining, casting, and forming of metals because of the strong nonisothermal nature of the arc-welding process. As a result, the understanding of the various chemical and physical reactions in arc welding is far less developed than it is for steelmaking, and, hence, predictions of weld microstructures and the resulting mechanical properties based on the consumables/parent plate and operational conditions can be subject at this point to no more than an incomplete theoretical treatment.

The main objective of this chapter is to give a survey of factors controlling the weld microstructures in arc welding of mild and low-alloy steels, stainless steels, and aluminum alloys. Consequently, the weld-metal solidification mode as well as the transformation behavior of the weld metal/base metal under various cooling conditions will be discussed in the light of information available in recent literature. To begin with, however, the specific features of the weldment that commonly need characterization will be briefly reviewed.

CHARACTERISTIC FEATURES OF WELDS

Single-Pass Weldments

It is convenient to begin a description of the various weld regions by using a single-pass bead as an example. The two types of weld, and the common terms applied to them, are shown in Fig. 12-1. As a rough classification, the weldment may be divided into two main regions: the fusion zone and the heat-affected zone (HAZ).

The Weld Metal Fusion Zone. Within the weld fusion zone, the peak temperature has exceeded the melting point of the metal, and the chemistry is usually different from that of the parent plate because of the chemical interactions between the liquid weld metal and its surroundings (arc atmosphere, slag). The chemical composition of the weld metal will in this case depend on the choice of welding consumables (i.e., the combination of filler metal, flux, and/or shielding gas), the base metal dilution ratio [see Fig. 12-1(b)], and the operational conditions applied, the first two being the most important.

During solidification of the weld metal, alloying and impurity elements segregate extensively to the center of the interdendritic or intercellular spaces and to the center parts of the weld under the conditions of rapid cooling, which, in the case of polyphase metals, will alter the kinetics of the subsequent solid-state transformation reactions. Accordingly, the transformation behavior of the weld metal is seen to be quite different from that of the base metal, even when the chemical composition is not significantly changed by the welding process.

The Heat-Affected Zone. In this region, the peak temperature has remained below the melting point of the parent plate and, hence, the chemical composition remains largely unchanged. Nevertheless, considerable microstructural changes take place within the HAZ during welding as a result of the rapid heating and cooling. In Fig. 12-2, the various HAZ regions of a single-pass mild steel butt weld are shown schematically, each zone being characterized by a specific peak temperature interval.

Multiple-Pass Weldments

In a multiple-pass weldment, the situation is much more complex than it is in a single-pass weld because of a partial refinement of the microstructure by subsequent weld passes, which to a large extent will increase the inhomogeneity of the various regions with respect to microstructure and mechanical properties.

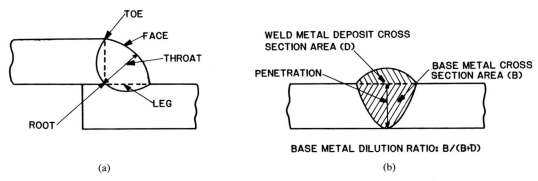

Fig. 12-1. Common terms applied to (a) a lap weld, and (b) a butt weld.

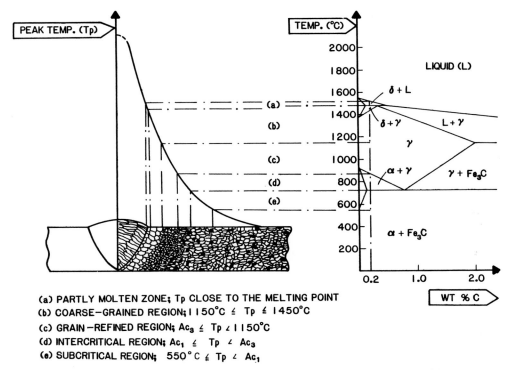

(a) PARTLY MOLTEN ZONE; T_p CLOSE TO THE MELTING POINT
(b) COARSE-GRAINED REGION; $1150°C \leq T_p \leq 1450°C$
(c) GRAIN-REFINED REGION; $Ac_3 \leq T_p < 1150°C$
(d) INTERCRITICAL REGION; $Ac_1 \leq T_p < Ac_3$
(e) SUBCRITICAL REGION; $550°C \leq T_p < Ac_1$

Fig. 12-2. The various HAZ regions in a 0.2-wt.% C single-pass steel butt weld.

The Weld Thermal Cycle

The weld thermal cycle is strongly influenced by the welding process and the operational conditions applied, the thermal conductivity of the parent plate, and the mode of heat flow, i.e., either two- or three-dimensional heat flow to correspond to the welding of thin and thick plates, respectively.

For welding of steels, the cooling time from 800° to 500°C, $\Delta t_{8/5}$, has been widely accepted as an adequate index for the weld cooling program. This index is a measure of the cooling rate that controls the austenite-to-ferrite transformation, determining the microstructures formed. In the case of three-dimensional heat flow, $\Delta t_{8/5}$ may be taken as a rough estimate directly proportional to the heat input, $E(kJ/mm)$,* where the proportionality constant is in the range of from 4 to 5 seconds per kJ/mm for mild and low alloy steel.

For a wider introduction to the metallurgy of arc welding, see the textbooks of Lancaster[1] or Easterling.[2]

SOLIDIFICATION MICROSTRUCTURE

Inherent to the welding process is the formation of a pool of molten metal directly below a moving heat source that effectively fuses together pieces of metal. The shape of this molten pool is determined by the flow of both

---

* The gross heat input can be expressed in terms of the amperage (I), the arc voltage (U), and the weld travel speed (v), as follows: $E(kJ/mm) = (IU)/(1000\ v)$.

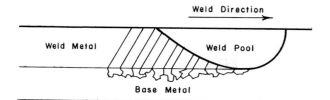

Fig. 12-3. Idealistic schematic representation of the grain structure in a longitudinal cross section taken through the center of a weld. Epitaxial columnar grains are shown to grow normal to the weld pool boundary.

heat and metal, with melting occurring ahead of the heat source and solidification behind it. Heat input determines both the volume of molten metal and, hence, dilution and weld metal composition, as well as the thermal conditions under which solidification takes place. Also important to solidification is the crystalline growth rate, which is geometrically related to weld travel speed and weld pool shape. Thus, weld pool shape, weld metal composition, cooling rate, and growth rate are all factors interrelated to heat input, which in turn will affect the solidification microstructure.[3-4] Some important points regarding interpretation of weld metal microstructure in terms of these four factors are summarized below.

Grain Morphology

Weld metal solidification proceeds by the growth of grains at the solid-liquid interface as defined by the shape of the weld pool boundary. Growth initiates at the outermost boundary of the weld pool, which constitutes the dividing line between regions of melting and solidification. Incipient melting at base-metal grain boundaries immediately adjacent to the fusion zone allows these grains to serve as seed crystals. Thus, weld metal grains tend to grow epitaxially, acquiring the size and crystalline identity of grains in the base metal as a result (shown schematically in Fig. 12-3). Growth of epitaxial grains is continuous until impingement occurs upon neighboring grains, with nucleation of new grains occurring only rarely. This continuous growth results in large columnar grains whose boundaries provide paths for easy crack propagation. Epitaxial columnar growth is particularly deleterious in multiple-pass welds where grains can extend continuously from one weld bead to another. Shown in Fig. 12-4 is a transverse cross section taken from a multiple-pass weld in a titanium alloy that exemplifies this problem (the grain structure was revealed using Kroll's etch). Since grains must grow normal to the liquid-solid interface, the shape of the weld pool determines the grain morphology.[5] A spherically shaped weld pool will yield a grain morphology much different from that of an elongated weld pool. When viewed from the top surface, as illustrated in Fig. 12-5, a spherical shape will yield curved and tapered columnar grains, whereas an elongated shape will yield straight and broad columnar grains. The latter condition promotes the formation of center-line cracking because of the mechanical entrapment of inclusions and enrichment of eutectic liquid at the trailing edge of the weld pool.

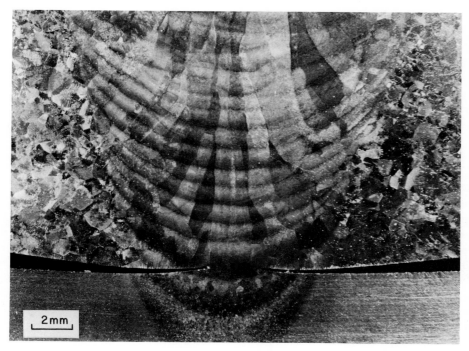

Fig. 12-4. Transverse cross section taken from a multiple-pass gas-metal-arc weld on a titanium–6-wt.% aluminum 4-wt.% vanadium alloy. Epitaxial columnar grains are seen to be continuous from one weld bead to another. (Courtesy, Martin-Marietta Corp.)

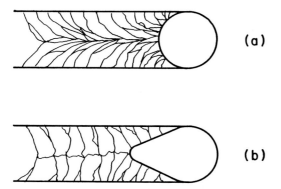

Fig. 12-5. Schematic comparison of the grain morphologies obtained for (a) spherically shaped weld pools versus (b) elongated weld pools, as viewed from the top surface.

Grain Substructure

Each individual grain in the weld metal has a substructure consisting of a parallel array of dendrites, or cells. This substructure may be masked by solid-state reactions as is the case with transformations from austenite in ferrous alloys. The crystallographic orientation of dendrites within a grain is determined by the orientation of the base-metal grain from which it originated, with dendrites aligned in the $\langle 100 \rangle$ direction for FCC and BCC metals. Grains with dendrites aligned in the direction of heat flow, normal to the interface, will tend to over-grow grains containing dendrites not having this preferred alignment. Competitive growth, early in the process, results in most grains having dendrites aligned nearly normal to the interface. The growth rate of dendrites is determined both by the position of the grain at the weld pool boundary as well as by the orientation of the dendrites within the grain. An expression for growth rate, R, can be obtained by relating pertinent growth directions to the direction of welding and weld travel speed, v.[6] Referring to Fig. 12-6, if φ represents the interface normal angle and θ the dendrite alignment angle, then

$$R = \frac{v \cos \varphi}{\cos (\varphi - \theta)}$$

It is clear that growth rate may increase as grains progress from initiation to impingement as determined by the curvature of the weld pool. This, together with a corresponding decrease in thermal gradient, may sometimes lead to a transition from cellular growth, at grain initiation, to dendritic growth at the trailing end of the weld pool.[7] The distinction between cells and dendrites lies primarily in their sensitivity to crystalline alignment. Cells do not necessarily have the $\langle 100 \rangle$ axis orientation; dendrites do. Often dendrites are observed to be free of secondary arms, a morphology referred to as *cellular-dendritic* and indicative of low undercooling between dendrites.

Redistribution of solute elements during solidification

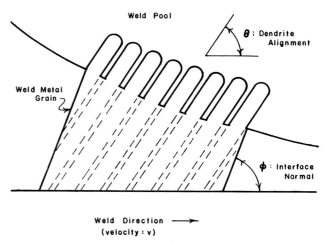

Fig. 12-6. Growth rate of dendrites within a grain is seen to be a function of both the dendrite alignment angle (θ) and the interface normal angle (ϕ) relative to the direction of welding.

of alloys is accomplished by partitioning of elements to the interdendritic liquid. Spacing between dendrites is controlled by the local solidification time available for diffusion of these solute atoms. The rapid cooling rates associated with welding result in short local solidification times which, in turn, yields microstructures much finer than those normally encountered in casting operations. Weld cooling rates can be easily approximated in terms of heat input[8] and are generally on the order of 100°C/s near the weld pool boundary. Dendrite spacing has been found to increase with heat input, being proportional to the square root of heat input for three-dimensional heat flow[9] and directly proportional to heat input for two-dimensional heat flow.[10] Dendrite spacing is generally on the order of 10–100 μm for a variety of metal alloys with typical welding heat inputs.

Interdendritic Eutectic

Microsegregation, which evolves from partitioning of solute elements to the interdendritic liquid, can be described quantitatively by using the Scheil equation.[11] Due to partitioning, a concentration gradient is established in the interdendritic liquid where, at the base of the dendrites, a maximum is reached at the eutectic composition. Thus, weld metal microstructures will typically consist of single-phase dendrites surrounded by a complex network of multi-phase eutectic constituents. An example of such a microstructure, taken from the weld metal of an aluminum–6-wt.% copper alloy, is shown in Fig. 12-7. Here, a 2-vol.% HF macroetch of a transverse cross section of an autogenous single-pass gas-tungsten-arc weld reveals the grain structure, fusion zone boundary, and heat-affected zone. Grain substructure was revealed using a double etch with the sample immersed in a 25-vol.% HNO$_3$ solution at 70°C for 60 sec followed by a 30-sec immersion in a solution of 0.5-g NaF + 1-ml HNO$_3$ +

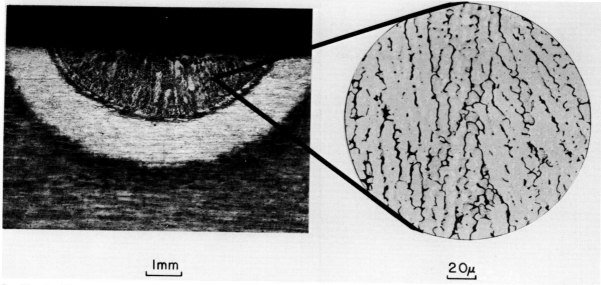

Fig. 12-7. The dendrite substructure within a grain as observed from a transverse cross section of an autogenous gas-tungsten-arc weld on an aluminum–6-wt.% copper alloy.

2-ml HCl + 97-ml H$_2$O. The elongated globular regions are aluminum dendrites, and the dark phase located between dendrites is the CuAl$_2$ constituent of the aluminum–copper eutectic. Solidification of interdendritic eutectics may be either coupled ("lamellar") or uncoupled ("divorced"). Rapid cooling rates and low solute concentrations have been found to favor formation of a divorced eutectic,[12,13] as they often do in welding. When eutectic growth is coupled, spacing between constituent phases will decrease with an increase in growth rate, varying inversely with the square root of the latter.[14] For the growth rates encountered in welding, eutectic spacing is usually on the order of 1 μm.[15,16] In addition to microsegregation, macrosegregation has also been observed in weld metal in the form of banding. Banding is thought to occur as a result of nonsteady advancement of the weld pool, where narrow bands extending across the width of the weld bead demark regions of different growth conditions.[17]

Solidification Defects

Weld metal defects associated with solidification include hot tearing, porosity, and inclusions, as shown in Figs. 12-8(a), (b), and (c). Hot tears originate near the liquid-solid interface when strains from solidification shrinkage and thermal contraction cause rupture of low-melting-point liquid films located at grain boundaries. Hot tears characteristically form along the centerline of the weldment and have a surface morphology that appears smooth and rounded when viewed with the scanning electron microscope (SEM). Hot tears are also found in the form of crater cracks, that is, cracks that develop in the weld pool when the arc is suddenly extinguished. An example of a crater crack is shown in Fig. 12-8(a), where a 2-vol.% HF etch reveals the intergranular nature of hot tearing in an aluminum–6-wt.% copper alloy. The susceptibility of an alloy to hot tearing appears to be related to its inability to accommodate strain through dendrite interlocking as well as the tendency of tears to back fill with eutectic liquid.[18,19] The time interval during which liquid films can exist in relation to the rate of strain generation may also play a role in hot tear susceptibility.[20] High-strength aluminum alloys are highly susceptible to this type of defect, particularly those containing combinations of copper, zinc, and magnesium.[21] When considering binary additions to aluminum, a peak susceptibility to hot tearing is normally observed where the susceptibility rises rapidly and then tapers off with increased solute additions. Pure metals and alloys of eutectic composition are least susceptible. Ferrous alloys can also be hot-tear sensitive depending upon the amount of phosphorus and sulfur impurities present. Some grades of stainless steel show high sensitivity, particularly when dendritic solidification proceeds as austenite rather than ferrite.[22] Austenite dendrites will partition chromium, and this will result in the formation of interdendritric eutectic ferrite. Ferrite dendrites will partition nickel, resulting in interdendritic austenite together with the retention of some delta ferrite at the dendrite core. Metallographic means have been developed, including color depositional etching[23] and magnetic etching,[24] to help identify austenitic and ferritic phase constituents. Depositional etching with a solution of 20-g NH$_4$HF$_2$ + 0.5-g K$_2$S$_2$O$_5$ in 100-ml H$_2$O will give nickel-rich regions a blue color and chromium-rich regions a yellow color, but retained delta ferrite will remain white. Magnetic etching, which involves coating the specimen with a colloidal solution containing fine iron

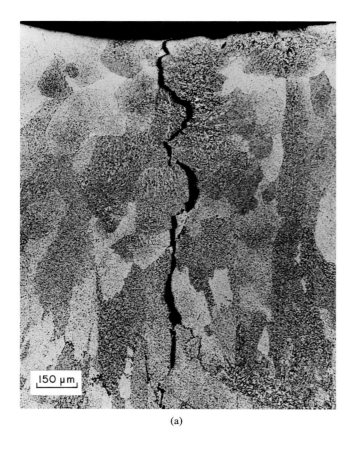

(a)

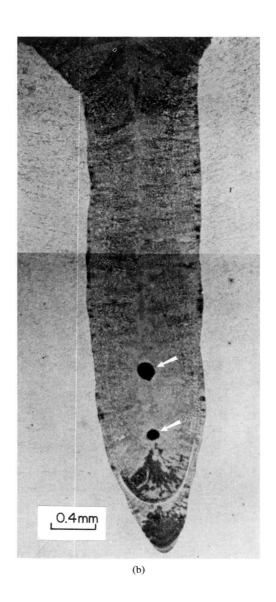

(b)

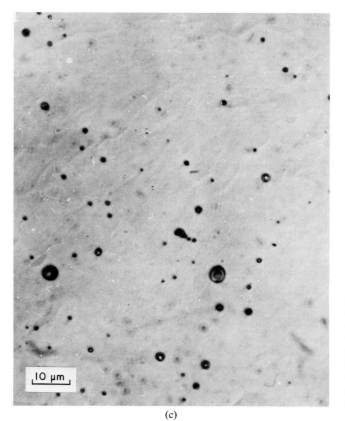

(c)

Fig. 12-8. Weld metal defects related to solidification include: (a) hot tears (shown is a crater crack in an autogenous gas-tungsten-arc weld on an aluminum–6-wt.% copper alloy); (b) porosity (shown is a transverse cross section of an autogenous electron-beam weld on AISI 304 stainless steel with macroporosity at the root of the weld; note the large depth-to-width ratio of the bead morphology, which is characteristic of electron-beam welds). (Courtesy, Rockwell International Corp.); and (c) inclusions (shown is a transverse cross section of a submerged-arc weld, polished and unetched, made on AISI 1020 steel).

particles (Ferrofluid*), will decorate regions of ferromagnetic ferrite when exposed to an externally applied magnetic field.

Porosity in weldments may be present either as interdendritic micropores or as macropores over 1 mm in diameter. Its origins are related to contamination of the weld joint, filler wire, or welding flux with organic material such as oil or water. Aluminum welds are particularly susceptible to porosity caused by hydrogen contamination.[25] Monatomic hydrogen is absorbed into the molten weld pool at superheat temperatures (T greater than liquidus) that allows high solubility. During solidification, hydrogen is partitioned at the liquid/solid interface as a result of the large decrease in solubility.

Once a critical concentration is reached, diatomic hydrogen gas can nucleate and coalesce into pores. Atmospheric pressure greatly influences the total volume of pores formed since high pressure suppresses pore formation, whereas a vacuum (such as in electron-beam welding) encourages pore formation, as shown in Fig. 12-8(b). Underwater welding of structural steel, while often performed at high pressures below the ocean surface, can produce large amounts of porosity (see Fig. 12-13). This situation arises from the extensive pick-up of hydrogen during welding as a result of the higher partial pressure of that gas in the arc atmosphere that prevails under those conditions. Macroporosity is sometimes associated with banding, and continuous strings of pores are found to traverse the weld metal. Although not as deleterious to mechanical properties as hot tears are, porosity does reduce the effective cross-sectional area of the weldment and can also provide a continuous path for leakage.

Inclusions are second-phase particles commonly found in weldments. They can be either exogenous or endogeneous, depending on their origin. The first type arises from the entrapment of welding slags and surface scales. If the inclusions are formed within the system as a result of deoxidation reactions (oxides) or solid-state precipitation reactions (carbides, nitrides), they are known as *endogeneous*. In steel-weld deposits, the volume fraction of inclusions will normally be considerably higher than that in cast-steel products because of the short time available for the growth and flotation of the particles. As a result, weld-metal inclusions are also significantly smaller in dimension (frequently less than 1 μm) and more finely dispersed [see Fig. 12-8(c)] than those in castings.

Inclusions affect the final properties of a weldment in two different ways. First of all, they may act as nucleation sites for various transformation products, resulting in preferential phases being associated with specific types of inclusions; consequently, different mechanical properties may be expected. Second, inclusions may act as crack initiation sites that lead to premature failing of the weld joint. For characterization of inclusions, properties such as composition, number density, volume fraction, and size distribution are used.

SOLID-STATE TRANSFORMATIONS IN WELD METAL

When the solidified weld metal cools down, solid-state transformation reactions may occur, resulting in modification of the solidification microstructure. Thus, the primary solidification pattern may be succeeded by secondary or tertiary structures according to the number of solid-state reactions in the system. Two main factors that determine the final microstructure are the chemical composition and thermal cycle of the weld metal. The micrograph in Fig. 12-9 shows the eutectoidal decomposition products of a low-carbon low-alloy steel weldment superimposed on the solidification structure revealed using a double etching technique.[26] This technique involves immersing the specimen in a saturated solution of picric acid with a few drops of Kodak Photo-Flo 200 (wetting agent) for 30 to 60 sec, followed by etching with a 2-vol.% nital solution. A water and methanol rinse is necessary between the two etching procedures. The 2-vol.% nital etch reveals the solid-state transformation products, while the saturated picric acid etch brings out the alignment of inclusions, thereby giving an indication of the prior solidification pattern. A rotation of the austenite grain with respect to the delta ferrite grain can also be observed, as shown schematically in Fig. 12-10. The fineness of the transformation products usually depends on the solidification grain structure as described by the rule of structural heredity.

Weld Metal Ferrite

In most structural steels, weld metal will solidify as delta ferrite. At the peritectic temperature, austenite will form from the reaction between liquid weld metal and delta ferrite, and subsequent cooling will lead to the formation of alpha ferrite. Prior to the austenite-ferrite transformation, however, austenite grain growth may occur because of the high temperature experienced. The extent of growth will depend on the number density and size distribution of inclusions present in the weld metal. During the austenite-ferrite transformation, proeutectoid ferrite first forms along the austenite grain boundaries, known as *grain-boundary ferrite* or *grain-boundary allotriomorphs*. Elongated or granulated, this ferrite grows into the austenite grain on one side of the boundary only. This reaction is also known as *ferrite veining* because of its branching aspect throughout the weld metal, delineating the prior-austenite grains (see Fig. 12-11). Subsequent to grain-boundary ferrite formation, ferrite sideplates (Widmanstätten ferrite) are developed in the form of long nee-

---
* A trademark of Ferrofluidics Corp., 40 Simon St., Nashua, NH 03061 USA.

Fig. 12-9. The eutectoidal decomposition structure of a low-carbon, low-alloy steel weldment superimposed on the solidification pattern, using a double etching technique (see text).

dle-like ferrite laths protruding from the allotriomorphs. The Kurdjumov-Sachs orientation relationship between sideplates and parent austenite grains has been verified.[27] A coarse austenite grain size and a low carbon content in combination with a relatively high degree of supercooling are found to promote ferrite sideplate formation, with inclusions and precipitates providing favorable nucleation sites. Weldments with a high inclusion density generally show large volume fractions of grain-boundary ferrite and ferrite sideplates. These needle-like laths can be properly characterized by their length-to-width aspect ratio, where values of 10:1 are commonly found and values as high as 20:1 have been observed.[28]

As the temperature continues to drop (i.e., approaching the bainite start temperature), intragranular acicular ferrite will nucleate and grow in the form of short laths separated by high angle boundaries. The inclination between orientations of adjacent acicular ferrite laths is usually larger than 20 degrees (see Fig. 12-11). The random orientation of these laths provides good resistance to crack propagation. An acicular ferrite lath is approximately two micrometers thick, with an aspect ratio varying from 3:1 to 10:1.[28] Ferrite veining, Widmanstätten ferrite, and acicular ferrite are easily revealed by etching the weldment with a 2%-nital solution, which delineates ferrite grain boundaries.

During proeutectoid ferrite formation, carbon is rejected continuously from the ferrite phase, enriching the remaining austenite. This carbon-rich austenite transforms later to a variety of constituents such as martensite (both lath and twinned), bainite, pearlite, and retained austenite (see Fig. 12-12) dependent on the cooling rate and alloy composition. Because of the acicular nature of the bainite laths, they can also be described by their

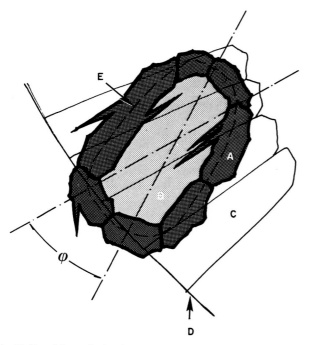

Fig. 12-10. Schematic showing the angle of rotation, ϕ, of an austenite grain with respect to delta ferrite grains. The following phases are indicated: (A) grain boundary ferrite, (B) intragranular products, (C) delta ferrite grains, (D) fusion line, and (E) austenite grain boundary.

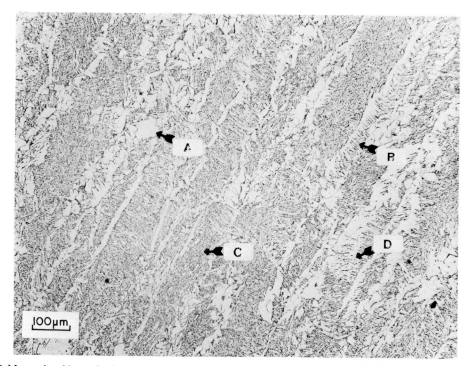

Fig. 12-11. Typical C–Mn steel weld metal microstructure showing: (A) ferrite veining, (B) ferrite sideplates, (C) acicular ferrite, and (D) bainite.

aspect ratio, with values similar to those of Widmanstätten sideplates.[28] More frequently, however, bainite laths occur in the form of packets and not as individual needles. Packets of bainite can often be seen associated with grain boundaries. For observation of such constituents, use of picral as a etching solution may be more appropriate than nital, since picral is a slower acting etchant that will not attack ferrite grain boundaries. Picral is also preferred for revealing microstructures consisting of ferrite and cementite since the etching response is more uniform. Nital is preferred for revealing martensitic microstructures and for revealing ferrite grain boundaries.

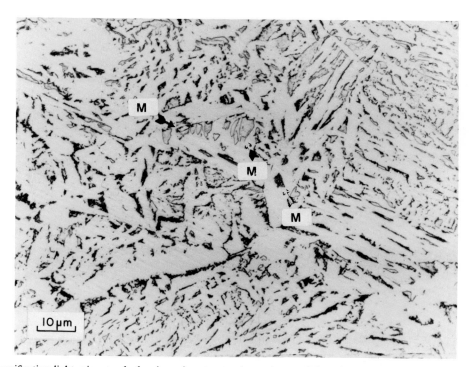

Fig. 12-12. High-magnification light micrograph showing microstructural constituents (M) such as martensite, pearlite, and retained austenite.

(a)

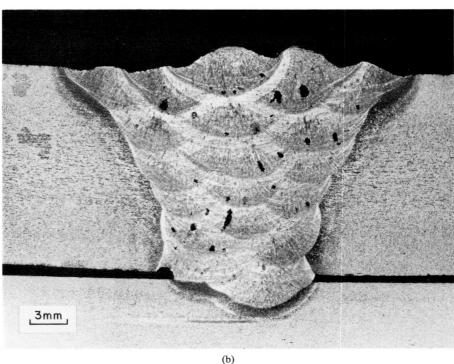

(b)

Fig. 12-13. Macrostructure of a multiple-pass, shielded-metal-arc weld, made under water on ASTM A-36 steel and showing multiple layers of heat-affected zones; note macro-porosity in the fusion zone, attributed to the presence of water: (a) macroetched with aqueous 15-vol.% nitric acid + 5-vol.% methanol, and (b) macroetched same as (a) followed by light grinding.

Multiple-Pass Welds

In a multiple-pass weld, a partial refinement of the microstructure by subsequent weld passes will take place in regions where the peak temperature has exceeded the Ac_3 temperature. By swabbing a ground transverse cross section of the weldment with a macroetchant (e.g., an aqueous 10-% ammonium persulfate solution or an aqueous 15-vol.% nitric acid + 5-vol.% methanol solution), the HAZ turns dark, thereby delineating successive passes (see Fig. 12-13). Deep etching may be necessary to give a macrograph of good quality. To avoid staining of the sample around porous areas, fissures, and cracks (caused by bleeding of the trapped etching solution), pores may be filled with parafin or silicone prior to etching. In order to bring out certain structural aspects, light grinding after etching reverses the coloration of the HAZ, an example of which is shown in Fig. 12-13(b).

As a result of the reheating of individual weld beads, carbon atoms diffuse out of any martensite previously formed, resulting in a tempered martensite microstructure with superior toughness. Besides its minor effect on the gross microstructure, preheating from the previous bead may lead to a general decrease in the hydrogen content and the stress level of the weldment, thereby reducing the tendency for hydrogen-induced cold cracking.

For steels that are not transformable (e.g. austenitic stainless steel), the primary solidification structure will not be altered. The epitaxial, columnar growth will continue through the beads and extend from the base plate cross the whole weld joint. In the case of austenitic stainless steels, this is not particularly harmful because of the characteristic high toughness of the FCC structure at low temperatures.

Weld Metal Precipitation

Prior to, or simultaneously with, the weld-metal transformation reactions, precipitation reactions may also occur. Carbides, nitrides, and carbonitrides of transition elements are the most common precipitates found in steels, their formation being dependent upon weld metal composition, temperature, and cooling rate. Chemical composition will affect the solubility of the precipitates and, indirectly, the temperature of precipitation. Because of the fast cooling rate of the welding process, precipitation may be suppressed, or if not, the resulting restricted growth of the particles will lead to the formation of submicroscopic particles (approximately 0.01 μm), and these will remain unresolved under a light microscope. High-magnification electron micrographs of carbon extraction replicas of weldments are customarily used to study precipitation phenomena.[29]

Fine precipitates will usually pin grain boundaries, restricting grain growth. At high temperatures, however, precipitates may coarsen or dissolve and lose their effectiveness for grain size control. In addition to the grain-boundary pinning effect, precipitates may also strengthen the matrix by dispersion hardening, which in turn will produce a marked decrease in toughness. These effects are particularly important in the heat-affected zone of multiple-pass weldments.

HEAT-AFFECTED ZONE

The following discussion will be concentrated on the HAZ transformation behavior of C–Mn and high-strength, low-

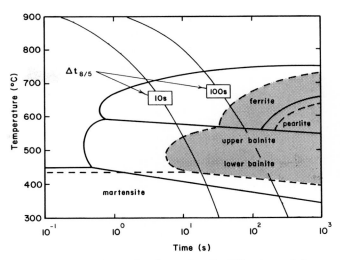

Fig. 12-14. CCT-diagram for a low-carbon Cu–Ni bearing steel. Superimposed on the CCT-diagram are two cooling curves corresponding to $\Delta t_{8/5}$ equal to 10 and 100s, with heavy solid line representing an austenitizing temperature of 900°C for 5 min and heavy broken line, an austenitizing temperature of 1300°C for 5 s.

alloy (HSLA) steels. With reference to Fig. 12-2, the various microstructures formed within a single-pass HAZ can be estimated by means of a continuous cooling transformation (CCT) diagram, as indicated in Fig. 12-14. For parts of the HAZ heated above the Ac_3 temperature, the reaction product may either be martensite, lower and upper bainite, ferrite/pearlite, or a mixture of these different constituents depending on the steel chemical composition and the cooling time through the critical transformation temperature range ($\Delta t_{8/5}$). Moreover, it can be seen from Fig. 12-14 that the phase transformation is strongly affected by the peak temperature or, more correctly, the prior-austenite grain size (i.e., the coarse-grained region close to the fusion line will be more sluggish to transform).

PARTLY MOLTEN ZONE

Depending on the system under consideration, the partly molten zone will usually exhibit a finer grain size compared to that of the coarse-grained region in spite of the higher peak temperatures experienced. This effect can probably be related to the segregation of certain alloying and impurity elements from the bulk of the delta-ferrite grains to the solid/liquid interface, which on subsequent transformation, will reduce the austenite grain growth as the result of an increased boundary drag.[2] However, an extensive segregation of impurity elements may, on the other hand, result in the precipitation of low-melting-point iron-manganese sulphides, leading to the formation of liquation cracks in this region and thereby affecting the mechanical integrity of the weldment.

Coarse-Grained Region

Embrittlement in the HAZ is often located in the coarse-grained region, where the peak temperature has exceeded about 1200°C, as a result of the formation of low-toughness microstructures. The prior-austenite grain size in this area is approximately 100 μm at low and medium heat inputs (see Fig. 12-15) and over 300 μm for high-heat input welds, in the absence of effective grain-growth pinning precipitates such as TiN. The microstructure formed within each austenite grain will normally not be uniform but consist of a complex mixture of two or more of the following constituents,[30] arranged in decreasing order of transformation temperature:

1. Grain-boundary ferrite and ferrite side plates at austenite grain boundaries.
2. Areas of high carbon content, ranging from pearlite to other forms in which the carbides precipitate as rods or spheroids.
3. Bainite colonies where the ferrite plates have grown in a side-by-side manner resembling upper bainite.
4. Lower bainite and martensite.

The constituents, their colony size, and their proportion will depend on the steel chemical composition and the operational conditions applied.

Grain-Refined Region

In the grain-refined region, the peak temperature has been so low that the precipitates have not been fully coarsened nor dissolved and, hence, austenite grain growth is largely restrained. Accordingly, on subsequent transformation, the austenite tends to decompose to a fine ferrite/pearlite microstructure, which will often be finer than that of the base metal (see Fig. 12-15).

Intercritical Region

In the partly transformed, intercritical region, a wide variety of microstructures can be obtained, varying from pearlite to upper bainite and martensite in a matrix of ferrite. However, the tendency for formation of low-toughness microstructures within the ferrite, arising from transformation of carbon-enriched austenite, will be less pronounced in HSLA steels then in C–Mn steels because of their lower carbon content and the presence of strong carbide formers such as niobium, vanadium, or titanium.

Subcritical Region

In the subcritical region, no significant microstructural changes take place. Nevertheless, the toughness may be strongly affected by tempering or strain aging reactions occuring at these temperatures.[31] As shown in Fig. 12-16, spheroidization of carbides arising from tempering is most likely to occur at the grain boundaries and triple-point grain junctions.

WELDING OF DISSIMILAR METALS

Joining of dissimilar metals comprises both metals that are chemically different (Al, Cu, Ni) and alloys of a partic-

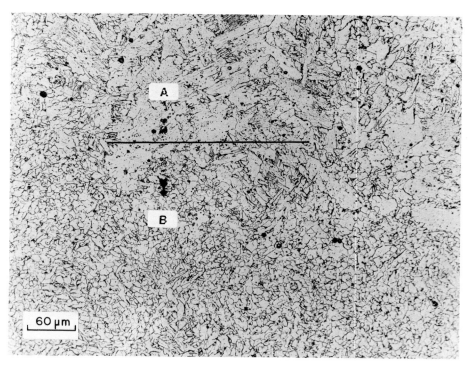

Fig. 12-15. Representative microstructures of the heat-affected zone showing the distinction between (A) coarse-grained region, and (B) fine-grained region.

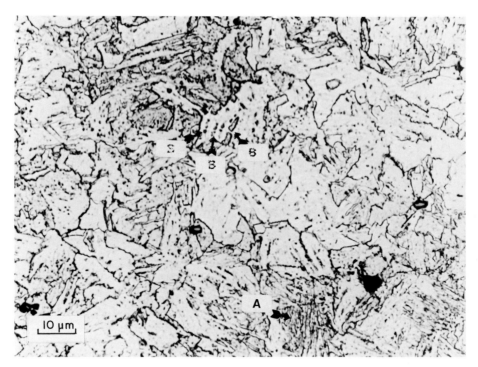

Fig. 12-16. Heat-affected zone exhibiting a wide variety of microstructures in intercritical and subcritical regions: (A) bainite and martensite, and (B) spheroidized carbides.

ular metal that are significantly different from a metallurgical standpoint (austenitic stainless steel to low-alloy ferritic steel, for example); it also includes the use of solid-state welding techniques, and brazing and soldering techniques as well as fusion-welding processes.[32] In the case of fusion welding of dissimilar metals, the most important consideration is the weld-metal chemistry and its mechanical and physical properties. Filler metals must be selected with an eye to avoiding hot tearing and formation of brittle phases within the weld metal on solidification and subsequent cooling. Doing so often requires the use of filler metals of a completely different chemical composition from that of either of the two base metals.

From a metallographical point of view, a chemically inhomogeneous weldment of this kind is very complicated to deal with. First of all, large variations in hardness across the weld will make it difficult to obtain a smooth specimen surface by means of conventional polishing techniques. Moreover, because of the chemical differences between the various metals/alloys under consideration, it may be hard to find a proper etching solution for the weldment that will provide a clear image of both the weld metal and the heat-affected zone microstructures. These problems can be overcome to some extent by the use of masking techniques, together with selective etching of individual regions within the weldment.

For welding austenitic stainless steel to low-alloy steel, stainless-steel filler metals such as AISI types 309, 310, or 312 can be used so long as the structure is going to be exposed to little or no wear and moderate stresses.[33]

If higher wear resistance and yield strength are required, austenitic Mn–Cr electrodes are usually recommended. Under certain welding conditions, however, austenitic filler metals may be susceptible to cracking and precipitation hardening, especially in the case of excessive base-metal dilution.[32] Thus, for many applications, nickel-bearing filler metals are preferred since these alloys can tolerate dilution from a variety of base metals without becoming crack sensitive. Because of the large difference in thermal expansion between the different materials involved, the ferritic base metal may be susceptible to reheat cracking in the coarse-grained region close to the fusion line.[34]

Acknowledgement. The authors acknowledge the support of the Division of Materials Science of the United States Department of Energy. The authors also appreciate the review and comments of Professor D. L. Olson.

REFERENCES

1. Lancaster, F. L. *Metallurgy of Welding*. 3rd Ed. London: George Allen & Unwin Ltd., 1980.
2. Easterling, E. *Introduction to the Physical Metallurgy of Welding*. London: Butterworths & Co. Ltd., 1983.
3. Davies, G. J., and Garland, J. G. Solidification structures and properties of fusion welds. *Int. Met. Rev.* 20:83–106 (1975).
4. Matsuda, F.; Hashimoto, T.; and Senda, T. Fundamental investigations on solidification structure in weld metal. *Trans. Nat. Res. Inst. Metals* 11:43–58 (1969).
5. Savage, W. F.; and Aronson, A. H. Preferred orientation in the weld fusion zone. *Welding Journal* 45:85s–89s (1966).
6. Savage, W. F., and Hrubec, R. J. Synthesis of weld solidification

using crystalline organic materials. *Welding Journal* 51:260s–271s (1972).
7. Savage, W. F.; Lundin, C. D.; and Aronson, A. H. Weld metal solidification mechanics. *Welding Journal* 44:175s–181s (1965).
8. Adams, C. M. Cooling rates and peak temperatures in fusion welding. *Welding Journal* 37:210s–215s (1958).
9. Brown, P. E., and Adams, C. M. Fusion zone structures and properties in aluminum alloys. *Welding Journal* 39:520s–524s (1960).
10. Jarman, R. A., and Jordan, M. F. Relationship between heat input and dendritic structure of full-penetration weld beads in a commercial aluminum-copper alloy. *Journal Inst. Metals* 98:55–57 (1970).
11. Brody, H. D., and Flemings, M. C. Solute redistribution in dendritic solidification. *Trans. AIME* 236:615–624 (1966).
12. Michael, A. B., and Bever, M. B. Solidification of aluminum-rich aluminum copper alloys. *Trans. AIME* 200:47–56 (1954).
13. Murty, Y., and Kattamis, T. Z. Structure of highly undercool cobalt-tin eutectic alloy. *J. Crystal Growth.* 22:219–224 (1974).
14. Chadwick, G. A. Eutectic alloy solidification. *Prog. in Materials Science,* Pergamon, 12:114–121 (1963).
15. Cross, C. E., and Olson, D. L. Modification of eutectic weld metal microstructure. *Welding Journal* 61:381s–387s (1982).
16. Rohatgi, P., and Adams, C. M. Colony and dendritic structures produced on solidification of eutectic aluminum copper alloy. *Trans. AIME* 245:1609–1613 (1969).
17. D'Annessa, A. T. Characteristic redistribution of solute in fusion welding. *Welding Journal* 45:569s–576s (1966).
18. Borland, J. C. Generalized theory of super-solidus cracking in welds and castings. *British Welding Journal* 4:508–512 (1960).
19. Clyne, T. C., and Davies, G. J. Comparison between experimental data and theoretical predictions relating to dependence of solidification cracking on composition. *TMS Proc. Int. Conf. Solidification-Sheffield:* 275–278 (1977).
20. Senda, T., and Matsuda, F. Fundamental investigations on solidification crack susceptibility for weld metals with trans-varestraint test. *Trans. Jap. Weld. Soc.* 2:141–162 (1972).
21. Dudas, J. H., and Collins, F. R. Preventing weld cracks in high-strength aluminum alloys. *Welding Journal* 45:241s–249s (1966).
22. Brooks, J. A.; Thompson, A. W.; and Williams, J. C. A fundamental study of beneficial effects of delta ferrite in reducing weld cracking. *Welding Journal* 63:71s–83s (1984).
23. Beraha, E., and Shpigler, B. *Color Metallography.* Metals Park, OH: ASM, 1977.
24. Gray, R. J. Magnetic etching with Ferrofluid: *Metallographic Specimen Preparation.* New York: Plenum, 1973, pp. 155–177.
25. Devletian, J. H., and Wood, W. E. Factors affecing porosity in aluminum welds—a review. *WRC Bulletin* 290:1–18 (1983).
26. Indacochea, J. E. Effect of the manganese—silicate base submerged arc welding flux composition on the weld metal composition and microstructure. Colorado School of Mines Ph.D. Thesis T-2511 (1981).
27. Aaronson, H. I. Proeutectoid ferrite and the proeutectoid cementite reactions. *Proc. Symp. Decomp. Austenite by Diffusional Processes.* Philadelphia: AIME, 1960, pp. 387–548.
28. Committee of Welding Metallurgy of Japan Welding Society. *Classification of microstructures in low C-low alloy steel weld metal and terminology.* IIW DOC IX-1282-83.
29. Loberg, B.; Nordgren, A.; Strid, J.; and Easterling, K. E. The role of alloy composition on the stability of nitrides in Ti-microalloyed steels during weld thermal cycles. *Met. Trans.* 15A:33–41 (1984).
30. Dolby, R. E. Factors controlling the HAZ and weld metal toughness in C-Mn steels. *Proc. 1st National Conference on Fracture.* Johannesburg, 1979, pp. 123–140.
31. Dolby, R. E., and Saunders, G. G. Subcritical HAZ fracture toughness of C-Mn steels. *Met. Constr.* 4:185–190 (1972).
32. Matthews, S. J., et al. Dissimilar metals. *AWS Welding Handbook* 4:513–547 (1982).
33. Graville, B. A., et al. High alloy steels. *AWS Welding Handbook* 4:169–209 (1982).
34. Campbell, G. M., et al. Evaluation of fractors controlling high temperature service life of 2 ¼ Cr–1Mo steel to austenitic stainless steel weldments. *ASM Proc. Trends in Welding Research:* 443–470 (1982).

13
MICROSCOPY AND THE DEVELOPMENT OF FREE-MACHINING STEELS

J. D. Watson

The Broken Hill Proprietary Company Ltd.
Victoria, Australia

Machinability has been studied by engineers for over 75 years, beginning with the significant contributions of F. W. Taylor between 1881 and 1906.[1] Since then, the mechanics of metal cutting has been extensively studied, and considerable progress has been made in developing theories of machinability based on a knowledge of mechanical properties.[2] Scientists, and, in particular, metallurgists, have only more recently contributed to this understanding because for many years following Taylor's pioneering work, machinability drew little or no attention from the many eminent metallurgists of the time, and free-machining steels *per se* were unrecognized.[3]

This chapter describes how metallography has contributed to the development of an understanding of the character of free-machining steels, and, in so doing, it necessarily focusses attention on the strong influence of microstructure on machinability. Nevertheless, its purpose is not to discuss in detail how microstructure influences the mechanisms of chip formation, but rather how microstructure is developed and how it is qualitatively and quantitatively assessed by the metallographer. For a review of the affects of microstructure on machinability, the reader is referred to several excellent conferences that have been wholly or partly devoted to the subject and that are listed in the Bibliography at the end of this chapter.

As will be shown throughout this chapter, the metallurgists' contribution began with the use of the techniques of optical microscopy and has expanded significantly with the growing use of electron microscopy (particularly SEM) and methods of micro and surface analysis. These, together with the development of the theory of stereology and its application involving the use of automated, microprocessor-controlled image-analysis equipment have thrown a perspective on machinability uncontemplated at the time of F. W. Taylor.

It is a truism that all materials can be machined—albeit with differing degrees of difficulty—but free-machining steels are those designed specifically to promote ease of machining. This goal is achieved by deliberate additions of certain elements. These elements are listed in Table 13-1, from which it may be seen that only two, phosphorus and sulfur, are elements that are normally present in any significant amount in steels. The positions of the elements in the periodic table determine how they behave in steels, and a natural division can be made into categories on the basis of the form in which an element is present in steel as suggested by Iwata.[4]

1. Insoluble metals, of which lead and bismuth from the sixth period of Groups IVB and VB are the most important, are considered first. These two elements, which are partially soluble in liquid iron but virtually insoluble in all allotropes of solid iron,[5-8] form metallic inclusions during solidification. Lead alone, or in combination with sulfur, is the basis of a wide range of free-machining carbon and low-alloy steels.
2. Elements forming insoluble compounds, especially with manganese, are the next most important. From Group VI, sulfur and the metalloids selenium and tellurium exhibit this property. Despite some chemical similarities, the properties of sulfides, selenides, and tellurides are sufficiently different to produce different behavior when present in steels, and metallography has played an important role in identifying how these differences are manifested during machining.
3. Phosphorus and nitrogen are Group V elements, and both are soluble in ferrite in low-carbon steels, provided (in the case of nitrogen) that no strong nitride-forming elements like aluminium, titanium, or vandium are present. The effects of these elements on machining behavior are not well understood or even agreed upon,[9-13] despite the fact that Bessemer steels containing high phosphorus and nitrogen were the forerunners of modern free-machining steels.[3,14-17] Because these elements are in solid solu-

Table 13-1. Periodic Table Indicating Elements Added to Steels to Enhance Machinability.

1 H																	2 He
3 Li	4 Be											5 B	6 C	7 N	8 O	9 F	10 Ne
11 Na	12 Mg											13 Al	14 Si	15 P	16 S	17 Cl	18 Ar
19 K	20 Ca	21 Sc	22 Ti	23 V	24 Cr	25 Mn	26 Fe	27 Co	28 Ni	29 Cu	30 Zn	31 Ga	32 Ge	33 As	34 Se	35 Br	36 Kr
37 Rb	38 Sr	39 Y	40 Zr	41 Nb	42 Mo	43 Tc	44 Ru	45 Rh	46 Pd	47 Ag	48 Cd	49 In	50 Sn	51 Sb	52 Te	53 I	54 Xe
55 Cs	56 Ba	57– La	72 Hf	73 Ta	74 W	75 Re	76 Os	77 Ir	78 Pt	79 Au	80 Hg	81 Tl	82 Pb	83 Bi	84 Po	85 At	86 Rn
87 Fr	88 Ra	80– Ac	104														

58 Ce	59 Pr	60 Nd	61 Pm	62 Sm	63 Eu	64 Gd	65 Tb	66 Dy	67 Ho	68 Er	69 Tm	70 Yb	71 Lu
90 Th	91 Pa	92 U	93 Np	94 Pu	95 Am	96 Cm	97 Bk	98 Cf	99 Es	100 Fm	101 Md	102 No	103 Lw

tion, their presence is not directly observable metallographically, and they will therefore not be considered further in this chapter. Their effects on machinability have been well-documented by Aborn[18-21] and others.[9-13,22-24]

4. Calcium from Group IIA is an alkaline earth, and its role in promoting free-machining behavior is somewhat different from that of the more common free-machining elements listed above. Calcia has the ability to flux with certain oxides, including alumina and silica, and to combine with sulfur to form complex low-melting-point compounds that improve machining under certain circumstances.

5. Boron from Group IIIA is not a free-machining additive in the same sense as the above elements for it has not been used to produce large-scale commercial free-machining steels and little is known about its behavior. It has, however, been suggested as an alternative additive to lead for environmental reasons, as will be discussed later.

The following sections consider in detail how microstructure varies with composition and steel-processing variables. First, the general features of the microstructures of steels containing each of the free-machining additives are discussed; then the methods for characterizing these microstructures quantitatively are outlined. The final section examines the techniques available for observing either directly or indirectly how microstructure influences chip formation.

SPECIMEN PREPARATION METHODS

Metallographic preparation methods for carbon, alloy, or stainless steels containing free-machining additives do not differ significantly from the now standard methods of mechanical polishing described by Samuels.[25] These utilize silicon carbide papers for grinding and diamond laps for rough and intermediate polishing.

If problems with excessive relief or pull-out of second-phase particles are encountered, they may usually be overcome by means of a modified practice recommended for the retention of inclusions[25]; this procedure involves a single polishing stage with a 1–μm diamond-charged, short-napped polishing cloth, thus protecting the inclusions from excessive rubbing by the nap of the pads. Because sulfide-type inclusions found in free-machining steels are not particularly soft (compared to the matrix), they do not pose the same problem as graphitic cast irons, and normally it is unnecessary to resort to special polishing procedures for routine metallography.

When metallic inclusions are present, as in lead or bismuth-bearing steels, some special precautions are required in order to preserve their integrity. Excessively long polishing times with napped cloths must be avoided, since these tend to scour the soft phase and leave only an empty socket or void that may easily mislead the metallographer, particularly if the void occurs in studies of processes taking place near the tool tip in metal-cutting experiments. In such cases, void formation is one of the several possible "mechanisms" by which additives promote machinability, and good metallographic practice must be adhered to if false conclusions are to be avoided. This subject will be discussed at more length later.

Second, if final polishing on alumina is to be attempted, care must be taken to avoid etching caused by electrochemical differences between the inclusions and ferrite. As noted by Chalfant,[26] polishing in a solution with a pH of exactly 7.0 obviates these effects. Blank and Johnson[27] used this method and claimed it gave better results than the techniques of Bardgett and Lismer,[28] Gerds and Melton,[29] or the slightly alkaline polishing solutions suggested by Schofield.[30] The technique has also been successfully used by Garvey and Tata[10,11] and Samuels.[31]

For improved retention of lead, Volk[32] has suggested grinding with abrasive paper smeared with paraffin wax, as was already recommended for polishing pure lead.[25]

Forgeng and Lee[33] and others[34] have recently advocated automatic polishing of specimens, both for Automatic Inclusion Assessment (AIA) on image-analysing microscopes and for routine metallography. For AIA, automatic polishing is almost mandatory because of the large numbers of fields and specimens that must be examined in order to achieve statistical reliability.[35] Apart from this aspect, however, the additional advantages are retention of inclusions and lack of edge rounding and/or relief, features that are particularly appropriate to the preparation of free-machining steels.

MACRO EVIDENCE OF MICROSTRUCTURE

The distribution of sulfur and lead in free-machining steel can be observed by direct printing onto photographic papers. Because segregation of sulfur and, more particularly, of lead occurs in these steels, direct printing methods allow a rapid assessment of the degree of inhomogeneity that will be useful for routine quality control.

Sulfur Print

The method developed by Baumann[36] has been used as a standard metallographic technique for many years[37] and is still used routinely for revealing sulfides in steel.[38] The method consists of taking a finely ground (not necessarily polished) specimen and pressing it into the surface of a photographic paper moistened with a dilute solution of sulfuric acid (2 to 5 percent). Care must be taken to remove air bubbles, and 1 to 2 minutes contact is usually sufficient to allow dark silver sulfide to stain the paper, after which the print can be washed, fixed, and dried in the usual way. Sulfur prints, together with corresponding optical micrographs, from 25.4-mm bars of resulfurized steel are shown in Fig. 13-1.

Lead Print

The lead print is basically similar to the sulfur print method. As described by Volk,[32] it involves soaking a silver-bromide photographic paper in concentrated acetic acid for 3 minutes, removing the excess acid, and then laying the paper emulsion-side-up on a glass slide. The finely ground lead-containing sample is then pressed onto the paper and allowed to react for approximately 5 minutes (for Pb < 10 percent). The sample is then removed and the paper immersed for 2 to 3 minutes in a hydrogen-sulfide solution, a process that converts the lead-acetate spots on the paper to black lead sulfide. As a last step, the paper is washed and dried.

QUALITATIVE METALLOGRAPHY

Steels Containing Metallic Inclusions (Pb, Bi)

The dissolution and dispersion of lead in molten steel has been investigated by a number of workers,[5-7] and, although there is some disagreement on the exact figure, it has been found that at steelmaking temperatures, steel dissolves approximately 0.2 to 0.3 percent of the lead. Above this level, the binary Fe–Pb system has a miscibility gap in which two conjugate liquids (Fe–0.2%Pb and almost pure lead) exist in equilibrium.

Since lead is virtually insoluble in solid iron, the microstructure of lead-bearing steels at room temperature consists of any lead undissolved (emulsified) at the steelmaking temperature plus that rejected from solution during solidification and cooling. For the levels of lead normally found in free-machining steel (0.2 percent), the latter should predominate.

Thompson, Quinto, and Levy[39] and Thompson and Levy[40] studied the lead distribution in ingots and rolled product of AISI 12XX series steels using a scanning electron microscope operated in the BSE (Back-Scattered-Electron) mode. The excellent atomic number contrast exhibited by lead under these conditions allows the distribution to be clearly seen both on polished sections [Fig. 13-2(b)] and on surfaces prepared by ductile fracturing along a plane containing the rolling direction [Fig. 13-2(c)].

The general conclusions of this and other work[27,41] are that lead in the ingot exists as both spherical particles of free lead and as lead envelopes around manganese-sulfide inclusions. The microradiographic studies of Blank and Johnson[27] showed not only all the essential features of optical and scanning electron photomicrographs but also more macroscopic effects such as the presence of a lead-free skin near the ingot base, the exact location of which depended on the time at which the lead addition was made during the ingot teem.

The size of the lead particles, like that of manganese

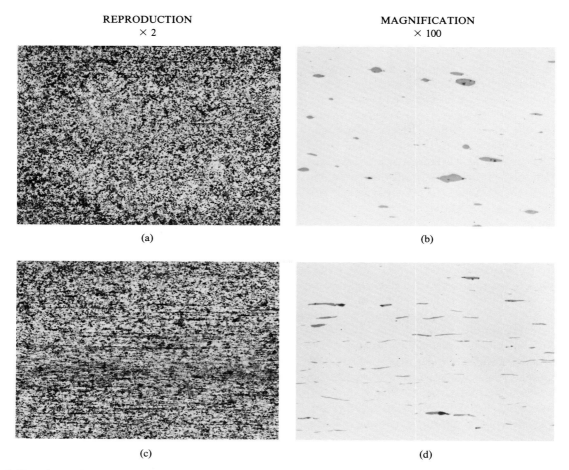

Fig. 13-1. Sulfur prints and corresponding optical micrographs from two 25.4-mm bar samples of low-carbon resulfurized steel showing: (a) and (b) even distribution of *globular* sulfides; and (c) and (d) segregated distribution of *stringer* sulfides. (a) and (c) reduction × 2. (b) and (d) magnification × 100.

sulfide, is found to vary with ingot cooling rate, becoming larger as the cooling rate decreases.

Bismuth, whether added in conjunction with lead[42-46] or in place of lead,[45-48] forms metallic inclusions that are metallographically identical to lead and can best be distinguished by X-ray analytical techniques.[47] Because of the lower density of bismuth, however, it is claimed that it has less tendency to segregate during ingot solidification.[49]

During hot rolling, low-melting-point metallic inclusions are in a liquid state, and, when present in the form of "envelopes" on sulfides, they tend to be squeezed to the extremities of the latter so that in longitudinal metallographic sections or microradiographs they usually appear as "tails,"[49-50] as seen in Fig. 13-2. As shown by Ramalingam et al.,[51] this "association" of lead and manganese sulfide has a profound influence on the way in which the sulfides deform during machining, and this is believed to contribute to the free-machining effect found in leaded, resulfurized steels.

Although lead-bearing alloy and stainless free-machining grades are available (see Table 13-2), they are not as extensively used as the low-carbon grades. According to Mills,[52] the tendency of lead to segregate has mitigated against its use in stainless steels; in high-strength steels, moreover, lead is known to reduce the elevated temperature ductility and have embrittling effects under certain conditions.[46,53-55] For these reasons, other additives such as selenium or tellurium are often used for these steels.

When the metallographer is forced to use optical methods (as opposed to electron metallography) to study lead distributions, a special etching method devised by Volk[32] may be of help. It has been successfully used by Levy et al.[56] and comprises etching in a solution of 50-ml alcohol, 3-ml glacial acetic acid, 20-ml water, and 2-g potassium iodide. Yellow lead iodide forms on the lead inclusions and can be viewed most effectively using polarized light.

Steels Containing Nonmetallic Inclusions

Sulfur-Bearing Steels. The classical free-cutting steels are low-carbon resulfurized steels belonging to the AISI 11XX and 12XX series (the latter also being rephosphorized). These steels, together with their leaded variants, are designed solely for ease of machining and are

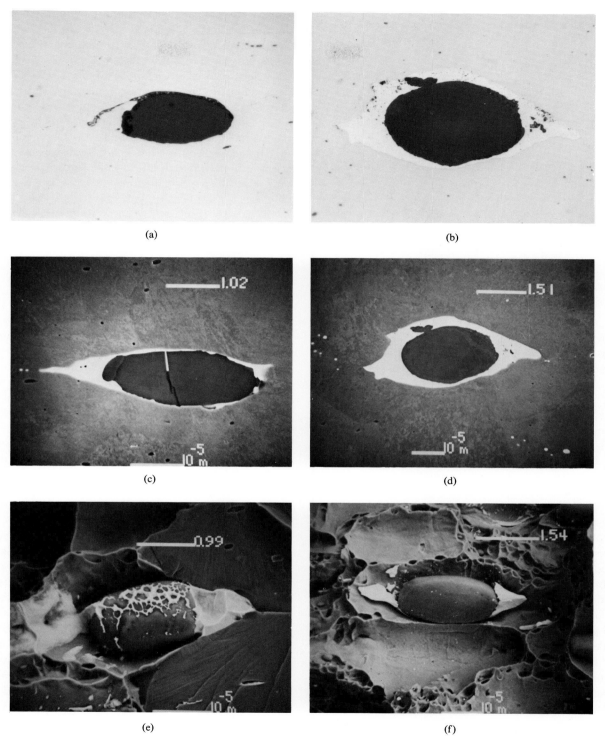

Fig. 13-2. Manganese-sulfide inclusions with associated metallic lead and bismuth in low-carbon, resulfurized free-machining steels revealed by: (a) and (b) optical bright field microscopy (unetched); (c) and (d) SEM of polished section (BSE image); and (e) and (f) SEM of fracture surface (BSE image).

Table 13-2. List of the Most Common Free-Machining Steels.

GRADE	ADDITIVES
Low Carbon Free Cutting:	
AISI 1200 Series	S + P (+ N optional)
AISI 12L00 Series	S + P + Pb
12X00 or 1200(X)	S + P + either Bi, Se, or Te
Low/Medium Carbon Free Cutting:	
AISI 1100 Series	S
AISI 11L00 Series	S + Pb
Carbon and Alloy Steels:	
E 4340	4340 Grade + Pb
E 51100	51100 Grade + Pb
E 52100	52100 Grade + Pb
10L00	Pb
Stainless Steels:	
303	302 Grade + S
303 Se	302 Grade + S + Se
430 F	430 Grade + S
430 F Se	430 Grade + S + Se
316 F	316 Grade + S + P

poorly suited to other operations (e.g., cold forming). When the final application demands better mechanical properties than can be achieved with these low-carbon steels, then medium-carbon grades (0.4-percent C) of the AISI 11XX series may be substituted, but with some loss of machinability as a result of the higher carbon content. For special applications, resulfurized alloy and stainless steels are available (see Table 13-2).

In all of these steels, sulfur is added up to a maximum of 0.4 percent, but the level is usually below this, being around 0.3 percent for the low-carbon and stainless grades and between 0.05 to 0.15 percent for the higher-carbon grades. Sufficient manganese is added to ensure the formation of the insoluble compound, manganese sulfide, that conveys the free-machining properties, the amount required being dependent on the alloy type. Thus, in low-carbon steels, the Mn:S ratio is generally kept above 3.5:1 in order to prevent hot shortness resulting from the formation of FeS. (FeS has a melting point of 1190°C when pure and 940°C when in eutectic form with iron[57].) In stainless steels, an even higher ratio is aimed for (generally > 6:1) in order to avoid the formation of the harder chromium-bearing sulfides.[58-60]

Methods for identifying the various metal sulfides using optical microscopy have been usefully summarized by Wells,[61] who points out that certain metal sulfides, because of substantial differences in their optical properties, can usually be distinguished from one another without having to resort to X-ray microanalysis. For example, manganese sulfide, which appears light gray in reflected light, is transparent with a greenish internal reflection. Iron and chromium sulfides are submetallic (i.e., they are opaque and have relatively high reflectivity) and may be distinguished from manganese sulfide by their anisotropy and color in reflected light; iron sulfide appears brassy yellow and chromium sulfide a creamy tan colour.[61] Similar differences are observed when manganese sulfide takes these or other elements into solid solution, as will be discussed later.

In commercial steels, the sizes and shapes of manganese-sulfide inclusions vary considerably, and this is believed to have a very important influence on machinability. The morphology of manganese sulfide, which is determined at the time of solidification, depends on both the liquid steel composition and the cooling rate. A fundamental understanding of the solidification processes taking place in melts containing iron, manganese, sulfur, and oxygen has been provided by the early work of Sims and Dahle,[62] Dahl et al.,[63] and Sims,[64] from which three "types" of sulfides were identified. Subsequent work[65-71] has allowed the phase equilibria and resulting microstructures to be explained in a consistent manner.

The so-called Type I manganese sulfide is now understood to form in oxygen-rich melts (i.e., in the absence of strong deoxidizers like aluminum and silicon). During solidification, iron-rich dendrites nucleate and grow, and the remaining liquid becomes enriched in manganese, sulfur, and oxygen to the degree that it eventually separates into two liquids. This separation occurs when the composition enters the region given by the miscibility gap in the pseudo-ternary Fe–MnO–MnS system,[63] from which it is clear that the composition of the minor phase is rich in manganese, sulfur, and oxygen. These small, solute-rich liquid droplets are trapped by the advancing iron dendrites and eventually solidify themselves as discrete spherical "inclusions" of manganese sulfide in conjunction with oxides of manganese or silicon. In rolled products, the silicate is often seen as a "tail" on the manganese sulfide, as shown in Fig. 13-3.

Type II sulfides form when the oxygen content is lower than that given by the boundaries of the miscibility gap, in which case the interdendritic sulfur-rich liquid eventually solidifies as a eutectic of iron and manganese sulfide, and the sulfides then have the characteristic fernlike morphology indicating their eutectic origin.

The third type of sulfide, Type III, is rarely expected to be found in free-machining steels since it requires an extremely low oxygen content and an excess of aluminum for its formation. For reasons that will become evident later, such strong deoxidation is never practiced in the production of free-cutting steels (specially deoxidized steels represent an exception, the reasons for which are to be discussed separately).

Qualitative metallographic evidence of the effects of cooling rate on sulfide size and shape have been given by Mohla and Beech[72]; the effects of oxygen content have been discussed by Paliwoda[73] and others.[74] The effects of oxygen are now recognized as being most important in determining not only the distribution of the sulfides, but also their composition and properties. Studying these

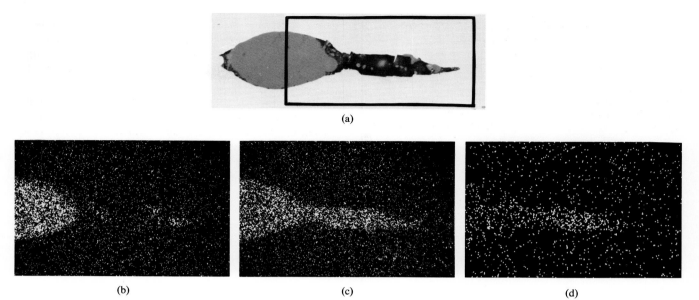

Fig. 13-3. Manganese-sulfide inclusion with eutectic, glassy manganese silicate tail in AISI 1215 steel: (a) optical micrograph ($\times$ 650); (b) corresponding S-K_α X-ray map from area shown; (c) corresponding Mn-K_α X-ray map from area shown; and (d) corresponding Si-K_α X-ray map from area shown.

differences in sulfide properties, Gaydos[75] showed that the room temperature hardness of sulfides which appeared reddish under polarized light was lower than that of those which appeared green and has a stringerlike morphology after rolling. Clayton and Brown[76] suggested that these observations could be directly related to the original oxygen content, since the reddish sulfides had associated manganese-oxide tails, whereas the green ones had silicate tails (like lead, silicates and oxide phases tend to be squeezed to the extremities of the duplex inclusions during hot rolling).

Kovach[58] pointed out that if the same relative hardness observed at room temperature is maintained at hot-rolling temperatures, the results are *not* consistent with the observed lower deformability of the red sulfides, and it was suggested that it is the liquid silicate tails rather than the sulfide composition that has a strong influence on the final shape of the sulfides. In this context, it is perhaps worthwhile noting that early attempts by Van Vlack[17] to relate machinability to sulfide size and shape via oxygen content are possibly misleading because the different oxygen levels used in that work were obtained by additions of silicon.

Riekels[77] has described the application of novel petrographic techniques for inclusion identification in low-carbon free-machining steels using thin sections and transmitted polarized light. In these studies, the reddish manganese sulfides were subsequently found to contain 9- to 18-percent iron, the yellow about 4-percent, and the green about 2-percent, using microprobe analysis. In no instance was manganese oxide found "dissolved" in the sulfide, and almost all the oxide phases were found as "tails" or envelopes around the sulfide inclusions.

Other polarized light microscopy studies of general application to steels have been described by Pickering,[78] and those specific to free-machining steels have been detailed by a number of others.[75,76,79] These techniques have been largely superceded by SEM–EDA, however, which allows for a more quantitative approach.

The microradiographic studies of Blank and Johnson[27] also confirm the results of optical metallography, their main conclusions being the following:

- Sulfide size increases as ingot cooling rate decreases (which occurs if the ingot size becomes larger or as distance from the chill surface increases).
- Sulfides in the columnar zone tend toward Type II.
- Sulfides in the equiaxed zone are sometimes associated with FeS.
- "A" type segregates contain "ropes" of sulfide inclusions.

From a metallographic point of view, superb evidence of the effects of oxygen on the morphology of manganese sulfides in ingots has been obtained by Steinmetz and Lindenberg,[80] who used scanning microscopy to examine specimens that were deep-etched in either methanol bromine or citrate solutions. Some of these are reproduced in Fig. 13-4.

The importance of these factors has been recognized in modern steelmaking practice for free-machining steels where it is usual to aim for large, "globular" Type I sulfides to promote machinability. These are achieved by deoxidation with manganese and the complete avoidance of all strong deoxidizers that not only adversely affect

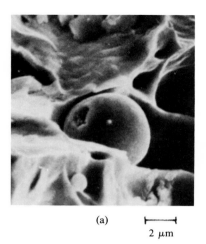 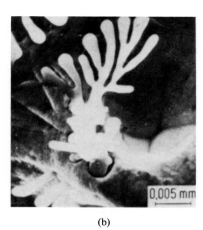 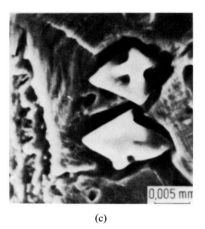

Fig. 13-4. The three morphologies of manganese sulfide revealed by scanning electron microscopy: (a) spherical Type 1 (Courtesy, *Met. Trans.*[67]); (b) dendritic Type 2 (Courtesy, *Arch. Eisen.*[80]); and (c) idiomorphic Type 3 (Courtesy, *Arch. Eisen.*[80]).

sulfide size and shape but also introduce harmful deoxidation products.

During hot rolling, the morphology of all soft inclusions is changed significantly. Because the plasticity of manganese sulfides is lower than that of steel, and because this ratio is temperature-dependent, the final size/shape distribution of sulfides in rolled products is affected by the following:

- Steel casting temperature
- Initial size and shape of the ingot, which determines the initial size/shape of the inclusions
- Rolling temperature and degree of deformation
- Presence of other inclusions (e.g., silicates)

Studies of the deformability of sulfides have been made by many researchers,[81-89] and the effects on machinability are reasonably well documented, albeit in a qualitative way. The greatest difficulties in quantifying such effects arise because "size" and "shape" are parameters that are

- Not single-valued (i.e., they are distributed in a statistical sense)
- Not easily measured once the shape deviates significantly from that of a sphere
- Not necessarily independent

Qualitatively, however, it may be said that sulfide deformability (relative to steel) decreases as

- Temperature increases above 900°C (see Fig. 13-5)
- Inclusion size decreases (see Fig. 13-6)

and the machinability increases as

- Sulfide size increases
- Sulfide aspect ratios decrease (i.e., spherical "shape")

Apart from oxygen, certain other additives have the ability to affect sulfide deformability and therefore machinability. Bellot and Gantois[89] suggest that both selenium and tellurium reduce the relative plasticity of manganese sulfide, thereby allowing a more "globular" shape to be retained during hot rolling, whereas Riekels has shown that dissolved iron increases sulfide hardness.[77] Bhattacharya[90] has demonstrated that cerium additions nucleate very nondeformable sulfides, and Yamaguchi et al.[91] and Philbrook[92] have found that zirconium and titanium sulfides also have very low deformability.

The final stage of processing for free-machining barstock is usually a cold drawing operation, the main purposes of which are to provide a close size tolerance on the bar, leave a good surface finish, and provide better chip-breaking characteristics. Apart from deforming the steel, however, this process also causes the sulfides to fracture, as shown in Fig. 13-7. This is somewhat unexpected in view of the fact that Jaffrey[93] demonstrated that sulfides deformed under uniaxial stress or strain do not fracture until strains in excess of 20 to 30 percent are reached.

It is also well known that sulfides may undergo significant shear deformation during the machining process itself,[94] so it must be concluded that the sulfide fracture observed in cold-drawn, resulfurized steels is due to their large size, complex states of stress, and/or an effect of temperature sensitivity. The effect of this cracking on machinability is unknown, but it is worth noting that "theories" of machinability that propose that inclusions enhance machinability by promoting void formation and fracture[79,95-97] rarely take into account that many of the sulfides in cold-drawn steels are fractured prior to participation in the machining operations and that even in hot-rolled steels it is possible to have cracklike defects in the form of the voids[81] or oxide tails[94] associated with sulfide inclusions.

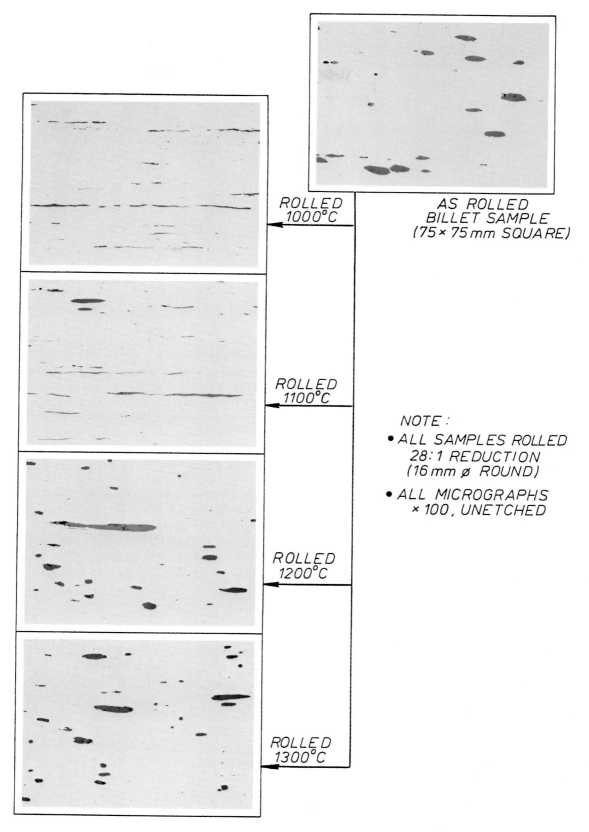

Fig. 13-5. Optical micrographs showing effect of rolling temperature on manganese-sulfide morphology in AISI 1215 steel.

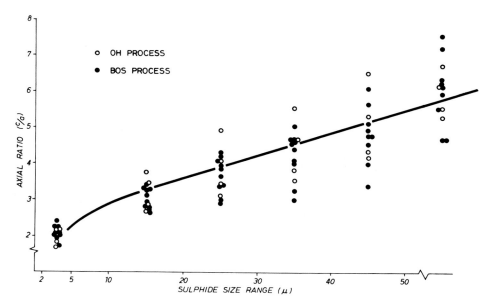

Fig. 13-6. Measured aspect ratios of manganese sulfides in hot-rolled billet sample of AISI 1215 as a function of size, showing the apparent size dependence of deformability.

Steels Containing Selenium and Tellurium. According to Kiessling et al.,[57,98,99] the monoselenide, MnSe, is cubic and isostructural with α manganese sulfide. The system MnS–MnSe accordingly shows complete solid solubility and inclusions in selenium-treated steels (with or without sulfur) that are very similar in appearance to manganese sulfides; the pure selenide, however, is slightly softer than the sulfide.[57,58]

Tellurium behaves in a more complex fashion; being only slightly soluble in manganese sulfide, it tends to form MnTe. In steels containing sulfur, the telluride forms envelopes around the manganese sulfides,[18-20,89,99,100] whereas in resulfurized steels containing lead, the compound PbTe forms in addition to MnTe.[3,18-19,101-103]

A photomicrograph of the complex inclusions in tellurium-treated steels is shown in Fig. 13-8. Manganese telluride has a melting point of only 1150°C,[95] which means that it may melt at hot-rolling temperatures and give rise to hot shortness. It should be noted, however, that at an Mn:Te ratio < 10, iron telluride may form, and Bellot and Gantois[89] suggest that this is the more likely cause of hot shortness. Although the machinability of tellurium-treated, low-carbon and medium-carbon steels is exceptionally good,[48,104] the tendency of tellurium to degrade hot ductility and its toxicity have limited its use in commercial steels.

Selenium, on the other hand, has been found to be the best additive for some alloy steels[45,46] from the point

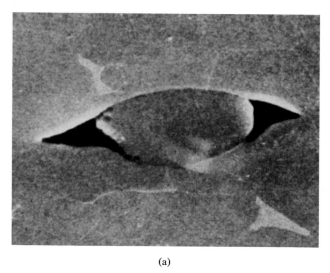

(a)

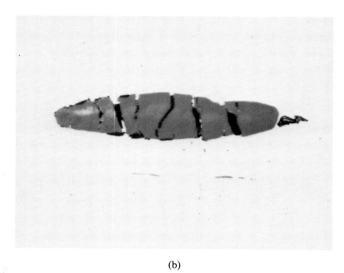

(b)

Fig. 13-7. Defects associated with manganese-sulfide inclusions in resulfurized steels that are present *prior* to machining: (a) voids produced by hot rolling (× 1500) (Courtesy, JISI[81]); and (b) cracks produced by bright drawing (× 800).

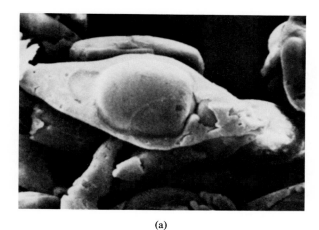

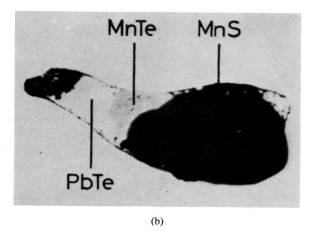

Fig. 13-8. Complex inclusions formed in tellurium-treated, resulfurized free-machining steels: (a) SEM of *extracted* inclusion showing metallic film on surface of manganese sulfide (× 1000); and (b) optical micrograph showing identity of tellurium-bearing phases in metallic film (× 2000). (Courtesy, *Jernkontornet Ann.*[103])

of view of machinability, and it is also useful in stainless free-machining grades[52,56,60] because the selenide inclusions do not have the same susceptibility to pitting corrosion as sulfides.

The use of selenium and tellurium in steel has been reviewed a number of times by Aborn,[18-21] and machinability studies have been reported by Tata and Sampsell[45,46] as well as others.[105-108]

Specially Deoxidized Steels and the Effects of Deoxidation Products. Oxides in free-machining steels can be divided into two categories based on their effects on machinability. One class contains the hard refractory oxides that find their way into steel from steelmaking slags, refractories, and deoxidation or reoxidation processes. Since these are undesirable, every effort is made to reduce them to the lowest possible level because of their detrimental effects on machinability. By comparison, deliberate and controlled additions of certain deoxidizers allow the formation of oxides that are physically and chemically quite different from the usual refractory oxides of silicon and aluminium. These form the second category of "specially deoxidized" steels, in which the machinability is enhanced rather than degraded by the presence of oxide phases. These two quite different effects of oxide phases will now be discussed.

Detrimental Oxides. The machinability of low-carbon free-machining steels has long been recognized as extremely sensitive to silicon content.[10,15] Quite apart from its indirect effect on sulfide shape through reduction of oxygen content, silicon has a direct effect on machining through the formation of hard oxide phases that rapidly wear cutting tools. It might be noted that the same detrimental effect is observed when aluminum is present as alumina or aluminates.[77,109] As discussed by Ramalingam and Watson,[110,111] abrasion is the most likely wear mechanism since the hardness of these oxide phases at metal-cutting temperatures satisfies the Kruschov criterion for significant wear to take place.

Ideally, free-machining steels should be free of oxides, but if any are present, they should be small and of low melting point. Oxides are usually found associated with the inclusions produced by the free-machining additives, most commonly as "tails" on the sulfides in resulfurized steels. The identity of these oxide phases depends on the chemical composition of the melt and the relative thermodynamic stabilities of the oxide phases. Type I sulfides in very low-silicon steels may dissove up to about 1.7-percent MnO,[112] above which the oxide appears as a separate phase. Even small amounts of silicon (~ 0.02 percent) give rise to the formation of silicate phases that solidify as a eutectic with manganese sulfide, giving rise to inclusions of the type shown in Fig. 13-9. As shown by Riekels,[77] manganese aluminate may also be found associated with sulfides, a condition extremely detrimental to machinability.

In stainless or other special steels, many other sources of hard inclusions are equally detrimental, including oxides, carbides, nitrides, and carbonitrides of chromium, molybdenum, titanium, zirconium, etc.[110]

Exogenous inclusions arising from physical entrapment of refractory materials are metallurgically unacceptable in almost all steel products and particularly deleterious to machinability. Studies of the erosion of various refractories by low-silicon, high-manganese steels of the 1214 type have been reported by Ramalingam and Watson[111] and indicate that manganese reduces silicates in the refractory, producing a soft glassy phase at the refractory/metal interface and releasing silicon to the metal. The soft glassy phase has a microstructure almost identical to that of subsurface exogenous inclusions [see Figs. 13-10(a) and (b)] often found in free-machining steels.[111] This fact strongly suggests that the glassy phase and/or other loosely adherent material may be swept into the steel by turbulent flow.[27,75] Furnace linings, ladles and nozzles, slags, holloware, and mould wash materials are

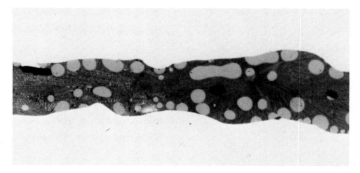

Fig. 13-9. Manganese sulfides associated in globular eutectic form with manganese silicate in high silicon heat (Si = 0.02 percent) of AISI 1215 ($\times$ 650).

all potential sources of the latter types of inclusions, whereas ferroalloy additions are a further possible source of silica and/or alumina unless special precautions are taken.

Specially Deoxidized Steels. The relatively recently developed medium-carbon steels deoxidized with "special" deoxidants (notably calcium) have been found to exhibit good machinability when cut with carbide tools. These steels were first discovered in Germany,[113] but since then enthusiastically developed in Japan.[114] Quite independently, and for different reasons, calcium deoxidation was also developed in the USA as a means of overcoming chronic machining problems with fine-grained gear steels contaminated with alumina inclusions.[3,115,116] During the last decade, calcium deoxidation has also been developed as one method for controlling sulfide shape in plate steels that are required to have good through-thickness properties.[117] There are, however, basic metallographic similarities in all of these steels, as follows:

- Oxides tend to be complex compounds either of calcia and alumina or of calcia, alumina, and silica and have relatively low melting points (compared to alumina).
- Calcium has a strong desulfurizing effect, and sulfides tend to be small, associated with the oxides, and relatively nondeformable.

Understanding the analogy that exists between low-carbon, resulfurized steels and medium-carbon, calcium-deoxidized steels is now possible.

Because medium-carbon steels require higher cutting forces than low-carbon steels, the cutting temperatures tend to be higher. Also, the advent of carbide cutting tools has allowed higher cutting speeds, which have pushed the cutting temperatures even higher. Under these conditions, the complex oxides of calcium can behave as deforming, nonabrasive inclusions that are capable of producing protective layers on the tool rake face.[113] The behavior, therefore, is entirely analogous to that of a resulfurized steel at a lower temperature.

Basically two types of calcium deoxidation practice have been developed,[116,118] both of which can be understood by reference to the phase diagram shown in Fig. 13-11. If deoxidation is effected by Ca–Al additions, the inclusions tend to be of the type, $x\text{CaO} \cdot y\text{Al}_2\text{O}_3$, with the values of x and y depending on the Ca:Al ratio.[119] As shown by Ramalingam et al.,[119,120] inclusions of the type, $\text{CaO} \cdot 6\text{Al}_2\text{O}_3$ (the first modification of alumina produced by calcium additions), form over a very small composition range, and the penalty for off-composition is the formation of alumina with resulting poor machinability. As reported by Tipnis et al.,[115,116] the Cal-Deox steels developed in the USA contained duplex inclusions with calcium-aluminate kernels surrounded by calcium-manganese sulfides.

The steels developed in Germany and Japan rely on calcium-silicon deoxidation, but the inclusions in the steels and the corresponding rake face smears (*Belag*) have a number of different compositions depending upon the technology used. As reviewed by Iwata,[4] the compositions of inclusions favored prior to 1970 were ternary oxides with the Gehlenite ($2\text{CaO} \cdot \text{Al}_2\text{O}_3 \cdot \text{SiO}_2$) and Anorthite ($2\text{CaO} \cdot \text{Al}_2\text{O}_3 \cdot 2\text{SiO}_2$) compositions, but since then work in Japan has demonstrated that other mixed oxide compositions are also capable of producing the desired machinability effects.[4,114]

Metallographic identification of inclusions in calcium-treated steels has been well-covered by Hilty[121] and Hilty and Farrell,[122,123] who have pointed out that identification of inclusions by conventional metallographic techniques is not simple and that X-ray analytical methods are almost indispensable for correct interpretation. Examples of the types of inclusions found in these steels are given in Fig. 13-12.

The identification difficulty arises because calcium aluminates in low-sulfur steels often appear glassy in reflected light and can easily be mistaken for silicates, even when viewed with polarized light where the typical silicate "optical cross" is observed. Quite apart from their X-ray spectra in electron optical instruments, calcium-bearing sulfides and aluminates may also be identified by their brilliant fluorescence under the electron beam.[123] Calcium aluminate appears as various shades of blue, whereas cal-

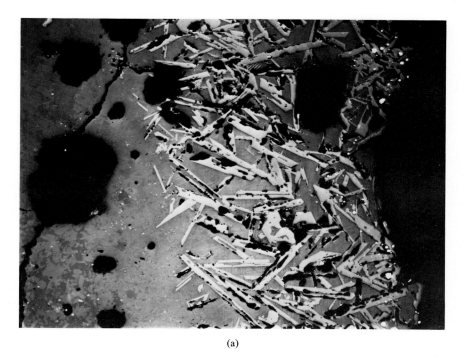

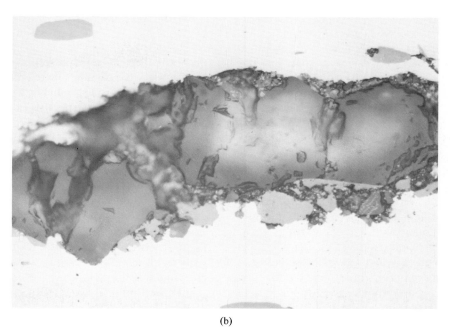

Fig. 13-10. Photomicrographs of (a) glassy phase alumina-silicates formed at refractory/metal interface of mullite crucible after 6 minutes contact with AISI 12L14 steel ($\times$ 200), and (b) complex subsurface inclusion in commercial AISI 12L14 and found to contain lead, manganese sulfide, and glassy manganese silicates and aluminates ($\times$ 800).

cium sulfide appears greenish-yellow. Mixed calcium/manganese sulfides fluoresce red-orange, and pure manganese sulfide does not fluoresce at all.

Boron-Treated Steels. Boron-treated steels will be discussed separately because very little is known about the role that boron plays in imparting machinability; historically, it has usually found use only as an element for improving hardenability. Although it may ultimately be shown that boron behaves simply as a deoxidant, until this is clearly established its chemical dissimilarity to the other free-machining additives is sufficient to warrant separate discussion here.

In studies of the effects of oxygen on the formation of manganese sulfides in resulfurized steels, Yeo[70] demonstrated that, although the machinabilities of heats treated

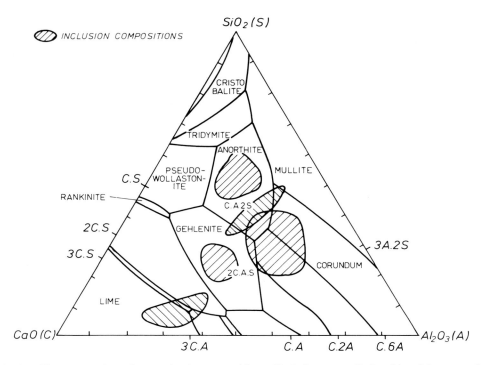

Fig. 13-11. Calcia-alumina-silica ternary phase diagram showing compositions of inclusions normally found in calcium-treated steels (After Iwata[4]).

with boron were lower than those of heats that were untreated, they were nevertheless considerably better than those of heats treated with strong deoxidizers like silicon and aluminum.

More recent studies by Reh[124] compared the performance of standard 1214 type steels with similar heats treated with lead or boron. These studies showed that the cutting forces for the boron-treated steels were intermediate between those of the leaded and nonleaded steels and that the tool lives were equivalent to those of the leaded steels.

Metallographic examination[124] has indicated that the boron is present as an oxide (as a tail on the sulfide in hot-rolled bar products) and that the sulfides appear to be "more globular" in the boron grade. This latter observation was also made by Yeo,[70] who claimed, in addition, that the degree of globularity was particularly favorable if the rolling temperature was high. Whether boron achieves its effect through modification of the sulfide shape by forming a nonabrasive oxide or by a combination of these effects is still not clear, and considerably more metallographic work seems to be required if its role in these types of steel is to be clarified.

QUANTITATIVE METALLOGRAPHY

Quantitative metallography can be divided into two parts: The first deals with the quantitative evaluation of discrete inclusions extracted from bulk samples, whereas the second considers the measurement of inclusion parameters on normal metallographic sections from which information about their spatial (three-dimensional) distribution can be deduced using probability theory applied to random sections of geometric shapes. This is the subject of stereology.[125-127]

ELECTROLYTIC INCLUSION EXTRACTION (EIE)

Sometimes referred to as the examination of "slimes," "residues," or "isolates," the technique of Electrolytic Inclusion Extraction (EIE) was originally devised by Klinger and Koch to separate inclusions and precipitates from their steel matrix.[128] It has subsequently been developed by others[129-131] to allow separation of extracted inclusions according to their type and has been further refined and applied specifically to free-machining steels by Müller et al.,[132] Kubinova,[133] and Wright.[134] A comprehensive bibliography of the subject is contained in the review by Vander Voort.[135]

As applied to free-machining steels, the advantages of the EIE method are that it provides the following type of sample:

- One amenable to gravimetric determination of inclusion content by type
- One ideal for assessing inclusion morphology and composition using SEM and EDA
- One that may provide information on the source of undesirable nonmetallic inclusions.

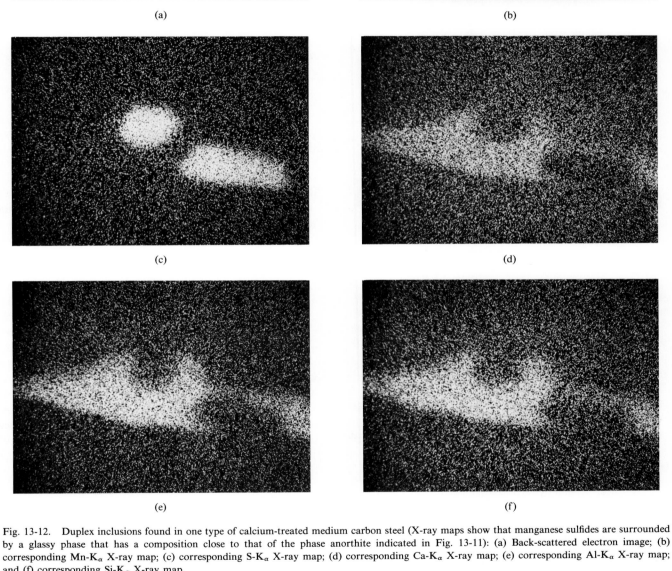

Fig. 13-12. Duplex inclusions found in one type of calcium-treated medium carbon steel (X-ray maps show that manganese sulfides are surrounded by a glassy phase that has a composition close to that of the phase anorthite indicated in Fig. 13-11): (a) Back-scattered electron image; (b) corresponding Mn-K_α X-ray map; (c) corresponding S-K_α X-ray map; (d) corresponding Ca-K_α X-ray map; (e) corresponding Al-K_α X-ray map; and (f) corresponding Si-K_α X-ray map.

Fig. 13-13. Large electrolytic cell (13-l capacity) used for extracting nonmetallic inclusions. Electrolyte: sodium citrate/potassium bromide/potassium iodide. Voltage: 20 V. Current density: 0.05 Acm^{-2}.

Using a modified Strohlein cell, Wright[134] has developed a technique that is particularly useful for the examination of low-carbon free-machining steels. The original 3 l extraction cell utilized an electrolyte of 5-percent sodium citrate, 1.2-percent potassium bromide, and 0.6-percent potassium iodide in aqueous solution, as originally proposed by Klinger and Koch,[128] and was operated at 20 V (or a current density of 0.015 Acm^{-2}), giving a dissolution rate of about 18 g of sample in a 24-hr period. A larger cell holding 13 l of electrolyte was subsequently designed (see Fig. 13-13) with several other modifications to the procedure, including:

- A secondary cell to remove metallic iron from the residue
- Centrifugal separation of the inclusions from the electrolyte
- Increased specimen size (up to 250 g)
- Increased current density (to 0.05 Acm^{-2})

This cell is capable of dissolving about 60 g of steel in 18 hr, providing a much faster sampling rate for routine analysis. This much larger specimen not only reduces the sampling errors but also allows anodes cut from different parts of bars and billets to be selectively examined.

This feature is very useful for tracing variations in inclusion distribution/content that arise from segregation effects, and a series of thin, concentric cylinders has been found to be particularly appropriate for selectively sampling free-machining steel-bar products.

The total inclusion content is determined by weighing the cleaned and dried residue and expressing this figure as a percentage of the weight loss of the anode. Separation of the oxides is achieved by treating the residue with hot hydrochloric acid; this dissolves the sulfides and allows the respective fractions of sulfide and oxide to be determined by difference.

The EIE method has increased significantly in value as a tool for examining inclusions with the increasing availability of scanning electron microscopes, especially those fitted with X-ray analytical facilities that are capable of providing *quantitative* elemental analysis on micron-sized particles. Examples of extracted sulfides and oxides examined in this way are given in Fig. 13-14.

A method of sizing extracted particles using a Coulter Counter has been described by Flinchbauch,[136] but methods are now available for doing this metallographically using AIA techniques. The latter also allows the inclusions to be imaged optically or with SEM. With either method, the true shape of single Type I sulfides in wrought billet or bar products can be shown to approach closely that of a prolate spheroid (an ellipsoid of revolution with the long axis being the symmetry axis parallel to the rolling direction) provided that the rolling temperature is high and the degree of deformation not too severe (see Figs. 13-2 and 13-14).

For a sample containing large numbers of sulfides, the "sizes" of these spheroids (i.e., their volumes) are not constant but are distributed about some mean value. In common with systems containing distributions of spherical particles, the frequency/"size" distribution of sulfides is probably of log-normal form,[137,138] with the mean size being determined by the local conditions prevailing during solidification. Similarly, the shapes (i.e., axial ratios or eccentricities) are not constant but are also distributed about some value. At least part of this variation is due to the size-dependence of the inclusion plasticity shown in Fig. 13-6.

To obtain a complete quantitative description of the inclusion distribution requires determination of the spatial, bivariate distribution function in which "size" and "shape" are plotted as either discrete or continuous functions of frequency of occurrence,[125,139,140] as shown in Fig. 13-15. Although it is, in principle, possible to derive data from the discrete case by direct measurement of inclusions released from bulk samples, it is impractical for several reasons:

- Extracted inclusions are usually randomly oriented and overlapping (Fig. 13-14); measurement is therefore difficult, although not impossible.

Fig. 13-14. Scanning electron micrographs of inclusions extracted from low-carbon free-cutting steels: (a) *globular* manganese sulfides ($\times$ 400); (b) *elongated* manganese sulfides ($\times$ 360); (c) silica particle ($\times$ 1700); (d) manganese aluminate-silicate particle ($\times$ 1000); and (e) manganese sulfide with alumino-silicate *tail* ($\times$ 750).

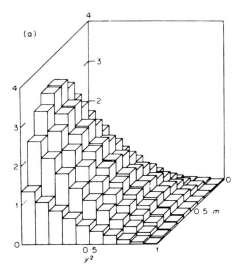 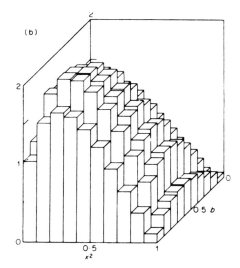

Fig. 13-15. Discrete size-shape distributions of prolate spheroids (b) and their sections (a) as given by Cruz-Orive.[139] The *size* parameter is given by b in the spatial distribution and by m in the section distribution, whereas the *shape* parameter is given by x^2 in the spatial distribution and y^2 in the section distribution. Note: The *size* parameter, b, is the length of the minor axis of an ellipsoid of major axis a. the *shape* parameter, x^2, is the square eccentricity of the ellipsoid defined by the equation, $x^2 = 1 - (b/a)^2$. (Courtesy, *J. Micros.*)

- Inclusions exposed using the fracture-surface method may present a biassed estimate of the distribution because the fracture path may be nonrandom, tending to follow the larger inclusions.

Combined image analysis by elemental composition, as well as by morphology, as has been described by Ekelund and Werlefors,[141] could be employed, but these methods are currently limited in their application to routine inspection because of their slow speed. Fröhlke[142] has shown that higher oxide volume fractions are often found in the surface regions of bar stock and that the EIE method gives results comparable with AIA carried out on metallographic sections, but, for the reasons discussed, size/shape distributions do not appear to have been derived from extracted inclusions.

For these and other historical reasons, it is more common to determine inclusion parameters using reflected light microscopy on longitudinal and transverse metallographic sections.

STEREOLOGICAL METHODS

The size, shape, and amount (and also the *distribution* of size and shape) of second-phase particles have long been recognized as factors influencing machinability. This is particularly true of resulfurized steels, for which published work often emphasizes, in a qualitative way, the beneficial effect of having large, "globular" manganese sulfides and the lowest possible amounts of refractory oxides.

As pointed out by Kovach[58] and Keane,[143] however, separation of the discrete and independent effects of sulfide size and sulfide shape has been satisfactorily demonstrated by few (if any) machinability studies. For example:

- Differences in machinability attributed to the small sulfide size in deoxidized steels may be due not to the size effect of the sulfide but to the presence of refractory oxides.
- Size and shape are almost certainly not independent parameters, as shown by Baker[81] and by Ramalingam and Watson.[111]
- The "true" (spatial) size and shape of an individual sulfide, or any other inclusion, can be revealed only by "releasing" it, using either the EIE technique or the ductile fracture method described by Thompson et al.[39,40]
- Large numbers of measurements are required for statistical accuracy.

Semiquantitative methods, which give inclusion distributions a numerical rating, are based on comparisons of representative photomicrographs with standard charts comprising a graded series of fields that contain inclusions of different size, shape, and amount. Such charts have been developed in several countries,[144-146] as reviewed by Vander Voort,[135] and a better-than-qualitative assessment of inclusion distribution can be obtained by comparing representative sample micrographs with these charts. Until fairly recently, these chart methods were the only alternative to carrying out quantitative metallography using the quite tedious methods of manual point counting and lineal or areal analysis. The latter do, however, generate section (area) information that is far less subjective than that provided by the chart methods.[147]

Volume Fraction

Effects. The beneficial effects of insoluble compounds in free-machining steels are usually found to saturate at some limiting volume fraction of inclusions. For example, in ingot-cast, low-carbon steels, machinability (as measured by flank wear on cutting tools) does not improve significantly once the sulfur content has reached ~ 0.3 wt.%.[109] Similarly, the effect of lead becomes less effective beyond levels of about 0.20 wt.%. These contents correspond to volume fractions of manganese sulfide and lead inclusions of approximately 0.015 and 0.0012, respectively, from which it can be seen that the effect of lead in particular is very large compared to the amount present.

This effect on machinability is closely paralleled by effects related to "friction" and the "smearing" of inclusions caused by the passage of the tool edge through the microstructure.[148] These observations provide strong circumstantial evidence that the enhancement of machinability comes about through a mechanism related to lubrication at the tool-chip interface,[57,97,103,138] but this evidence is not conclusive, and the opposing view—that inclusions act as stress concentrators providing sites for void nucleation and fracture—is still widely held.[79,95-97] Metallographic studies have yet to discriminate completely between these views, and it is an interesting field for further research.

Accuracy. The accuracy of volume-fraction determinations carried out by the methods of point counting and lineal and areal analysis have been assessed by a number of workers,[149,150] and these have been reviewed by Pickering,[151] who demonstrated the efficiency of the systematic, two-dimensional point count.

Because of their speed, modern AIA systems have the ability to measure large numbers of inclusions in relatively short times, thereby obviating some of the statistical problems of counting. Nevertheless, other major problems arise from inhomogeneity of samples,[35,152] and this is particularly a problem for free-machining steels since inclusions may readily segregate in large ingots.[27,74]

Comparison of metallographically determined volume fractions with those expected from considerations of bulk chemical analysis show that data from longitudinal metallographic sections give closer agreement with chemistry than data from transverse sections[153]; this is attributed to the inherent inaccuracies involved in measuring very small particles.

Size, Shape and Distribution

Size Effects. A large number of studies have attempted to show an effect of sulfide size on machinability,[17,76,154-156] and in all but one case[156] machinability was reported to improve with increasing inclusion size.

The majority of these investigations have not employed rigorous stereological analysis but indirect methods instead. For example, Pickering and Gladman[154] and Marston and Murray[155] have shown that machinability improves in resulfurized steels as the number of inclusions in a given area decreases (i.e., as average size increases). This approach is valid only if the steels being compared have equal volume fractions and similar shape factors, as demonstrated below using standard stereological relationships.[125-127]

Since

$$V_V = A_A \qquad (13\text{-}1)$$

and

$$N_A = A_A / \overline{A} \qquad (13\text{-}2)$$

where

V_V = volume fraction of inclusions
A_A = area fraction of inclusions on section
N_A = number of inclusions per unit area observed on section
$\overline{A}$ = mean intercepted area of inclusions

Then, substituting V_V for A_A in Eq. 13-2 indicates that if the volume fraction is increased at constant mean intercept area (i.e., constant size), then the number of inclusions observed per unit area increases. This is self-evident.

Consider, on the other hand, an ingot through which any random section plane contains N_A soft spherical inclusions. If this ingot is rolled, the originally spherical inclusions are deformed into ellipsoids, and mean intercept area $\overline{A}$ becomes progressively smaller in the transverse section and larger in the longitudinal section. The values of N_A vary correspondingly in the two sections. N_A is therefore shape- or deformation-dependent and is an accurate estimate of "size" only in very special circumstances.

Shape Effects. Measurement of the mean aspect ratios of spheroidal inclusions seems intuitively appealing as a way of determining shape, and this has been used in a number of studies on the effect of shape on machinability.[76,155]

The universal conclusion of these investigations has been that lower aspect ratios favor machinability. As pointed out by Fröhlke,[158] however, measuring "shape" using aspect ratios at different mean inclusion size is misleading for the same reason (but in reverse) as measuring "size" with different mean inclusion shape.

In a purely comparative and nonquantitative study of shape effects, Garvey and Tata[10] rolled two billets from the same cast through different hot-roll schedules and were able to demonstrate that the resulting materials had different machinability. On the assumption that this experiment obviated the effects of variable chemistry and inclusion size and distribution, the differences in machinability were attributed to the differences in sulfide shape. This appears to be the only study in which shape effects have been dissociated from size effects, but unfortunately the results were not reported quantitatively nor was the statistical reliability tested with repeat experiments.

Combined Size/Shape Effects. Fröhlke[158-160] has attempted to overcome some of the problems outlined above by developing a method of measuring inclusions and graphically plotting the data. Although this method is claimed to reflect changes in size and shape, it paradoxically employs only transverse metallographic sections of wrought products and requires measuring the "diameters" of all the sectioned inclusions. These are then plotted in size categories against the logarithm of frequency of occurrence. The data are generally found to fall on a straight line of slope $-m$ and intercept N, which Fröhlke calls the "manganese sulfide identification line" (or MIL) according to the equation:

$$\ln N_i = \ln N_o - m \cdot d_i \quad (13\text{-}3)$$

where

N_i = number of inclusions in i^{th} size category
d_i = maximum "diameter" in i^{th} size category
N_o = intercept at zero diameter
m = slope

Using measured machinability rankings and regression analysis, the following equation relating machinability to the inclusion parameters was also derived by Fröhlke[159]:

$$\ln V_{60} = A + B \cdot m + C \cdot \ln N_o - D \cdot \ln F_{ox} \quad (13\text{-}4)$$

where

V_{60} = sixty-minute tool life for a given wear criterion
F_{ox} = volume fraction of oxide inclusions
m, N_o = MIL parameters
A, B, C, D = constants

A similar equation has been theoretically derived by Ramalingam and von Turkovich.[161] Riekels[77] has also demonstrated the same general relationship using regression analysis of machining data. Yamaguchi[162] noted the apparent discrepancy in the sign of m given by Fröhlke's equations, and it seems that a transformation of m is necessary to obtain Eq. 13-4 from Eq. 13-3.

Other difficulties with the MIL are the covert limitations of its applicability. For example, Eq. 13-3 suffers from the following restrictions:

- It is strictly appropriate only to an exponential relationship between number (frequency) and size.
- If used to describe distributions that are log-normal, it fits only the tail of the distribution.
- It cannot treat bimodal distributions.

Nevertheless, Fröhlke's work has provided a significant contribution to the linking of microstructure and machinability, and it does appear to provide a relatively simple quantitative method for describing sulfide distributions in high-sulfur steels. Experiments with rolled bars, drawn wires, and tellurium treatment have shown that m and N_o appear to vary systematically in the expected fashion.[158,160]

Mathematically more rigorous methods for determining the true spatial size/shape distributions have been developed by Wicksell,[163] De Hoff,[164] Evans and Clarke,[165] and Cruz-Orive,[139,140] and these have been reviewed by Weibel.[125] The most general solution is that given by Cruz-Orive. This is capable of treating random and nonrandom dispersions of particles, the sizes and shapes of which are distributed in dependent or independent ways. Although potentially capable of overcoming many of the problems raised earlier in this section, these sophisticated stereological techniques do not yet appear to have found widespread application to studies of free-machining steels.

METALLOGRAPHIC TECHNIQUES APPLIED TO THE METAL CUTTING PROCESS

Quick-Stop Tests

Microscopy has made a significant contribution to understanding of machinability through studies of the processes taking place at the tool tip using so-called "quick-stop tests." In this technique, the metal-cutting operation is interrupted by extremely rapid removal of the cutting tool part way through the cut, so that the workpiece is left with an adhering chip that is effectively "frozen" at the moment the tool is retracted. The external features of the chip and chip root are best revealed by scanning electron microscopy, as shown in Fig. 13-16(a), whereas the metallographic features of the bulk deformation processes that were taking place at the last instant of cutting can be examined *post factum* by sectioning the sample in a plane normal to tool edge and using conventional metallographic procedures to prepare and examine the specimen, as shown in Fig. 13-16(b). Because fine feeds are ordinarily used in these tests, the chips are quite thin, and good practice is to electroplate the specimen prior to grinding and polishing to ensure good edge retention.[25]

Although it is possible to obtain stopped-chips using jigs attached to the crosshead of a tensile test machine,[166] deceleration effects cannot be eliminated, and much better methods using explosive devices with shear pin arrangements are available to impact the tool rapidly away from the workpiece.[167,168]

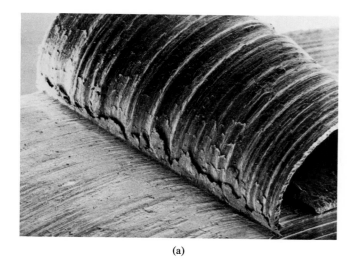

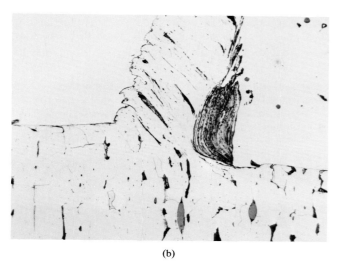

Fig. 13-16. Features of chip formation in AISI 12L14 steel revealed by quick-stop test at a cutting speed of 90 m/min: (a) SEM showing rear face of chip, as-machined workpiece surface, and the irregular nature of the built-up-edge ($\times$ 25). Note the effect on the surface finish of the rear face of the chip. (b) Optical micrograph of transverse section through chip root showing microstructural features of workpiece, chip, and built-up-edge ($\times$ 250). (Both courtesy, E. D. Doyle.)

Applied to free-machining steels, quick-stop tests have yielded considerable information with respect to the processes taking place in primary and secondary shear zones and the built-up-edge.[94,96,169-172]

The essential features revealed by these studies are the following:

- Severe shear takes place in a plane (or, more correctly, a zone) extending from the tool tip to the workpiece surface, called the "primary shear zone." As soft inclusions pass through this zone, they are deformed into thin sheets that lie approximately parallel to the shear plane. They are, therefore, disposed in just the correct orientation to promote chip fracture (chip breaking) when the curling chip is impeded in its motion. The analogy with lamellar tearing in plate steel has been noted by Watson and Brown.[94] Several studies have pointed to void formation around second-phase particles and suggested that this is a major contributor to the free-machining effect.[79]

- At the rear face of the chip, where it contacts the rake face of the tool, friction gives rise to further shearing forces in the sliding chip to form the so-called "secondary shear zone," in which free-machining inclusions are further deformed.[94,106,169] Exposure of these soft phases at the tool/chip interface occurs, and this is the basis of the "lubrication" theory that offers an alternative explanation of the free-machining effect of inclusions.[108]

- Under certain cutting conditions (speed/feed, etc.), free-machining steels form chips with a large dead zone or built-up-edge (BUE). This significantly affects the machined surface finish, cutting forces, chip thickness, cutting temperature, and, ultimately, tool life. As shown by Milovic and Wallbank,[170] the effect of free-machining additives is to cause the built-up-edge to be preserved at higher cutting speeds than in equivalent plain carbon steels. Although this shifts the "unstable" region of the force versus cutting speed to higher cutting-speed values, overall the cutting forces and cutting temperatures are lowered, with a beneficial effect on tool life.

Abeyama and Nakamura[173] have measured BUE heights and used empirical methods to develop relationships between cutting temperatures and surface finish, but the wavelike nature of the BUE caused by its transient instability (see Fig. 13-16) means that many measurements must be taken if statistical variations are to be overcome and the results meaningful.

Direct Observation of the Cutting Process

Direct evidence of the metallographic processes taking place during machining has been obtained by the use of cine-film techniques applied to a specially constructed optical microscope[174] and by *in situ* machining in the SEM.[4,175-179] Although both methods represent very elegant uses of modern metallographic tools and provide extremely fascinating insights into the behavior of microstructure under machining conditions, it is arguable whether all of the observed features are relevant to "real" metal cutting for the following reasons:

- Both techniques rely on observation of processes taking place at the free surface where the state of stress is almost certainly *not* representative of that of the bulk.
- In the case of the SEM studies, only very fine feeds and low speeds can be studied, and this has already been shown to cause quite different BUE conditions.

- When using the SEM, cutting can be carried out only in vacuum, which is likely to affect tool/chip interface characteristics.

For these reasons, the observations of cracking and void formation around inclusions in free-machining steels[96,176,178] that have been interpreted as mechanistic evidence of the role played by inclusions must still be of questionable significance.

Metallographic Studies of Cutting Tools

Wright and Trent[180] have demonstrated how metallography can be used to estimate the steady-state temperature gradient existing in high-speed steel tools during metal cutting by observing the state of temper in zones beneath the tool tip after cutting has ceased. This technique has been used by Milovic and Wallbank[170] to demonstrate that free-cutting steels machine at lower temperatures than their nonfree-cutting equivalents.

Microscopy has also been used extensively to study the worn surfaces of tools used to machine various grades of free-cutting steels. Tipnis and Joseph,[181] have shown, *inter alia,* that groove formation on the flank face is a feature that is dominant in the cutting of leaded steels and is ultimately responsible for tool failure (on either a part-growth or surface-roughness criterion).

Identification of surface films on the rake face crater attributed to deposition of the free-cutting additives has been carried out using electron optical methods of analysis (EDA or WDA). The presence of lead, sulfur, and, in particular, calcium-containing deposits from calcium deoxidized steels has been well documented[4,39,106,113] using these techniques, again pointing to the surface-related effects of the free-machining additions. These effects vary according to the additive and the tool material, and the question as to whether the rake face deposits act as diffusion barriers, friction-lowering solid lubricants, or both is not yet clear.

CONCLUSION

Metallography has provided valuable insights into the deformation processes taking place during metal cutting and, in particular, has identified the micro-mechanisms by which certain additives produce beneficial effects on machining behavior. It has also allowed the deleterious effect of refractory oxides to be explained in a plausible manner, thus paving the way for development of improved steelmaking practices. These contributions to an understanding of *machinability* are unique and could not have been made by the traditional engineering approach based on the mechanics of metal cutting.

The two approaches are complementary, however, and a total picture will emerge only as workers in these respective science and engineering disciplines learn to understand the theory and practices of each other. The microscopist committed to understanding machinability should therefore be prepared to devote time to understanding the machining process.

Acknowledgements. The author is grateful to the many colleagues who contributed material for this publication and assisted with preparation of the manuscript. The permission of The Broken Hill Proprietary Co. Ltd. to publish this work is also acknowledged.

BIBLIOGRAPHY

Sulfide Inclusions in Steel. ASM Materials/Metalworking Series No. 6 (1975).

Influence of Metallurgy on Machinability. ASM Materials/Metalworking Series No. 7 (1975).

Proc. Int. Symp. on Influence of Metallurgy on Machinability of Steel. Joint ISIJ/ASM Publication (1977).

The Machinability of Engineering Materials. Metals Park, OH: ASM (1983).

Machinability. ISI Special Report No. 94. London: Iron and Steel Institute (1967).

Non-Metallic Inclusions in Steel. ISI Special Report No. 90. London: Iron and Steel Institute (1964).

Non-Metallic Inclusions in Steel, Part II. ISI Special Report No. 100. London: Iron and Steel Institute (1966).

Non-Metallic Inclusions in Steel, Part III. ISI Special Report No. 115. London: Iron and Steel Institute (1968).

REFERENCES

1. Thomsen, E. G. "F. W. Taylor—A historical perspective." In *On the Art of Metal Cutting—75 Years Later.* New York: ASME, 1982.
2. Oxley, P. L. B. Machinability—A mechanics of machining approach. In *On the Art of Metal Cutting—75 Years Later.* New York: ASME, 1982.
3. Tipnis, V. A. Influence of Metallurgy on Machining—Free Machining Steels. In *On the Art of Metal Cutting—75 Years Later.* New York: ASME, 1982.
4. Iwata, K. The effect on machinability of infinitesimal amounts of elements in steel. *Proc. Int. Conf. on Production Technology.* Inst. of Eng. (Aust.), Nat. Conf. Publ. 74/3:183 (1974).
5. Ikeda, T. and Ichihashi, H. Solubility of lead in molten steel and its segregation during solidification. *Trans. I.S.I.J.* 20(9):B-372 (1980).
6. Araki, T. Observations on the dissolution and dispersion phenomena of lead in molten steel. *Trans. Nat. Res. Inst. for Metals* 5(3):14 (1963).
7. Morozov, A. N., and Gol'dshtein, Y. A. Steels Containing Lead. *Met. Sci. and Heat Treatment* 22(11–12):779 (1980).
8. Hansen, M. *Constitution of Binary Alloys.* New York: McGraw-Hill, 1958.
9. Watson, J. D., and Davies, R. H. The effect of nitrogen on the machinability of low carbon free machining steels. *J. App. Metalworking* 3, No. 2:110 (1984).
10. Garvey, T. M., and Tata, H. J. Factors affecting the machinability of low-carbon free-machining steels. *Mechanical Working of Steel II.* AIME Met. Soc. Conf. No. 26:99 (1965).
11. Garvey, T. M., and Tata, H. J. Machinability and metallurgy of resulfurized low-carbon free-machining steels. ASTME Conf. Paper EM66-180 (1966).
12. Aylward, P. T. Machinability as influenced by the composition,

microstructure, and cold extrusion of steel. SME Int. Auto. Eng. Conf. Paper 730113 (1973).
13. Murphy, D. W., and Aylward, P. T. *Machinability of Steel.* Bethlehem Steel Corp., 1965.
14. Feigenbaum, S. Open hearth screw steels replacing Bessemer. *J. Metals* 15:796 (1963).
15. Boulger, F. W.; Moorhead, H. A.; and Garvey, T. M. Superior Machinability of MX Explained. *Iron Age* 167:90 (1951).
16. Paliwoda, E. The influence of chemical composition on the machinability of rephosphorized open hearth screw steel. *Trans. ASM* 47:680 (1955).
17. Van Vlack, L. H. Correlation of machinability with inclusion characteristics in resulfurized Bessemer steels. *Trans. ASM* 45:741 (1953).
18. Aborn, R. H. *The Role of Metallurgy, Particularly Bismuth, Selenium, and Tellurium on the Machinability of Steels.* New York: American Smelting and Refining Company, 1968.
19. Aborn, R. H. Role of additives in the machinability of steel. *Proc. Int. Symp. on Influence of Metallurgy on Machinability.* Joint ISIJ/ASM Publication: 381 (1977).
20. Aborn, R. H. Trace additives: Introduction to bismuth, selenium and tellurium in iron and steel. In *Trace Additives: Bismuth, Selenium, and Tellurium in Iron and Steel.* Metals Park, OH: ASM, 1970.
21. Aborn, R. H. The role of selenium and tellurium in ferrous metals. In *Selenium and Tellurium in Iron and Steel.* Stockholm: Institutet für Metallforskning, 1969.
22. Leslie, R. T. and Lorenz, G. Tool life exponents in light of regression analysis. *Annals of the C.I.R.P.* 12:266 (1963–64).
23. Bunge, G. Effect of nitrogen on the machinability of free cutting steels. *Neue Hütte* 15, No. 7:420 (1970).
24. Ohno, T.; Takahashi, M.; Sakai, T.; and Hamahata, S. Simplified measurement of tool life by continuous acceleration of cutting speed. In *Machinability Testing and Utilization of Machining Data*, p. 164. ASM Materials Metalworking Technology Series, 1979.
25. Samuels, L. E. *Metallographic Polishing by Mechanical Methods.* 3rd ed. Metals Park, OH: ASM, 1982.
26. Chalfant, G. M. Revealing lead inclusions in leaded steel. *Metal Progr.* 78, No. 3:77 (1960).
27. Blank, J. R., and Johnson, W. Nature and distribution of inclusions in leaded steels, Parts I–III. *Steel Times,* July 23:110; July 30:148; Aug. 6:176 (1965).
28. Bardgett, W. E., and Lismer, R. E. Mode of occurrence of lead in lead-bearing steels and the mechanism of the exudation test. *JISI* 151, No. 1:281 (1945).
29. Gerds, A. E., and Melton, C. W. New etch spots leaded steels. *Iron Age* 178, No. 9:86 (1956).
30. Schofield, T. H. The microscopical examination of samples of lead-bearing and lead-free steels and ingot irons. *JISI* 151, No. 1:277 (1945).
31. Samuels, L. E. *Optical Microscopy of Carbon Steels.* Metals Park, OH: ASM, 1980.
32. Volk, K. E. The metallographic determination of lead in steels. *Arch. Eisen.* 16:81 (1942).
33. Forgeng, W. D., and Lee, A. G. Applications of automatic specimen preparation techniques in the metallography of steel. In *Metallography As a Quality Control Tool,* p. 89. New York: Plenum Press, 1980.
34. Martensson, H., and Persson, S. An automatic apparatus for the grinding and polishing of metallographic specimens. *Pract. Metallog.* 7:691 (1970).
35. Allmand, T. R., and Coleman, D. S. Technical problems in assessing non-metallic inclusions in steel for quality control. *Microscope* 20:57 (1972).
36. Baumann, R. Sulfur in iron. *Metallurgie* 3:416 (1906).
37. Desch, C. H. *Metallography.* 4th ed. Longmans Green and Company, 1937.
38. *Metallographic Structure and Phase Diagrams. Metals Handbook,* vol. 8. Metals Park, OH: ASM, 1973.
39. Thompson, R. W.; Quinto, D. T.; and Levy, B. S. Scanning electron microscopy and Auger electron spectroscopy observations on the role of lead in the machining of steel. *Proc. 2nd North American Metalworking Research Conf.,* p. 545, Madison, WI (1974).
40. Thompson, R. W., and Levy, B. S. On the morphology of lead in as-cast and rolled steel. *Mechanical Working and Steel Processing, XIV,* p. 291. AIME Met. Soc. Conf. (1976).
41. Dench, W. A.; Pugh, D. J.; and Stoddart, C. T. H. The metallography of lead and manganese sulfide inclusions in free cutting steel. *Proc. Modern Metallography Conference and Exhibition.* Metals Society (1976).
42. *Making, Shaping and Treating of Steel.* US Steel Corporation, 1971.
43. *USS Mach-5 Free Machining Steel.* US Steel Corporation, 1962.
44. Tata, H. J. The effect of special additives on the machinability of steel. In *Trace Additives: Bismuth, Selenium, and Tellurium in Iron and Steel,* p. 99. Metals Park, OH: ASM, 1970.
45. Tata, H. J., and Sampsell, R. E. Machinability of AISI 4140 Type steels maximized by selenium. *Mechanical Working and Steel Processing, X,* p. 378 AIME Met. Soc. Conf., 1972.
46. Tata, H. J., and Sampsell, R. E. Effect of additives on the machinability and properties of alloy steel bars. SAE Int. Automotive Engineering Congress, Paper No. 730114 (1973).
47. Bhattacharya, D. New free machining steels with bismuth—Pt. I Development. *Mechanical Working and Steel Processing, XVIII,* p. 153. AIME Met. Soc. Conf., 1980.
48. Schmitt, W. B. New free machining steels with bismuth—Pt. II Evaluation. *Mechanical Working and Steel Processing, XVIII,* p. 168. AIME Met. Soc. Conf., 1980.
49. Schrader, C. F. and Geiger, L. J. Statistical study on the effect of chemical composition on the machinability of 12L14. *Mechanical Working of Steel, II,* p. 83. AIME Met. Soc. Conf. No. 26, 1965.
50. Podgornik, A., and Smolej, A. Selective microradiography and electron microanalysis of free-cutting steels. *Mining and Metallurgy Quarterly,* Nos. 2 and 3:109 (1972).
51. Ramalingam, S.; Basu, K.; and Hazra, J. The role of lead and its effect on MnS inclusions leaded free cutting steels. *Proc. 3rd North American Metalworking Research Conf.,* p. 374. Pittsburgh, Pa., 1975.
52. Mills, B. Machinability of stainless steels. *Metallurgist and Materials Technologist* 6, No. 6:259 (1974).
53. Mostovoy, S., and Breyer, N. N. The effect of lead on the mechanical properties of 4145 steel. *Trans. ASM* 61:219 (1968).
54. Swinden, T. Leaded manganese molybdenum steel. *JISI* 148, No. 2:441 (1943).
55. Woolman, J., and Jacques, A. Influence of lead additives on the mechanical properties and machinability of some alloy steels. *JISI* 165:257 (1950).
56. Levy, B. S.; Thompson, R. W.; Henger, G. W., and Balakrishnan, M. V. A comparison of the machinability and inclusion morphology of AISI 12L14 steel produced by an ingot or ladle lead addition technique. In *Influence of Metallurgy on Machinability,* p. 89. ASM Materials/Metalworking Technology Series No. 7, 1975.
57. Kiessling, R., and Lange, N. *Non-metallic Inclusions in Steel, Part II.* Iron and Steel Institute Publication No. 100, 1966.
58. Kovach, C. W. Sulfide inclusions and the machinability of steel. In *Sulfide Inclusions in Steel,* p. 459. ASM Materials/Metalworking Technology Series No. 6, 1975.
59. Kovach, C. W., and Eckenrod, J. J. Free machining austenitic stainless steels. *Mechanical Working and Steel Processing, IX,* p. 300. AIME Met. Soc. Conf., 1971.
60. Kovach, C. W., and Moskowitz, A. Effects of manganese and sulfur on the machinability of martensitic stainless steels. *Trans. AIME* 245:2157 (1969).

61. Wells, R. G. Metallographic techniques in the identification of sulfide inclusions in steel. In *Sulfide Inclusions in Steel,* p. 123. ASM Materials/Metalworking Technology Series No. 6, 1975.
62. Sims, C. E., and Dahle, F. B. Effect of aluminium on the properties of medium carbon cast steel. *Trans. Amer. Foundrymen's Assoc.* 46:65 (1938).
63. Dahl, W. H.; Hengstenberg, H.; and Düren, C. Conditions of formation of various types of sulfide inclusions. *Stahl und Eisen* 86:782 (1966). (Brutcher Translation HB 6828.)
64. Sims, C. E. The nonmetallic constitutents of steel. *Trans. AIME* 215:367 (1959).
65. Hilty, D. C., and Crafts, W. Liquidus surface of the Fe–S–O system. *J. Metals* 5, No. 12:1307 (1952).
66. Yarwood, J. C.; Flemings, M. C.; and Elliot, J. F. Inclusion formation in the Fe–O–S system. *Met. Trans.* 2:2573 (1971).
67. Turkdogan, E. T., and Kor, G. J. W. Sulfides and oxides in Fe–Mn alloys: Part I, Phase relations in Fe–Mn–S–O system. *Met. Trans.* 2:1561 (1971).
68. Fredriksson, H., and Hillert, M. Entectic and monotectic formations of MnS in steel and cast iron. *JISI* 209:109–113 (Feb. 1971).
69. Turkdogan, E. T. Deoxidation of steel. *JISI* 210, No. 1:21 (1972).
70. Yeo, R. B. G. The effect of oxygen in resulfurized steels—Parts I and II. *J. Metals,* June: 29; July: 3 (1967).
71. Fredriksson, H., and Hillert, M. On the formation of manganese sulfide inclusions in steel. *Scand. J. Met.* 2:125 (1973).
72. Mohla, P. P. and Beech, J. Effect of cooling rate on the morphology of sulfide inclusions. *JISI* 207:177 (1969).
73. Paliwoda, E. J. The role of oxygen in free cutting steels. *Proc. 6th Mechanical Working and Steel Processing Conf.,* p. 27. AIME Met. Soc. Conf., 1964.
74. Carney, D. J. and Rudolphy, E. C. Examination of a high sulfur free-machining ingot, bloom and billet sections. *J. Metals* 6, No. 8:999 (1953).
75. Gaydos, R. Free machining steels and the effects of sulfide and siliceous nonmetallics. *J. Metals* 16, No. 12:972 (1964).
76. Clayton, D. B., and Brown, J. R. Sulfides in free-machining steel. *Iron and Steel* 42, No. 4:219 (1969).
77. Riekels, L. P. Inclusion characteristics in resulfurized AISI 1215 free machining steels with varying manganese contents. In *The Machinability of Engineering Materials,* p. 42. Metals Park, OH: ASM 1983.
78. Pickering, F. B. Identification of inclusions. *Iron and Steel* 30, No. 1:3 (1957).
79. Tipnis, V. A., and Cook, N. H. Influence of MnS—Bearing inclusions on flow and fracture in a machining shear zone. In *Mechanical Working and Steel Processing, IV,* p. 13. AIME Met. Soc. Conf. No. 44, 1965.
80. Steinmetz, E., and Lindenberg, H. V. Morphology of iron-manganese sulfides and oxysulfides. *Arch. Eisen.* 47, No. 9:521 (1976). (BISIT No. 15898.)
81. Baker, T. J., and Charles, J. A. Deformation of MnS inclusions in steel. *JISI* 210:680 (1972).
82. Banks, T. M., and Gladman, T. Sulfide shape control. *Metals Technology* 6, No. 3:81 (1979).
83. Maunder, P. J. H., and Charles, J. A. Behaviour of non-metallic inclusions in a 0.2% carbon steel ingot during hot rolling. *JISI* 206:705 (1968).
84. Vodopivec, F., and Gabrovsek, M. Relative plasticity of manganese sulfide inclusions during hot rolling of some industrial steels. *Metals Technology* 7, No. 5:186 (1980).
85. Malkiewicz, T., and Rudnik, S. Deformation of non-metallic inclusions during rolling of steel. *JISI* 201:33 (1963).
86. Segal, A., and Charles, J. A. Influence of particle size on deformation characteristics of manganese sulfide inclusions in steel. *Metals Technology* 4, No. 4:177 (1977).
87. Brunet, J. C., and Bellot, J. Deformation of MnS inclusions in steel. *JISI* 211:511 (1973).
88. Di Gianfrancesco, E., and Filippi, P. Use of automatic image analysis to characterize morphological parameters of manganese sulfides in free machining steels. *La Metallurgia Italiana* 9:434 (1976). (BISIT No. 17370.)
89. Bellot, J., and Gantois, M. The influence of sulfur and tellurium compounds on the hot deformability and mechanical properties of steels. *Trans. ISIJ* 18:536 (1978).
90. Bhattacharya, D. The effect of cerium on the machinability and mechanical properties of AISI 1144 steels. In *The Machinability of Engineering Materials,* p. 65. Metals Park, OH: ASM, 1983.
91. Yamaguchi, Y.; Shimohata, T.; Kaneda, T.; and Furasawa, S. Zirconium-treated resulfurized steels with improved machinability and cold forgeability. *Proc. Int. Symp. on Influence of Metallurgy on Machinability of Steel,* p. 289. Joint ISIJ/ASM Publication, 1977.
92. Philbrook, W. O. Oxygen reactions with liquid steel. *Int. Met. Rev.* 22:187 (1977).
93. Jaffrey, D. The failure of sulfide inclusions in steel under uniaxial stress or strain. *Metals Forum,* 5, No. 4:217 (1982).
94. Watson, J. D., and Brown, G. G. Effects of steel chemistry on chip formation in metal cutting. *Proc. Fourth Tewksbury Symposium on Fracture.* University of Melbourne, 1979.
95. Brown, R. H., and Luong, H. S. Observations of deformation in the orthogonal machining of mild steel. *Metals Technology* 2:2 (1975).
96. Brown, R. H., and Luong, H. S. The influence of microstructure on discontinuous chip formation. *Annals. of the CIRP* 25/1:49 (1976).
97. Poirer, D. R., and Kieras, A. P. The anisotropic cutting behaviour of free-machining steels. *J. Engng. Ind.* (*Trans. ASME Ser. B*) 97:1094 (1975).
98. Kiessling, R.; Hässler, B.; and Westman, C. Selenide-sulfide inclusions and synthetic compounds of the (Mn, Me)(S, Se)–Type. *JISI* 205:531 (1967).
99. Kiessling, R. Basic studies on the effect of Se and Te on the machinability of steel. In *Selenium and Tellurium in Iron and Steel.* Institutet für Metallforskning, 1969.
100. Malmberg, T.; Runnsjo, G.; and Aronsson, B. The addition of selenium and tellurium to carbon steels: Their recovery and effect on inclusions and machinability. *Scand. J. Met.* 3:169 (1974).
101. Thompson, R. W.; Coward, M. D.; and Levy, B. S. An assessment of the variability in machining behaviour in low-carbon resulfurized free machining steels. Society Automotive Engineers Paper No. 730116. Int. Automotive Eng. Congress, Detroit, Mich., 1973.
102. Ramalingam, S.; Thomann, B.; Basu, K.; and Hazra, J. The role of sulfide type and of refractory inclusions on the machinability of free cutting steels. In *Influence of Metallurgy on Machinability,* p. 111. ASM Materials/Metalworking Technology Series No. 7, 1975.
103. Gustafsson, J., and Carlsson, T. Isolation of inclusions in free machining steel. *Jernkont. Ann.* 154:285 (1970).
104. Schmitt, L. B. Tellurium: Rx for machinability. *American Machinist* (June 7, 1965).
105. Araki, T.; Yamamoto, S.; and Uchinaka, Y. Microscopic observations on the machining behaviour of mild steels containing telluride, sulfide, and selenide. *Tetsu-to-Hagane* 54, No. 4:444 (1968).
106. Tipnis, V. A., and Poole, S. W. The effect of selenium additions on the machinability of low carbon resulfurized steels. In *Trace Additives: Bismuth, Selenium and Tellurium in Iron and Steel* Metals Park, OH: ASM, 1969.
107. Becker, G. Free-cutting steels for serial production. *Thyssen Technische Berichte,* No. 2:59 (1977).
108. Kishi, K., and Eda, H. The lubrication and deformation mechanism of MnTe, MnS, MnSe, and Pb inclusions in various steels during wear and cutting processes. *Wear* 38:29 (1976).
109. Blank, J. R.; Naylor, D. J.; and Wannell, P. H. Improved and more consistent steels for machining. *Proc. Int. Symp. on Influence*

of Metallurgy on Machinability, p. 397. Joint ISIJ/ASM Publication, 1977.
110. Ramalingam, S., and Watson, J. D. On machinability variations, tool wear and steel making practice. *Proc. Int. Symp. on Influence of Metallurgy on Machinability,* p. 67. Joint ISIJ/ASM Publication, 1977.
111. Ramalingam, S., and Watson, J. D. Steelmaking, microstructure, and machinability. *Proc. 22nd Mechanical Working and Steel Processing Conf.,* p. 94. AIME, 1980.
112. Chao, H. C.; Smith, Y. E.; and Van Vlack, L. H. The MnO–MnS Phase Diagram. *Trans. Met. Soc. AIME* 227:796 (1963).
113. Opitz, H. and Konig, W. Machinability differences between various heats of grade Ck 45 steels. *Arch. Eisen.* 33, No. 12:831 (1962).
114. Sata, T.; Murata, R.; and Akasawa, T. Review of recent research and development on machinability of steel in Japan. *Proc. Int. Symp. on Influence of Metallurgy on Machinability,* p. 357. Joint ISIJ/ASM Publication, 1977.
115. Tipnis, V. A.; Joseph, R. A.; and Doubrava, J. H. Improved machining steel through calcium deoxidation. *Metals Engineering Quarterly* (Nov. 1973).
116. Tipnis, V. A. Calcium treatment for improved machinability. In *Sulfide inclusions in steel,* p. 480. ASM Materials/Metalworking Technology Series No. 6, 1975.
117. Gray, J. M., and Wilson, W. G. Effect of processing variables on the properties of low-carbon niobium steel proposed for arctic pipelines. *Mechanical Working and Steel Processing, X,* p. 199. AIME Met. Soc. Conf., 1972.
118. Bellot, J. Steels with improved machinability. *Met. Sci. and Heat Treatment* 22, No. 11–12:794 (1981).
119. Faulring, G. M., and Ramalingam, S. Inclusion precipitation diagram for the Fe–O–Ca–Al system. *Met. Trans.* 11B:125 (1980).
120. Ramalingam, S., and Watson, J. D. Inclusion chemistry control for machinability enhancement in steels. *Mat. Sci. and Eng.* 43:101 (1980).
121. Hilty, D. C. Mechanisms and modification of inclusion formation in steel. *Proc. Int. Symp. on Influence of Metallurgy on Machinability of Steel,* p. 199. Joint ISIJ/ASM Publication, 1977.
122. Farrell, J. W., and Hilty, D. C. Minimizing indigenous inclusion stringers with calcium. *Iron and Steelmaker,* ISS-AIME (Aug. 1976).
123. Hilty, D. C., and Farrell, J. W. Modification of inclusions by calcium. *Iron and Steelmaker,* ISS-AIME (May 1975).
124. Reh, B. Development of a micro-alloyed free-cutting steel. *Neue Hütte* 24 No. 12:451 (1979). (BISIT No. 20140.)
125. Weibel, E. R. *Stereological Methods.* Vols. 1 and 2. London: Academic Press Inc., 1980.
126. Underwood, E. E. *Quantitative Stereology.* Reading, Mass: Addison-Wesley, 1970.
127. DeHoff, R. T., and Rhines, F. N. *Quantitative Microscopy.* New York: McGraw-Hill, 1968.
128. Klinger, P., and Koch, W. Progress in the isolation of inclusions and structural constituents in alloyed and unalloyed steels. *Stahl und Eisen* 68 No. 9:321 (1948).
129. Plöckinger, E., and Randak, A. Formation and separation of sulfide inclusions in steel and cast iron. *Radex. Rund.* No. 5–6:768 (1957).
130. Tulepova, I. V.; Orlova, E. M.; and Ryutina, T. V. Removal and analysis of sulfide inclusions in boiler steel. *Zavod. Lab.* 38 No. 11:1319 (1972).
131. Kawamura, K. Phase analysis of sulfides of rare earth elements in steel. *Trans. ISIJ* 18:212 (1978).
132. Müller, C. A.; Stetter, A.; and Zimmermen, E. The effect of various deoxidation processes of mild, plain carbon free-cutting steels. *Arch. Eisen.* 37 No. 1:27 (1966).
133. Kubinova, A. Isolation and determination of manganese sulfides in free-cutting steel. *Hutn. Listy* 1:50 (1972).
134. Wright, P. W. The extraction of inclusions. *BHP Tech. Bull.* 23, No. 1:33 (1979).
135. Vander Voort, G. F. Inclusion measurement. In *Metallography as a Quality Control Tool.* New York: Plenum Press, 1980.
136. Flinchbauch, D. A. Use of a modified Coulter counter for determining size distributions of macroinclusions extracted from plain carbon steels. *Anal. Chem.* 43:178 (1971).
137. Simpson, I. D., and Standish, N. Derivation of the log-normal form of both section and spatial distributions of particles. *Metallography* 10:149 (1977).
138. Simpson, I. D., and Standish, N. Determination of the size-frequency distributions of non-metallic inclusions in as-cast metals. *Metallography* 10:433 (1977).
139. Cruz-Orive, L. M. Particle size-Sharpe distributions: The general spheroid problem II. Stochastic model and practical guide. *J. Micros.* 112:153 (1978).
140. Cruz-Orive, L. M. Stereological analysis of spheroidal particles of variable size and shape. *Practical Metallography Special Issue* No. 8:350 (1978).
141. Ekelund, E., and Werlefors, T. A system for the quantitative characterization of microstructures by combined image analysis and X-ray discrimination in the scanning electron microscope. *Scanning Electron Microscopy:* 417 (1976).
142. Fröhkle, M. Determination of the oxide purity degree in free cutting steels with the aid of quantitative image analysis. *Arch. Eisen.* 45, No. 12:899 (1974).
143. Keane, D. M. The influence of inclusions on machinability. In *Inclusions and their Effects on Steel Properties,* p. 19/1. British Steel Corporation, 1974.
144. ASTM E45–81. Standard Recommended Practice for Determining the Inclusion Content of Steel.
145. German Standard 1572. Microscopic Testing of Free Cutting Steels for Non-Metallic Sulfide Inclusions by Means of a Series of Pictures (1977). (BISIT No. 15913.)
146. The "Fox" Inclusion Count. A Quantitative Method of Expressing the Cleanness of Steel. S. Fox & Co. Ltd.
147. Blank, J. R., and Allmand, T. R. Evaluation of operator errors occurring in assessment of non-metallic inclusions by conventional metallographic methods. In *Automatic Cleanness Assessment of Steel,* p. 1. Iron and Steel Institute Publication No. 112, 1968.
148. Yamaguchi, K., and Kato, T. Friction reducing actions of inclusions in metal cutting. *J. Eng. Ind.* (*Trans. ASME B*) 102, No. 3:221 (1980).
149. Hilliard, J. E., and Cahn, J. W. An evaluation of procedures in quantitative metallography for volume fraction analysis. *Trans. Met. Soc. AIME* 221:344 (1961).
150. Gladman, T. The accuracy of lineal analysis in metallographic investigations. *JISI* 201:1044 (1963).
151. Pickering, F. B. *The Basis of Quantitative Metallography.* Inst. of Metallurgical Technicians, Monograph No. 1, 1976.
152. Ruddleston, R.; Thornton, J. R.; and Bowers, M. D. Assessment of steel cleanliness using Quantimet 360. *Metals Technology* 3, No. 9:422 (1976).
153. Blank, J. R. Evaluation of the Quantimet image analysing computer for quantitative metallographic observations. In *Automatic Cleanness Assessment of Steel,* p. 63. Iron and Steel Institute Publication No. 112, 1968.
154. Pickering, F. B., and Gladman, T. Modern approach to metallography. *Steel and Coal* (Dec. 1962).
155. Marston, G. J., and Murray, J. D. Machinability of low-carbon free-cutting steel. *JISI* 208:568 (1970).
156. Riekels, L. P. The influence of microstructure on the metal cutting behaviour of AISI 1215 resulfurized free machining steels. Ph.D. Dissertation, University of Illinois (June 1979).
157. Radtke, D., and Schreiber, D. Relationship between sulfide formation and machinability in free cutting steel. *Steel Times* 193:246 (1977).

158. Fröhlke, M. Determination of characteristic values for manganese sulfides in free-cutting steels by means of quantitative image analysis. *Microscope* 19:403 (1971).
159. Fröhlke, M. The influence of non-metallic inclusions on machinability of low-carbon free cutting steels. *Int. Symposium on Influence of Metallurgy on Machinability of Steel,* p. 41. Joint ISIJ/ASM Publication, 1977.
160. Bartholome, W.; Fröhlke, M.; and Köster, H. H. Determination of the manganese sulfide content in free-machining steels by means of quantitative image analysis. *Radex-Rund.* 4:514 (1971).
161. Ramalingam, S., and Von Turkovich, B. F. Structure property relations in free machining steels. *J. Eng. Ind. (Trans. ASME B)* 102, No. 1:91 (1980).
162. Yamaguchi, K. *Supplementary Volume to Proc. Int. Symposium on Influence of Metallurgy on Machinability of Steel,* p. 14. Joint ISIJ/ASM Publication, 1977.
163. Wicksell, S. D. The corpuscle problem second memoir. Case of ellipsoidal corpuscles. *Biometrika* 18:151 (1926).
164. De Hoff, R. T. The determination of the size distributions of ellipsoidal particles from measurements made on random plane sections. *Trans. Met. Soc. AIME* 224:474 (1965).
165. Evans, D. A., and Clarke, K. R. Estimation of embedded particle properties from plane section intercepts. *Adv. App. Prob.* 7:542 (1975).
166. Quinto, D. T.; Bhattacharya, D.; and Thompson, R. W. An investigation of cutting energy and metallurgical properties in leaded and non-leaded AISI 1215 steels. *Proc. International Symposium of the Influence of Metallurgy on Machinability,* p. 433. Joint ISIJ/ASM Publication, 1977.
167. Brown, R. H. A double shear-pin quick-stop device for very rapid disengagement of a cutting tool. *Int. J. Mach. Tool Des. Res.* 16:115 (1976).
168. Williams, E. F.; Smart, J. E.; and Milner, D. R. The metallurgy of machining II: The cutting of single-phase, two-phase, and some free machining alloys. *Metallurgia* 81:51 (1970).
169. Trent, E. M. *Metal Cutting.* London: Butterworths, 1977.
170. Milovic, R., and Wallbank, J. The machinability of low carbon free cutting steels with high speed steel tools. *J. App. Metalworking* 2, No. 4:249 (1983).
171. Wallbank, J. Structure of built-up edge in metal cutting. *Metals Technology* 6:145 (1979).
172. Form, G. W., and Beglinger, H. The mechanism of chip formation—Part II: Experimental verification and general discussion. *Z. Metallkude.* 64:8 (1973).
173. Abeyama, S., and Nakamura, S. Machinability evaluation of steel in an automatic screw machine. In *Machinability Testing and Utilization of Machining Data,* p. 184. ASM Materials/Metalworking Technology Series, 1979.
174. Warnake, G. A new method of visualizing the cutting process. *Proc. 5th North American Metalworking Research Conf.,* p. 229. Buffalo, NY, 1977.
175. Ramalingam, S., and Bell, A. C. A scanning electron microscope stage for the observation of chip formation. *Rev. Sci. Instrum.* 44, No. 5:573 (1973).
176. Hazra, J.; Caffarrelli, D.; and Ramalingam, S. Free machining steels—The behavior of Type I MnS inclusions in machining. *J. Eng. Ind. (Trans. ASME B)* 96:1230 (1974).
177. Iwata, K., and Ueda, K. Dynamic behaviour of manganese sulfide inclusions in machining under scanning electron microscope observation. *Proc. 1st Int. Conf. on Production Engineering,* p. 516. Tokyo JSPE, 1974.
178. Iwata, K., and Ueda, K. The mechanism of fracturing in micromachining. *Proc. 3rd Int. Conf. on Production Engineering,* p. 266. Kyoto JSPE, 1977.
179. Doyle, E. D., and Silva, V. M. The dynamics of metal machining: A study in the SEM. *Proc. 4th Australian Conf. on Electron Microscopy,* p. 62. Sydney, 1976.
180. Wright, P. K., and Trent, E. M. Metallographic methods for determining temperature gradients in cutting tools. *JISI* 211:364 (1973).
181. Tipnis, V. A., and Joseph, R. A. A study of plunge (or form) machining of low carbon resulfurized steel on a multispindle automatic screw machine—Part 2: Influence of speed, feed, and duration of cutting on worn tool geometry. *J. Eng. Ind. (Trans. ASME B)* 93:571 (1971).

14
MICROSCOPY AND TITANIUM ALLOY DEVELOPMENT

C. G. Rhodes
Rockwell International Science Center

Although the Kroll process for refining titanium was invented in 1937, significant production in the U.S. did not begin until 1948. It was the advent of the aircraft jet engine that stimulated growth of the titanium industry, primarily because of titanium's excellent strength-to-weight ratio. A wide variety of alloy compositions evolved for various aircraft applications, not only for engines, but for airframe structures as well.[1] As the demand for these applications ebbed and flowed in the years following 1948, the industry began to cultivate other areas for use. The corrosion resistance of Ti made it a viable replacement for stainless steels in many applications in the chemical processing industry.[2] More recently, Ti has been used in various types of prostheses for surgical implant in humans.[3]

Titanium undergoes an allotropic transformation at 882°C, having a body-centered-cubic structure (beta phase) above the transformation temperature and hexagonal-close-packed structure (alpha phase) below 882°C. It is this allotropic nature of Ti that provides the basis for the development of a large spectrum of alloy compositions. Various alloying elements can raise or lower the transformation temperature, resulting in stabilization of either the alpha or beta phase. Many of the alloys developed for structural applications contain a mixture of alpha and beta phases, whose morphologies and distributions can be altered by processing or heat treatment. In addition, depending upon alloy composition, the metastable beta phase can decompose on cooling to mixtures of alpha and beta, to martensite, or to a metastable transition phase. Microstructures can therefore be adjusted by alloying, as well as by heat treatment and processing.

In the early years of Ti alloy development, microstructures were examined exclusively by light microscopy of polished and etched surfaces.[4] Special metallographic techniques had to be developed to reveal the underlying microstructures, and the techniques frequently varied with alloy composition.[4] As the transmission electron microscope became available, surface replica techniques were used to study microstructural features that could not be resolved by the conventional light microscope.[5] Omega phase, for instance, was first observed by replica transmission electron microscopy (TEM) after having been discovered by X-ray diffraction.[5,6] Thin-foil TEM followed replica TEM and revealed many microstructural features that could not be resolved by the other techniques. Light and electron microscopy have led to an understanding of the metallurgy of Ti alloys and the ways in which microstructure influences mechanical properties.

The interpretation of Ti alloy microstructures has assisted in the design of alloys for particular applications by manipulating the parameters of composition, processing, and heat treatment to develop microstructures that correspond to the desired mechanical properties. For example, lenticular alpha phase is known to provide excellent elevated-temperature creep resistance;[7] hence, alloys are specially formulated to stabilize the alpha phase. Similarly, properties such as corrosion resistance,[8] fracture toughness,[9-11] and fatigue life[12] can be optimized by means of microstructure manipulation. Processing and heat treatments that alter microstructure morphology have been utilized to optimize properties such as superplasticity[13] and fatigue crack propagation resistance.[14-15]

METALLOGRAPHIC TECHNIQUES

Some of the earliest concentrated efforts in microstructure analysis of Ti alloys were conducted at the Columbus Laboratories of Battelle Memorial Institute. In 1958, H. R. Ogden and F. C. Holden[4] of Battelle published a handbook of Ti microstructures, entitled *Metallography of Titanium Alloys,* that included a collection of metallographic techniques. At that time, the primary technique for preparing Ti samples required the use of acid wheel polishing because of the propensity of the alpha phase to smear when polished with fine alumina. Metallographers could easily recognize smeared metal by examining a polished surface under polarized light. Because of its hexagonal crystal structure, alpha Ti is quite active under

Fig. 14-1. Polarized light photomicrograph of electropolished Ti–5Al–2.5Sn. (Courtesy of M. Calabrese and J. Chesnutt.)

polarized light, as shown in Fig. 14-1. This polarized light photomicrograph of Ti–5Al–2.5Sn reveals the grain structure and absence of smeared metal.

In the quest for ways to prepare metallographic surfaces

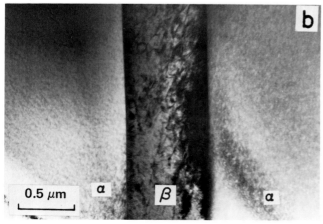

Fig. 14-2. Transmission electron micrographs of Ti–6Al–4V demonstrating the effects of thinning technique on the appearance of the alpha/beta interface phase: (a) electrolytically thinned, and (b) ion milled.

so as to reveal the true microstructure, electrolytic polishing solutions were developed at several laboratories involved in the study of Ti alloys.[4] Unfortunately, many of the most effective solutions contained the highly unstable perchloric acid. Because of the number of documented cases of explosions resulting from the use of this acid, electropolishing did not become a widely accepted technique for the metallographic preparation of Ti. In the mid 1960s, it was discovered[16] that Ti and its alloys could be electropolished in the relatively safe solution of 1%-HF/1.5%-H_2SO_4/97.5%-methanol that had been developed for refractory metals.[17] Since mechanically ground surfaces could be satisfactorily electropolished in this solution, the need for mechanical polishing was eliminated. Electropolishing has since emerged as the most efficient method for metallographic sample preparation of Ti alloys.

Frequently, however, the phases present in the microstructure of Ti alloys are of such a size that they cannot be resolved by light microscopy of polished and etched surfaces. Even scanning electron microscopy (SEM) of metallographically prepared surfaces cannot reveal many of the submicroscopic phases, such as omega phase, some martensites, or beta prime. Thin-foil TEM, therefore, is required to resolve these finely divided phases, as well as deformation components such as dislocations and twins.

Both chemical and electrolytic solutions have been used for the thinning of Ti alloys[18-20] for TEM examinations. Most of the chemical polishing solutions contain nitric and hydrofluoric acids, whereas the electrolytic thinning solutions include mixtures of perchloric acid and alcohol or sulfuric acid and alcohol.[20] Titanium readily absorbs hydrogen at room temperature, and the amount of hydrogen generated in a chemical or electrolytic polishing bath can result in the formation of hydride precipitates in the alpha phase during thinning.[19] Consequently, all solutions are usually used at temperatures in the range of −20° to −30°C to minimize absorption of hydrogen into the Ti sample. In spite of these precautions, however, recent studies have shown that Ti microstructures may be altered during electrolytic thinning.[21]

An alternate thinning technique for TEM that is emerging with the advent of commercially available instruments is ion milling.[21] This approach precludes the absorption of hydrogen into the alpha and beta phases and minimizes or eliminates spontaneous relaxation in the beta phase. Figure 14-2 compares the effects of electrolytic thinning and ion milling on the alpha/beta interface region in Ti–6Al–4V. There is currently some debate over whether the alpha/beta interface phase is a bulk microstructural feature or a thin-foil artifact.[22,23] The problem lies in the fact that an interface phase is present in thin-foils prepared by electrolytic polishing and absent in thin-foils prepared by ion milling (see Fig. 14-2). It has not yet been unequivocally resolved whether the interface phase is being intro-

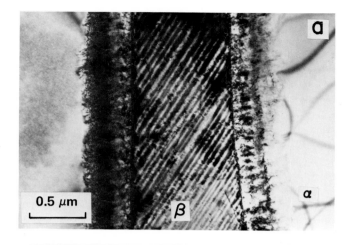
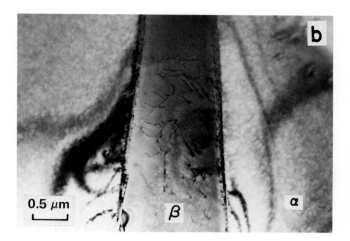

Fig. 14-3. Transmission electron micrographs of Ti–6Al–4V demonstrating the effects of thinning technique on the beta phase: (a) electrolytically thinned, (b) ion milled, and (c) ion damage.

duced by hydrogen generation during electrolytic thinning or whether it is being dissolved by the ion bombardment and attendant thermal excursion during ion milling. Current studies should resolve the question.

Figure 14-3 illustrates the results of the two thinning techniques on the beta phase in Ti–6Al–4V. It can be seen that the heavy density of linear features present in the sample thinned by electropolishing in Fig. 14-3(a) is absent in the ion-milled microstructure in Fig. 14-3(b). Ion milling does alter the microstructure somewhat, however, by the introduction of ion damage, which is manifested, in Fig. 14-3(c), by the formation of vacancy and/or interstitial loops. Metallographers must be keenly aware, therefore, that even though the TEM provides a view of the microstructure that can be obtained in no other way, improper sample preparation can lead to artifacts.

MICROSTRUCTURE DEVELOPMENT

A variety of microstructures is available to Ti alloy designers for tailoring alloys to particular applications. The variety is made possible by alloying additions that can either strengthen the alpha phase, which is stable at low temperature, or retain the beta phase, which is the high-temperature form of Ti. These manipulations can be accomplished because some alloying elements lower the beta transus, thereby stabilizing the beta phase, whereas others raise the beta transus, resulting in stabilization of the alpha phase.

Although most of the important Ti alloys are multielement alloys, a schematic binary phase diagram such as that shown in Fig. 14-4 can be used to describe them. Alpha alloys are the result of small amounts of beta stabilizer additions; alpha+beta alloys form with increased beta stabilizer additions; beta alloys are produced when compositions push the alloy into the beta-phase field. Typical examples of microstructures of these types of alloys are shown in Figs. 14-5 through 14-7.

The alpha alloy shown in Fig. 14-5 contains alpha grains having a lenticular morphology that form on cooling from the beta-phase region. The transformation from body-centered-cubic beta phase to hexagonal-close-packed alpha phase is controlled by the crystallographic relationship first described by Burgers[24] and results in six equally possible variants of alpha phase. In general, the Burgers orientation relationship between alpha and beta phases holds whether the alpha phase forms martensitically or by nucleation and growth. In either case, the resulting alpha phase has an elongated grain structure

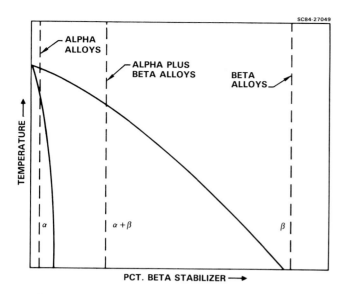

Fig. 14-4. Schematic binary phase diagram for Ti-X systems.

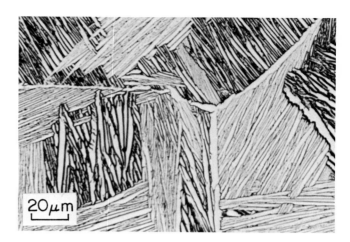

Fig. 14-6. Lenticular alpha-phase grain morphology in another alpha-beta Ti alloy.

with [0001] parallel to the long axis. The aspect ratio of the alpha grains can be controlled by adjusting the cooling rate from the beta-phase field.

Microstructures of the alpha+beta alloys consist of a mixture of the two phases, with volume fractions being a function of alloy composition and thermal treatment. The beta phase can be retained below the beta/alpha+beta transus temperature because partitioning of the beta stabilizing elements during cooling results in continued enrichment of the beta phase to the point that the composition lies in the beta-phase field. The alloy shown in Fig. 14-6 has approximately 90-percent alpha phase and 10-percent beta phase, and, as was the case for the alpha alloy in Fig. 14-5, the alpha grains exhibit an acicular morphology separated by thin strips of beta phase.

The commercially important beta alloys are more properly designated "metastable" beta alloys because, although they can be quenched from high temperature to retain 100-percent beta phase (see Fig. 14-7), they can be aged at intermediate temperatures to precipitate alpha phase. Since the body-centered-cubic beta phase is more easily worked than the hcp alpha phase, the beta alloys are used in applications where cold forming is required.[25] Because the beta alloys are age-hardenable, they can be used in thick cross-sectional components, such as heavy forgings, that can be strengthened through thickness. After forming, aging treatments that precipitate alpha phase can improve mechanical properties.

Depending upon the specific alloying element and its concentration, decomposition of the beta phase, either on cooling or during isothermal subtransus treatment, can take one of several paths. For instance, rapid cooling of lean beta phase results in martensitic transformation, with the product being either hcp alpha prime or orthorhombic alpha double prime. At slower cooling rates, decomposition of a lean beta phase occurs by nucleation and growth of hcp alpha phase. Beta phase with higher levels of beta-stabilizing elements can decompose through

Fig. 14-5. Lenticular alpha-phase grain morphology in an alpha-phase Ti alloy.

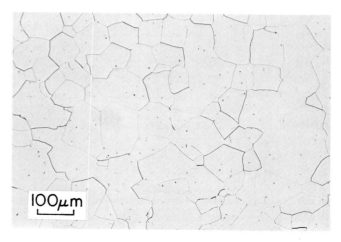

Fig. 14-7. Equiaxed grain morphology in a metastable beta-phase Ti alloy. (Courtesy of R. Spurling and C. H. Hamilton.)

the formation of metastable phases, omega or beta prime. Since alloy designers usually concentrate on compositions that avoid the formation of these metastable phases, only the martensites and the equilibrium alpha precipitation are important in commercial Ti alloys.

Microstructures of the alpha and alpha+beta alloys can also be adjusted by thermomechanical treatments. The acicular morphology of alpha-phase grains, as described in the preceding paragraphs, can be altered to a more equiaxed shape by deformation at elevated temperatures. Processes, such as rolling or forging, that are carried out at a temperature below the beta transus, but high in the alpha+beta phase field, will produce equiaxed alpha-phase grains. Figure 14-8 is a typical example of an alpha-beta Ti alloy that has been processed in this manner.

This very brief synopsis of Ti alloy microstructures has been intended to show the wide variety of microstructures available to the alloy designer. Over the years, researchers have demonstrated the influence of each of these microstructures on many important mechanical properties.[25-27] The metallurgist who wishes to develop an alloy for a particular application, then, has a large body of knowledge to draw upon and can tailor alloy composition to produce the desired microstructure. Historically, light microscopy played an important role in the development of the first commercial Ti alloys, and the concept of microstructure manipulation to impart specific mechanical properties has grown through the following years. Thin-foil TEM opened up new vistas of microstructural analyses that have led to the explanation of many phenomena in the areas of phase transformations and deformation and fracture. A better understanding of these phenomena has improved the performance of Ti alloys in general through the refinement of compositions or adjustments in processing that create the required microstructures. The remainder of this chapter will describe the contributions of light microscopy, SEM, and TEM to the development and refinement of commercial Ti alloys.

ALPHA AND NEAR-ALPHA PHASE ALLOYS

Early work on Ti alloy development revealed that for long-term, elevated-temperature applications, alloys with little or no beta phase performed better than alpha+beta alloys. This behavior was explained as being a consequence of the thermal instabilities of beta phase;[28] that is, thermal decomposition of beta during long-term exposures results in alteration of mechanical properties. Alloys designed for applications where components are creep-limited, therefore, evolved to be comprised primarily of alpha phase. Using light microscopy or replica TEM, metallurgists observed that microstructures consisting of an acicular alpha-grain morphology exhibited better elevated temperature tensile and creep properties than did those with an equiaxed alpha-grain morphology.[7] It was further discovered that a coarse Widmanstatten alpha, as seen in Fig. 14-6, has better creep resistance than a fine martensitic alpha structure,[7] as seen in Fig. 14-9. Although the reasons for this behavior were not explained until much later, the design of commercial Ti alloys was based, at least partially, on being able to develop a lenticular morphology of the alpha grains.

This knowledge was taken into account in the development of Ti–8-wt.%Al–1-wt.%V–1-wt.%Mo (8–1–1). The alloy was designed to be a high-temperature, creep-resistant, near-alpha-phase alloy, in which Al was added for solid-solution strengthening and stabilization of the alpha phase, and Mo and V were added to eliminate the "third phase formed in binary Ti–Al alloys."[29] The latter was subsequently shown to be the ordered Ti_3Al.[30] Recommended processing included initial deformation, such as forging or rolling, to be started above the beta transus temperature, thereby ensuring a basketweave microstruc-

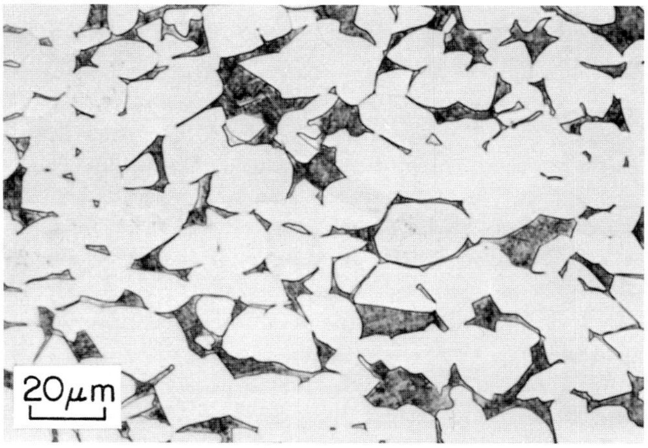

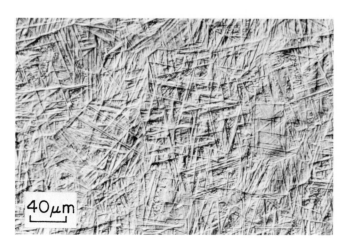

Fig. 14-8. Equiaxed alpha-grain morphology in an alpha-beta Ti alloy (Courtesy of M. Calabrese and J. Chesnutt.)

Fig. 14-9. Martensite in an alpha-phase Ti alloy. (Courtesy of R. Spurling and C. H. Hamilton.)

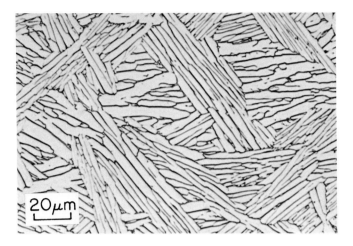

Fig. 14-10. Basketweave microstructure in an alpha-phase Ti alloy.

ture, such as that shown in Fig. 14-10. Table 14-1 lists data reported by Bohanek and Kessler[29] that compares the amount of creep deformation at 538°C (1000°F) after 150 hr at various stress levels for the two types of microstructure. The superiority of the basketweave structure over the equiaxed structure is quite pronounced.

Some years after the introduction of 8-1-1, it was discovered to be highly susceptible to stress-corrosion cracking, especially in a sea water environment.[31,32] After further metallographic studies of 8-1-1 revealed the existence of ordered Ti_3Al phase after long-term elevated temperature exposures,[30] it was concluded that the presence of ordered domains contributed to the susceptibility of the alloy to sea-water stress corrosion.[33] The 8-1-1 experience convinced alloy designers to avoid compositions with more than 6-wt.% Al.

Titanium metallurgists, recognizing the importance of microstructure as well as Al concentration, later designed creep-resistant alloys with additional alpha-phase solid-solution strengtheners, such as Sn and Zr, and low levels of beta stabilizers. These alloys, such as Ti–6Al–2Sn–1.5Zr–1Mo–0.35Bi–0.1Si (Ti-11), have been designated "super alpha" alloys because of their improved properties and the presence of beta phase in the microstructure. This second generation of high-temperature Ti alloys has retained the concept of the basketweave alpha structure but with the addition of Si to improve creep resistance.

Alloy designers anticipated that Si would generate increased creep strength because its limited solubility in the alpha phase would result in its precipitation as an intermetallic compound.[34] Experimental alloys demonstrated that Si additions did indeed improve creep resistance, although there appeared to be a limit to the amount of Si that was effective. It was initially assumed that the presence of Ti silicide particles along grain boundaries inhibited the grain-boundary sliding component of creep.[35] High-resolution microscopy was later to demonstrate, however, that the mechanism by which Si enhanced creep resistance was quite different from that first suspected.

Paton and Mahoney[36] showed that grain-boundary sliding contributed less than 10 percent of total creep strain at 538°C (1000°F) in both experimental and commercial alloy compositions. Using thin-foil TEM, they concluded that the primary mechanism by which Si enhanced creep resistance was not the inhibition of grain boundary sliding but rather the precipitation of silicides on mobile dislocations during creep exposure. It can be inferred from their work that the presence of grain-boundary silicides not only does not enhance creep resistance but may actually reduce post-creep tensile ductility. Paton and Mahoney have thus explained the optimum levels of Si (0.2–0.5 wt.%) in creep-resistant Ti alloys in terms of microstructure.

The latest generation of creep-resistant Ti alloys is typified by Alloy 829 (Ti–5.5Al–3.5Sn–3Zr–1Nb–0.25Mo–0.3Si), which is designed for use at temperatures up to 550°C.[37] Based on the microstructural data developed earlier, this near-alpha alloy relies on a basketweave microstructure with the addition of Si to obtain creep properties. In addition, it was determined during the development of this alloy that the prior-beta grain size is significant in controlling creep properties; hence, the manufacturer's recommended processing results in a beta grain size that falls in the range of 0.5 to 0.75 mm for optimum properties.[38]

The desire to raise the upper temperature limit at which Ti alloys may effectively operate has led to research in several areas, most of which centers around microstructural effects and consequently depends upon microscopy for its successful completion. One of these studies, and probably the most conventional, is directed toward the modification of Alloy 829 by composition and processing to achieve a prior-beta grain size on the order of 0.3 mm.[38] The producers of Alloy 829 feel that this reduced beta grain size will help bring about a significant improvement in creep resistance.[38]

A rather more unconventional approach to high-temperature Ti alloy design has been taken in the research on alloys strengthened by an ordered Ti_3Al phase.[39] TEM revealed that aging the alloy Ti–8Al–5Nb–5Zr–0.25Si at 600°C produces a fine dispersion of Ti_3Al that improves

Table 14-1. Total Creep Deformation of Ti–8Al–1V–1Mo at 538°C (1000°F) (from Ref. 29).

	CREEP DEFORMATION (%)	
TEST PARAMETERS	BASKETWEAVE ALPHA	EQUIAXED ALPHA
150 hr, 103 MPa	0.06	0.24
150 hr, 207 MPa	0.16	1.33
150 hr, 276 MPa	0.28	4.03

the elevated-temperature tensile strength and creep resistance. The room-temperature brittleness generally associated with Ti$_3$Al is ameliorated by the addition of Nb. The well-established acicular alpha phase morphology, as well as the addition of Si, is used in this alloy to optimize creep resistance.

The so-called "conventional wisdom" that now exists among Ti-alloy producers regarding optimum parameters for high-temperature alloy design is based in no small part on microstructures. The basketweave alpha-grain morphology and the minimum prior beta grain size are essential for all creep-resistant Ti alloys, including the new departure alloys containing high concentrations of Al. It has been light microscopy, SEM, and TEM that have led to the establishment of the current "conventional wisdom" surrounding high-temperature Ti alloys.

Alpha-phase alloys are also used for cryogenic applications, primarily because they are less sensitive to notch brittleness than the alpha-beta or beta-phase alloys. Ti–5Al–2.5Sn is a commercial alpha alloy that has been used extensively in subambient temperature applications since its appearance in the late 1950s. Some early studies on this alloy revealed that trace interstitial impurities of oxygen and nitrogen had a negative effect on the ability of the alloy to resist notch embrittlement at temperatures down to −190°C (−320°F).[40] No note was apparently made of the microstructure, however, and it was not until a few years later that the importance of the morphology of alpha-phase grains was discovered.[41] This latter work demonstrated through microstructural analysis that not only was the level of interstitial impurities important, but that the equiaxed alpha grain shape resulted in a reduction of notch sensitivity as compared to Ti–5Al–2.5Sn with an acicular alpha grain shape. Although the effect was more pronounced in the alpha-beta alloys than in the all-alpha alloy, it was nevertheless sufficiently significant to warrant the use of equiaxed microstructures in cryogenic applications of Ti–5Al–2.5Sn.

One of the advantages of using alpha phase alloys is their response to welding. In general, alpha-phase alloys are more amenable to welding than alpha/beta or beta alloys because the problem with segregation of beta-stabilizing elements in the fusion zone is absent in the former.[42] Welding of Ti alloys results in exposure to temperatures above the beta transus not only in the fusion zone, but also in a heat-affected zone surrounding the weld itself. During cooling, then, an acicular microstructure develops in all areas above the beta transus as a result of the welding. If the cooling rate is sufficiently rapid, a fine acicular or martensitic structure will result, whereas if the cooling rate is slower, a coarser acicular structure formed by nucleation and growth will be produced. When the starting condition of the workpiece includes an acicular alpha-grain morphology, the variation in mechanical properties attendant with variations in microstructure as a result of welding will be minimized. Just the opposite is the case for components having an equiaxed alpha microstructure prior to welding.

ALPHA+BETA ALLOYS

Ti alloy designers and producers recognized early that mechanical properties such as tensile strength, fracture toughness, and fatigue resistance could be improved by the addition of beta stabilizers to dilute Ti–Al alloys. The result of the inclusion of elements such as V, Fe, Mo, Cr, or Ta was the retention of beta phase in the microstructure at room temperature after elevated-temperature processing.

The first structurally significant alpha+beta alloy was Ti–6Al–4V, which was developed in the early 1950s. The addition of Al for strength and V for beta-phase stabilization produced a mixture of alpha and beta phases that resulted in reasonable ductility accompanying the higher strength. This alloy can be solution-treated to produce retained beta phase that is subsequently aged to improve the alloy's properties. The aging treatment results in decomposition of the retained metastable beta phase, producing a finely divided mixture of alpha and beta phases, as shown in Fig. 14-11. With the outstanding performance of Ti–6Al–4V, developers searched for new alloys of this type to improve upon its properties.

Among the second generation of alpha+beta alloys was Ti–6Al–6V–2Sn, which was developed about ten years after the advent of Ti–6Al–4V. Because 6-percent Al was felt to be the maximum allowable (based on experience with 8–1–1), increased strength was achieved by the addition of Sn, which, like Al, is a strong solid-solution strengthener of alpha phase. It was recognized from the microscopy of Ti–6Al–4V that an equiaxed primary alpha phase mixed with aged beta phase constituted an optimum compromise of tensile strength and ductility. The beta stabilizing element V was known to improve the strength of Ti,[43] as was demonstrated by Ti–6Al–4V. Increasing V from 4 to 6 percent results in an increase in volume fraction of beta phase that can be retained at room temperature after solution treatment. Subsequent low-temperature aging then produces a larger volume fraction of transformed beta, and the expected increases in strength associated with transformed beta are realized. The differences in distributions of primary alpha phase and volume fractions of transformed beta phase between the two alloys are shown in Fig. 14-12. Once again, microstructure has been a primary factor in the development of a new Ti alloy.

The latest alpha+beta Ti alloy to be developed is Ti–4.5Al–5Mo–1.5Cr (CORONA 5), which was designed for high fracture toughness and moderate tensile strength.[44] The properties goal for this experimental alloy included a tensile strength at least as high as Ti–6Al–4V and fracture toughness of at least 110 MPa $\sqrt{m}$ (Ti–6Al–4V in the recrystallization annealed condition has fracture

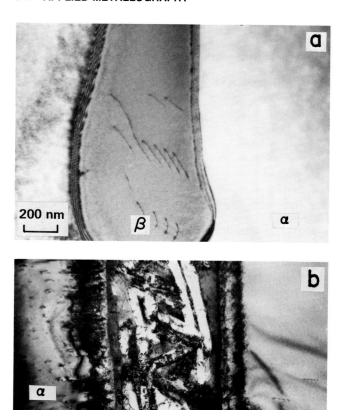

Fig. 14-11. Transmission electron micrographs illustrating the effect of aging treatment on beta phase in Ti–6Al–4V: (a) as-solution treated, retained beta phase (courtesy of M. Calabrese and M. Mitchell), and (b) aged, decomposed beta phase.

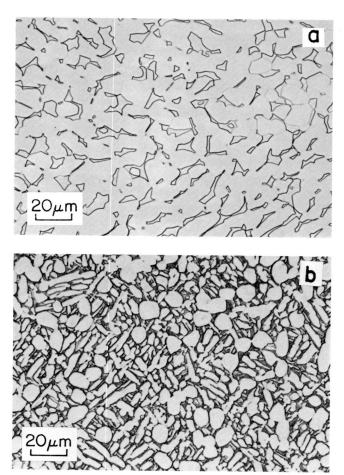

Fig. 14-12. Demonstrating the effect of concentration of beta-stabilizing elements on equilibrium volume fraction of alpha phase in alpha-beta Ti alloys: (a) Ti–6Al–4V having ~0.85 alpha phase, and (b) Ti–6Al–6V–2Sn having ~0.60 alpha phase.

toughness of around 82 MPa $\sqrt{m}$). To achieve this goal, the alloy designers based their initial concepts on pre-existing microstructure/property data. As we have seen, microstructures of Ti alloys can be modified by alloying additions and by processing; both of these factors were to play a significant role in the development of CORONA 5.[45]

The desired tensile strength level could be easily reached by solid-solution strengthening, primarily with Al, and the toughness goal was addressed by adding beta stabilizers that not only increased the equilibrium volume fraction of beta phase but also provided an easy means for modifying the morphology of the microstructure. The selection of Mo as the primary beta stabilizer was based on its beneficial influence on stress-corrosion resistance, whereas Cr was added as a partial replacement for Mo to decrease the density of the alloy.[44] One of the side effects of Mo was to reduce the scale of the microstructure compared to Ti–6Al–4V. The result of this finer mixture of alpha and beta phases in the microstructure was to generate more alpha-beta boundaries that could contribute to crack branching and improved fracture toughness.

Prior studies had shown that a Widmanstätten alpha morphology was usually associated with higher fracture toughness than that obtained with an equiaxed primary alpha morphology for any particular alloy.[46] For example, Table 14-2 lists results reported in Ref. 46 for Ti–6Al–6V–2Sn and Ti–6Al–2Sn–4Zr–6Mo showing the improvement in fracture toughness of the Widmanstätten microstructure over the equiaxed structure. When CORONA 5 was processed above the beta transus to produce a Widmanstätten microstructure, however, the resulting fracture toughness was not significantly greater than that obtained for the equiaxed structure. The reason for this behavior appears to be a combination of factors. First, the fracture toughness of the alloy with an equiaxed primary alpha structure is relatively high to begin with, so that further increases would not be realistically expected. Second, the beta processing treatment results in formation of alpha phase at prior-beta grain boundaries along which the fracture progresses in such a way that the Widmanstätten structure probably contributes less to the fracture

Table 14-2. Fracture Toughness of Two Alpha+Beta Titanium Alloys (from Ref. 46).

ALLOY	MICROSTRUCTURE	FRACTURE TOUGHNESS
Ti–6Al–6V–2Sn	Equiaxed primary alpha	55 MPa $\sqrt{m}$
Ti–6Al–6V–2Sn	Widmanstätten alpha	85 MPa $\sqrt{m}$
Ti–6Al–2Sn–4Zr–6Mo	Equiaxed primary alpha	26 MPa $\sqrt{m}$
Ti–6Al–2Zn–4Zr–6Mo	Widmanstätten alpha	57 MPa $\sqrt{m}$

Table 14-3. Superplastic Properties at a Strain Rate of 2×10^{-4} s^{-1} (from Ref. 49).

ALLOY	TEMP. (°C)	m*	FLOW STRESS (MPa)	TOTAL ELONGATION (%)
Ti–6Al–4V	927	0.70	9.6	600
Ti–6Al–4V + 2Ni	816	0.85	13	720
Ti–6Al–4V + Fe + 1Ni	816	0.68	14	550
Ti–6Al–4V + 1Co + 1Ni	816	0.82	19	550

* m is a measure (slope of a log-log plot of flow stress versus strain rate) of the superplastic response of a material and falls in the range of 0.3 to 1.0.

toughness than does the grain-boundary alpha. Third, the plate material being evaluated exhibited directionality in properties, apparently as a result of texture effects. The variations in directional properties of the equiaxed alpha material were sometimes greater than the variations caused by microstructure.[44,45] Directionality of properties was reduced in the alloy when beta-processed.

CORONA 5 was developed from a basic understanding of the influence of alloying elements on microstructure and of the influence of microstructure on mechanical properties. Particular mechanical properties can be achieved in the alloy by selecting processing to develop the microstructure that controls that property.

Superplasticity is a property of Ti alloys that has recently been developed into an important industrial forming process.[47] The propensity of a material to sustain unusually large amounts of deformation without necking (superplasticity) is very much dependent upon microstructure.[48] The requisite microstructure is two-phased and fine-grained, with both the grain size and phase composition being stable at the superplastic forming temperature.[13,48] The alpha+beta Ti alloys fall into this category of materials. Once it had been found to exhibit superplastic properties, Ti–6Al–4V became the subject of research to develop processing techniques that would exploit this remarkable property.[47]

One of the disadvantages of superplastic forming of Ti–6Al–4V is the expense incurred in time and equipment necessary to reach and sustain the elevated temperature required for optimum processing, 927°C (1700°F). This drawback motivated recent research aimed at alloy modifications that would reduce the optimum superplastic forming temperature without degrading other mechanical properties.[49] This research, conducted by Drs. J. Wert and N. Paton of the Rockwell International Science Center, is a classic example of the use of microstructure manipulation to develop an alloy with special properties for a particular application. Their goal was to modify Ti–6Al–4V so that its superplastic properties at 816°C (1500°F) would be equivalent to those of conventional Ti–6Al–4V at 927°C, while the room-temperature tensile properties would remain unchanged.

Wert and Paton reasoned that, since the superplastic forming temperature should correspond to that temperature at which the alloy contained equal volume fractions of alpha and beta phases, the alloying addition should be an element that lowered the alpha+beta/beta transus of Ti–6Al–4V. They further concluded that the alloying addition should be an element that diffuses rapidly, since it is well known that high diffusion rates in the beta phase are associated with improved superplastic flow properties of alpha-beta Ti alloys.[50] With this premise, Ni, Fe, and Co were selected for addition to Ti–6Al–4V, with the amount of each quaternary addition based on calculations and experimental determinations of the beta transus. Some of their results are presented in Table 14-3.

The results of Wert and Paton proved that their premise was correct and that an alloy could be designed from microstructural principles to exhibit special properties. As this section has demonstrated, the alpha+beta Ti alloys are especially amenable to microstructural manipulation, and metallurgists have used this characteristic to design a variety of alloys. We have touched on only a few of the special properties—viz. creep resistance, fracture toughness, and superplasticity—but there are others such as corrosion resistance,[51] hydrogenation forming,[52] and high-temperature tensile strength[53] that could also have been explored.

BETA ALLOYS

Titanium alloys in which the beta phase is stable at room temperature contain significant amounts of beta-stabilizing elements such as V, Mo, Fe, or Cr. Adding these elements to Ti results in alloys with increased density, which is a definite disadvantage for aerospace applications. Furthermore, beta-phase alloys generally exhibit poor high-temperature properties, especially creep resistance. Grain growth in beta alloys during heat treatment can result in the degradation of fracture properties common in materials of large grain size.[54] For these reasons, there are fewer commercial beta-phase Ti alloys than alpha or alpha+beta alloys, and the beta alloys are certainly less well developed than the other types.

One important advantage of the body-centered-cubic beta-phase alloys is that they are more easily formable than their hexagonal-close-packed alpha-phase counterparts. The commercial beta-phase alloys that have been

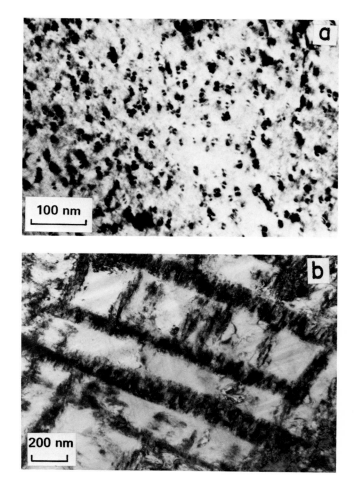

Fig. 14-13. Transmission electron micrographs of Ti–8V–8Mo–2Fe–3Al demonstrating the effect of aging temperature on precipitation: (a) coherent metastable beta prime phase formed during low-temperature aging, and (b) equilibrium alpha precipitation formed during intermediate temperature aging.

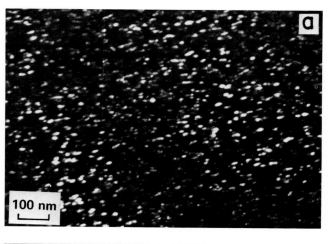

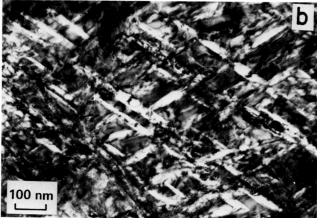

Fig. 14-14. Transmission electron micrographs of Ti–11.5Mo–4.5Sn–6Zr demonstrating the effect of aging temperature on precipitation: (a) metastable omega phase formed during low-temperature aging, dark field image; and (b) equilibrium alpha phase formed during intermediate temperature aging.

developed were designed for specific applications such as deep drawing and sheet and foil forming. Because they are all age-hardenable, they would be more correctly designated as metastable beta alloys.

The first beta alloy developed, Ti–13V–11Cr–3Al,[55] could be quenched from a solution temperature of 774°C (1425°F) to retain 100-percent beta phase and subsequently aged at 482°C (900°F) to precipitate the alpha phase that provided the increased strength. Although designed to be a stable, age-hardenable alloy, Ti–13V–11Cr–3Al proved to be unstable when exposed to elevated temperatures for long periods of time because of the formation of the embrittling intermetallic compound, $TiCr_2$. Cr is one of the beta-stabilizing elements that forms a eutectoid with Ti, and the presence of 11-wt.% Cr in the alloy results in extensive intermetallic compound formation through the eutectoid decomposition reaction. This development and the experimental observations of Hunter and Arnold,[56] who originally designed metastable beta alloys to be precipitation-hardened by intermetallic compound formation in the manner of Ni-base superalloys, led designers toward alloys that could be strengthened by alpha-phase precipitation without the formation of intermetallics.

The second generation of metastable beta alloys included Ti–8V–8Mo–2Fe–3Al (8823),[56,57] Ti–11.5Mo–4.5Sn–6Zr (Beta III),[58] and Ti–3Al–8V–6Cr–4Mo–4Zr (Beta C). They were designed to avoid intermetallic compound formation, although 8823 does contain a small amount of Fe and Beta C contains Cr. These alloys were developed for their deep hardenability as designers took advantage of the favorable deformation and age-hardening characteristics of the metastable beta phase. Aging treatments that were recommended for developing optimum properties[57,58] resulted in the precipitation of equilibrium alpha phase,[59-62] thereby avoiding a beta-phase separation reaction in 8823[59] (Fig. 14-13) and in Beta C[62] and omega-phase formation in Beta III[60] (Fig. 14-14).

Subsequent beta alloys include Ti–10V–2Fe–3Al (10–

2–3) and Ti–15V–3Cr–3Al–3Sn (15–3), which were also designed for specific applications. The former was developed to have a lower density and lower production costs than earlier metastable beta alloys.[63] There are several aerospace applications for this alloy.[63] Since it has excellent cold fabrication characteristics,[64,65] 15-3 is currently being characterized for strip and foil applications.

SUMMARY

There is great latitude in the design of Ti alloys because of the variety of microstructures made possible by the phenomena of allotropic transformation and the stabilization of beta phase to room temperature. Variations in microstructures are significant for structural applications because mechanical properties depend upon them.

Analysis of microstructures relies on the ability of the metallographer to prepare specimens devoid of artifacts. Techniques for doing so have developed over the years to the point where electrolytic polishing is the simplest and the most reliable now available, whether microstructures are to be studied by light microscopy or electron microscopy.

Elements such as Al and Sn stabilize the alpha phase to higher temperatures than are possible for unalloyed Ti, whereas elements such as V, Mo, Cr, and Fe stabilize the beta phase to lower temperatures than those provided by unalloyed Ti. As a result, microstructures, and consequently mechanical properties, can be altered by alloying additions. Alpha- and beta-phase morphologies and volume fractions in any particular alloy can also be controlled by processing and heat treatment.

Alpha- and near-alpha-phase alloys have been designed for high-temperature creep resistance by introducing processing steps that result in an acicular alpha-phase morphology with little or no retained beta phase. On the other hand, this class of alloy is more effective and less notch sensitive in cryogenic applications when the microstructure consists of an equiaxed alpha-phase morphology. Alpha-phase alloys are more amenable to welding than other types of Ti alloys because the problem with segregation of beta-stabilizing elements is absent.

Alpha+beta phase alloys have been developed for improved tensile, fatigue, and fracture toughness properties compared to those of the alpha-phase alloys. A Widmanstätten alpha-phase morphology provides better resistance to fatigue-crack growth rates and fracture-toughness failures than an equiaxed morphology. Conversely, superplastic forming of Ti alloys is most efficient when the microstructure consists of equal volume fractions of equiaxed alpha and beta phases.

Metastable beta-phase alloys are more easily formed than the alpha or alpha+beta alloys. Because they are precipitation-hardenable, the beta alloys can be processed in the BCC state and subsequently aged to high strength. Applications include massive components that require a deep hardening capability.

Acknowledgements. The metallographic specimens and thin-foil specimens, as well as the light micrographs presented in this paper, were prepared either by Robert Spurling or Michael Calabrese of the Rockwell International Science Center. I am also grateful to J. C. Chesnutt, C. H. Hamilton, M. R. Mitchell and J. C. Williams for permission to publish photomicrographs of their research specimens.

REFERENCES

1. Jaffee, R. I., and Promisel, N. E., eds. *The Science, Technology and Application of Titanium.* New York: Pergamon Press, Ltd., 1970.
2. Inomata, S.; Goto, A.; Yano, K.; Tsuchimoto, M.; Shibata, S.; Fujii, T.; Sakurai, T.; and Kanamoto, M. On the explosive bonding and forming of titanium. Ibid., pp. 1065–1080.
3. Zwicker, U.; Buhler, K.; Muller, R.; Beck, H.; Schmid, H. J.; and Ferstl, J. Mechanical properties and tissue reactions of a titanium alloy for implant material. In *Titanium '80 Science and Technology,* edited by Kimura, H., and Izumi, O., pp. 505–518. Warrendale, PA: Met. Soc. AIME, 1980.
4. Ogden, H. R., and Holden, F. C. Metallography of titanium alloys. *TML Report No. 103.* Columbus OH: Battelle Memorial Inst., 1958.
5. Frost, P. D.; Parris, W. M.; Hirsch, L. L.; Doig, J. R.; and Schwartz, C. M. Isothermal transformation of titanium-chromium alloys. *Trans. ASM* 46:231–256 (1954).
6. Holden, F. C., and Young, A. D. Electron micrographic study of aging in a beta titanium alloy. *Trans. Met. Soc. AIME* 212:287–288 (1958).
7. Fentiman, W. P.; Goosey, R. E.; Hubbard, R. T. J.; and Smith, M. D. Exploitation of a simple alpha titanium alloy base in the development of alloys of diverse mechanical properties. In *The Science, Technology and Application of Titanium,* edited by Jaffee, R. I., and Promisel, N. E., pp. 987–999. New York: Pergamon Press, Ltd., 1970.
8. Zwicker, U., and Katsch, E. Cracking of titanium alloys under stress during oxidation in air. Ibid., pp. 299–306.
9. Coyne, J. R. The beta forging of titanium alloys. Ibid., pp. 97–110.
10. Green, T. E., and Minton, C. D. T. The effect of beta processing on properties of titanium alloys. Ibid., pp. 111–119.
11. Frederick, S. F., and Hanna, W. D. Fracture toughness and deformation of titanium alloys at low temperatures. *Met. Trans.* 1:347–352 (1970).
12. Kennedy, J. Fatigue behavior of solution treated and quenched Ti–6Al–4V. *Grumman Res. Dept. Report RE-630.* 1981.
13. Lee, D., and Backofen, W. A. Superplasticity in some titanium and zirconium alloys. *Trans. TMS-AIME* 239:1034–1040 (1967).
14. Hall, I. W., and Hammond, C. The relationship between crack propagation characteristics and fracture toughness in alpha+beta alloys, In *Titanium Science and Technology,* edited by Jaffee, R. I. and Burte, H. M., pp. 1365–1376. New York: Plenum Press, 1973.
15. Thompson, A. W.; Williams, J. C.; Frandsen, J. D.; and Chesnutt, J. C. The effect of microstructure on fatigue crack propagation rate in Ti–6Al–4V. In *Titanium and Titanium Alloys,* edited by Williams, J. C., and Belov, A. F., pp. 691–704, New York; Plenum Press, 1982.
16. Spurling, R. A. Rockwell International Science Center, Thousand Oaks, CA. Unpublished research.

17. Cortes, F. R. Electrolytic polishing of refractory metals. *Metals Progress* 88:97–100 (1961).
18. Rice, L.; Hinesley, C. P.; and Conrad, H. Techniques for optical and electron microscopy of titanium. *Metallog.* 4:257–268 (1971).
19. Blackburn, M. J.; and Williams, J. C. The preparation of thin foils of titanium alloys. *Trans. TMS-AIME* 239:287–288 (1967).
20. Spurling, R. A. A technique for preparing thin foils of Ti and Ti alloys for transmission electron microscopy. *Met. Trans.* 6A:1660–1661 (1975).
21. Spurling, R. A.; Rhodes, C. G.; and Williams, J. C. The microstructure of Ti alloys as influenced by thin foil artifacts. *Met. Trans.* 5:2597–2600 (1974).
22. Rhodes, C. G., and Paton, N. E. Formation characteristics of the alpha/beta interface phase in Ti–6Al–4V. *Met. Trans.* A 10A:209–216 (1979).
23. Banerjee, D., and Williams, J. C. The effect of foil preparation technique on interface phase formation in Ti alloys. *Scr. Met.* 17:1125–1128 (1983).
24. Burgers, W. G. On the process of transition of the cubic-body-centered modification into the hexagonal-close-packed modification of zirconium. *Physica* 1:561–586 (1934).
25. Jaffee, R. I. Metallurgical synthesis. In *Titanium Science and Technology*, edited by Jaffee, R. I., and Burte, H. M., pp. 1665–1693. New York: Plenum Press, 1973.
26. Williams, J. C., and Belov, A. F., eds. *Titanium and Titanium Alloys*. New York: Plenum Press, 1982.
27. Kimura, H., and Izumi, O., eds. *Titanium '80 Science and Technology*. Warrendale, PA: Metallurgical Society of AIME, 1980.
28. Rosenberg, H. W. Titanium alloying in theory and practice. In *The Science, Technology and Applications of Titanium*, edited by Jaffee, R. I., and Promisel, N. E., pp. 851–859. New York; Pergammon Press, Ltd., 1970.
29. Bohanek, E., and Kessler, H. D. An advanced titanium alloy for service at temperatures in excess of 800°F. In *Reactive Metals*, vol. 2, edited by Clough, W. R., pp. 23–41. NY: Interscience Publishers, (1959).
30. Blackburn, M. J. Relationship of microstructure to some mechanical properties of Ti–8Al–1Mo–1V. *Trans. ASM* 59:694–708 (1966).
31. Brown, B. F. ASTM Annual Meeting, cited in Ref. 32.
32. Beck, T. R. Stress corrosion cracking of titanium alloys. *J. Electr. Soc.* 114:551–556 (1967).
33. Jackson, J. D., and Boyd, W. K. Stress corrosion cracking in titanium and titanium alloys. In *The Science, Technology and Application of Titanium*, edited by Jaffee, R. I., and Promisel, N. E., pp. 267–281. New York: Pergamon Press, Ltd., 1970.
34. Erdeman, V. J., and Ross, E. W. Long time stability of Ti–679 after creep exposure for times to 15,000 hour. Ibid., pp. 829–837.
35. Rosenberg, H. W. High temperature alloys. In *Titanium Science and Technology*, edited by Jaffee, R. I., and Burte, H. M., pp. 2127–2140. New York: Plenum Press, 1973.
36. Paton, N. E., and Mahoney, M. W. Creep of titanium-silicon alloys. *Met. Trans.* A 7A:1685–1694 (1976).
37. Neil, D. F. and Blenkinsop, P. A. Effect of heat treatment on structure and properties of IMI829. In *Titanium '80 Science and Technology*, edited by Kimura, A., and Izumi, O., pp. 1287–1297. Warrendale, PA: Metallurgical Society of AIME, 1980.
38. Blenkinsop, P. A. IMI Titanium Limited, Birmingham, England. Private communication.
39. Rhodes, C. G.; Paton N. E.; and Mahoney, M. W. Creep properties of Ti–8Al–5Nb–5Zr–0.25Si. In *Titanium Science and Technology*, edited by Lütjering, G.; Zwicker, U.; and Bunk, W., pp. 2355–2361. Oberursel, W. Germany: Deutsche Gesellschaft für Metallkunde V., 1985.
40. Klier, E. P., and Feola, N. J. Notch tensile properties of selected titanium alloys. *Trans. TMS-AIME* 209:1271–1277 (1957).
41. Ogden, H. R.; Douglass, R. W.; Holden, F. C.; and Jaffee, R. I. The notch sensitivity of Ti–5Al–2.5Sn, Ti–6Al–4V and Ti–2Fe–2Cr–2Mo titanium alloys. *Trans. Met. Soc.* AIME 221:1235–1240 (1961).
42. Becker, D. W.; Messler, R. W. Jr.; and Baeslack, W. A., III. Titanium welding. In *Titanium '80 Science and Technology*, edited by Kimura, A., and Izumi, O., pp. 255–275. Warrendale, PA: Metallurgical Society of AIME, 1980.
43. Craighead, C. M.; Simmon, O. W.; and Eastwood, L. W. Ternary alloys of titanium. *Trans. AIME* 188:514–538 (1950).
44. Berryman, R. G.; Froes, F. H.; Chesnutt, J. C.; Rhodes, C. G.; Williams, J. C.; and Malone, R. F. High toughness titanium alloy development. *Technical Report TFD-74–657*, Naval Air Systems Command, Washington, D.C., 1974.
45. Williams, J. C.; Froes, F. H.; Chesnutt, J. C.; Rhodes, C. G.; and Berryman, R. G. Development of high-fracture toughness titanium alloy. In *Toughness and Fracture Behavior of Titanium*, STP 651, pp. 64–114. Philadelphia: Am. Soc. for Testing and Materials, 1978.
46. Chesnutt, J. C.; Rhodes, C. G.; and Williams, J. C. Relationship between mechanical properties, microstructure and fracture topography in alpha+beta titanium alloys. In *Fractography—Microscopic Cracking Processes*, ASTM STP 600, pp. 99–138. Philadelphia: Am. Soc. for Testing and Materials, 1976.
47. Hamilton, C. H.; Stacher, G. W.; Mills, J. A.; and Li, H. Superplastic forming of titanium structures. AFML-TR-76-62, Wright-Patterson AFB, OH, April 1976.
48. Edington, J. W. Physical metallurgy of superplasticity. *Met. Tech.* 3:138–151 (1976).
49. Wert, J. A., and Paton, N. E. Enhanced superplasticity and strength in modified Ti–6Al–4V alloys. *Met. Trans.* 14A:2535–2544 (1983).
50. Hammond, C. Superplasticity in titanium base alloys. In *Superplastic Forming of Structural Alloys*, edited by Paton, N. E., and Hamilton, C. H., pp. 131–145. Warrendale, PA: TMS-AIME, 1982.
51. Blackburn, M. J.; and Smyrl, W. H. Stress corrosion and hydrogen embrittlement. In *Titanium Science and Technology*, edited by Jaffee, R. I., and Burte, H. M., pp. 2577–2609. New York: Plenum Press, 1973.
52. Kerr, W. R.; Smith, P. R.; Rosenblum, M. E.; Gurney, F. J.; Mahajan, Y. R.; and Bidwell, L. R. Hydrogen as an alloying element in titanium (Hydrovac). In *Titanium '80 Science and Technology*, edited by Kimura, A., and Izumi, O., pp. 2477–2846. Warrendale, PA: Metallurgical Society of AIME, 1980.
53. Molinier, R.; Moulin, J.; and Syre, R. A. A study of the metallurgical characteristics of Ti–6Al–6V–2Sn alloy. In *The Science, Technology and Application of Titanium*, edited by Jaffee, R. I., and Promisel, N. E., pp. 979–982. New York: Pergamon Press, Ltd., 1970.
54. McLean, D. *Mechanical Properties of Metals*: New York: John Wiley & Sons, Inc., 1962.
55. Petersen, V. C.; Bomberger, H. B.; and Vordahl, M. B. An age hardening titanium alloy. *Metal Progress* 76:119–122 (1959).
56. Hunter, D. B.; and Arnold, S. V. Metallurgical characteristics and structural properties of Ti–8Mo–8V–2Fe–3Al sheet, plate and forgings. In *The Science, Technology and Application of Titanium*, edited by Jaffee, R. I., and Promisel, N. E., pp. 959–967. New York: Pergamon Press, Ltd., 1970.
57. Bohanek, E. Deep hardenable titanium alloys for large airframe elements. In *Titanium Science and Technology*, edited by Jaffee, R. I., and Burte, H. M., pp. 1993–2007. New York: Plenum Press, 1973.
58. Petersen, V. C.; Froes, F. H.; and Malone, R. F. Metallurgical characteristics and mechanical properties of Beta III, a heat treatable titanium alloy. Ibid., pp. 1969–1980.
59. Williams, J. C., and Rhodes, C. G. Rockwell International Science Center, Thousand Oaks, CA. Unpublished research.
60. Williams, J. C.; Froes, F. H.; and Yolton, C. F. Some observations on the structure of Ti–11.5Mo–6Zr–4.5Sn (Beta III). *Met. Trans.* 11A:356–358 (1980).
61. Hagemeyer, J. W., and Gordon, D. E. Properties of two beta titanium alloys after aging at several different temperatures. In *The*

Science, Technology and Application of Titanium, edited by Jaffee, R. I., and Promisel, N. E., pp. 1957–1968. New York: Pergamon Press, Ltd., 1973.
62. Rhodes, C. G., and Paton, N. E. The influence of microstructure on mechanical properties in Ti–3Al–8V–6Cr–4Mo–4Zr (Beta C). *Met. Trans.* 8A:1749–1861 (1977).
63. Chen, C. C.; Hall, J. A.; and Boyer, R. R. High strength beta titanium alloy forgings for aircraft structural applications. In *Titanium '80 Science and Technology*, edited by Kimura, A., and Izumi, O., pp. 459–466. Warrendale, PA: Metallurgical Society of AIME, 1980.
64. Hicks, A. G.; Nelson, G. W.; and Rosenberg, H. W. Beta titanium foil. *J. Met.* 34 (12):A35 (1982).
65. Rosenberg, H. W. Ti–15–3: A new cold-formable sheet titanium alloy. *J. Met.* 35 (11):30–34 (1983).

15
USE OF MICROSCOPY IN FAILURE ANALYSIS

Iain Le May

Metallurgical Consulting Services Ltd.

Microscopy is one of the basic tools employed in conducting a failure analysis, its purpose being to examine fracture surfaces (the science of fractography) to determine their morphology and, hence, the mechanisms of fracture. In addition, it is used to examine the microstructure of failed components to determine the phases present, heat treatment, and processing history, as well as the path of the primary and any secondary cracks. Thus, microscopic examination can provide information pinpointing the origin, the path, and the nature of the fracture and whether improper processing, for example, contributed to it.

Depending on the nature of the failure, microscopy may be utilized at various levels of magnification, ranging from low-power binocular viewing of the fracture surface to high-resolution transmission electron microscope (TEM) examination of detailed microstructural features. It is important that fractures be examined on a macroscopic level first, before being examined at higher magnification using optical, scanning electron microscope (SEM) or TEM techniques (utilizing replicas of the surface in the case of the latter). Specimens for metallographic examination of the microstructure should also be taken only after the relevant parts of the component and its surfaces have been examined and recorded photographically.

As general sources of information on the use of microscopy in failure analysis, the reader is referred to volume 10 of the *Metals Handbook*[1] and to the symposium proceedings edited by McCall and French.[2] More detailed discussion of metallographic methods will be found elsewhere,[3] and detailed examples of fracture surfaces and fractography are also available for comparison.[4]

In this chapter, it is proposed to discuss the various failure mechanisms that arise, indicating some of the distinctive fractographic and metallographic features to be observed with the use of microscopy techniques and providing some guidance to the metallographer in identifying the causes of failure of a component or structure. It must be emphasized, however, that there is no substitute for practical "hands on" experience in failure analysis. It is all too easy to overemphasize details without attaining a sufficiently broad overall view—a classic case of not seeing the wood for the trees.

FAILURE MECHANISMS

Overload

The overload failures of greatest familiarity are probably those arising from laboratory testing in which the whole fracture surface is rapidly formed. In few service situations does the complete fracture take place as a result of overloading; in most cases, the overload failure occurs in the material remaining after a crack has developed with time or under repeated loading, sometimes from a preexisting defect. The final fracture is almost always of an overload type.

Ductile Fracture. Ductile fracture involves extensive plastic deformation during separation of the two parts of a fracture surface. Various types of ductile fracture may be identified, but ductile overload fracture is normally characterized by the formation of many voids at inclusions in the material, leading to the development of a fracture surface displaying dimples where these voids have linked up. Figure 15-1 shows an example of such a fracture surface, the inclusions from which the voids nucleated being clearly seen.

Brittle Fracture. Brittle fracture involves only very localized deformation during separation of the two fracture surfaces; the surfaces thus have the appearance of brittle separation, somewhat similar to the appearance of broken glass. In an ideally brittle material, the atoms separate along particular crystallographic planes with no plastic deformation, and the energy dissipated during fracture is only that required to separate the planes, i.e., that needed to form the new surfaces. In practice, even in the most brittle of materials, some plastic deformation occurs on a localized scale, and the energy required to

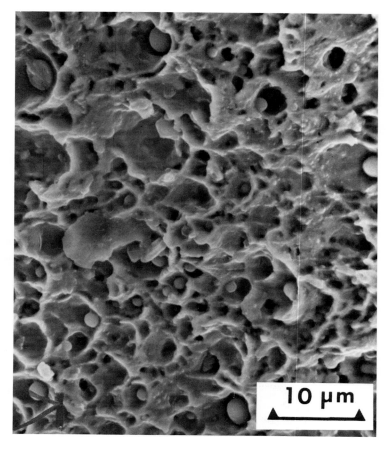

Fig. 15-1. Ductile fracture in the welded region of a structural steel; note the inclusions within the voids, from which the latter nucleated (SEM view).

cause fracture is significantly greater than that for an ideally brittle solid.

Brittle fractures are promoted by low-temperature conditions and by the presence of triaxial tensile stresses such as those that may be developed at notches and sharp changes in section. They are more commonly found in body-centered-cubic metals and alloys such as iron and structural steel and hexagonal-close-packed metals such as titanium and zinc, but not normally observed in face-centered-cubic metals such as aluminum or copper.

The term *cleavage* is used to characterize separation along crystallographic planes; because of its nature, it leads to the formation of a fracture surface consisting of flat, shiny facets. The crack changes its orientation in crossing a grain boundary because of the different grain orientations and because of the nature of crack propagation on specific families of planes (e.g., on {100} in α-iron). Many small cracks may be formed when the crack crosses a twist subgrain or grain boundary, with fine cleavage steps between them. The parallel cracks join together gradually, the fine steps coalescing to form a larger one, and thus the river patterns characteristic of cleavage fracture are produced. Figure 15-2 shows an example of this.

Another characteristic feature of cleavage cracks is the presence of tongues, as shown in Fig. 15-3. These are produced by local fracture along the interfaces between the matrix and a twin, the mechanism being indicated in more detail elsewhere.[5]

When the fracture is less crystallographic, the term *quasi-cleavage* is often used to describe the resulting brittle fracture. An example is shown in Fig. 15-4, the brittle fracture having been promoted in this case by a high

Fig. 15-2. Cleavage fracture in a mild steel bolt, showing characteristic river patterns (two-stage, plastic-carbon replica viewed in the TEM).

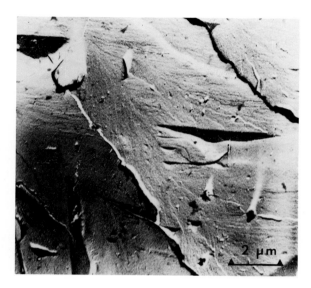

Fig. 15-3. Tongues on the surface of cleaved mild steel (two-stage, plastic-carbon replica viewed in the TEM).

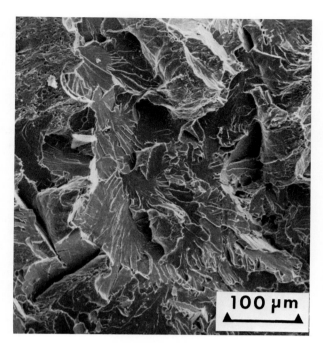

Fig. 15-4. Quasi-cleavage in a structural steel (SEM view).

degree of triaxiality and a rapid overload, allowing little time for plastic flow to occur.

A very useful feature of many brittle fractures, so far as the failure analyst is concerned, is that running brittle cracks often produce a series of chevron marks on the fracture surface, pointing back arrowlike to the fracture origin. These can be recognized macroscopically and enable the metallographer to concentrate on the particularly important region at which the fracture initiated.

Fatigue

Fatigue failure, which involves crack initiation and growth under cyclic or repeated loading, is probably the most common mechanism responsible for structural failure. In many cases, it is possible to see evidence of fatigue fracture almost at a glance, since distinctive "clam shell" or beach marks may be present on the surface. These are formed either when the crack stops growing for a period because of a reduction in load and then restarts or when the load level changes markedly. Figure 15-5 gives an example of this.

In other cases, it may be impossible to determine if the fracture was of a fatigue nature by macroscopic viewing, and more detailed examination may be required. Because of the localized nature of the failure at the tip of the crack and the fact that the crack front moves forward by an incremental amount during each tensile load cycle, it is frequently found that striations are left on the fracture surface, lying at right angles to the direction of crack growth. These are shown in Fig. 15-6, and their presence is an almost certain indication that the failure took place by fatigue.

Another indicator that is an absolutely certain sign of fatigue fracture is the presence of "tire tracks." These are caused by the entrapment of particles or inclusions between the fracture surfaces, leaving a series of indentations as the crack advances and the particles move forward. An example is shown in Fig. 15-7.

It is worth pointing out that when fatigue occurs at high levels of mean stress, the process may be effectively one of progressive ductile rupture, and there may be little evidence of striations on the fracture surfaces. Similarly, with fatigue at high temperature, the presence of striations is by no means assured, and operating factors and load history may be of much greater importance than surface fracture appearance in determining the failure mode.

Fig. 15-5. Beach marks formed as a fatigue crack grew from the outside of a crankshaft. The fracture origin is arrowed.

Fig. 15-6. Fatigue striations on the fracture surface of AISI 4140 steel (Fe–0.4%C–1%Mn–1%Cr–0.2%Mo) in the quenched and tempered condition (two-stage, plastic-carbon replica viewed in the TEM).

Environmentally Assisted Fracture

A number of different situations may be identified in which the combined action of stress and environment can lead to failure. These include stress-corrosion cracking (SCC), corrosion fatigue, and fretting, which will be dealt with in turn.

Stress-Corrosion Cracking. This is an important failure mechanism, particularly in petrochemical and chemical plant, but it can occur in many situations where the alloy in contact with a liquid would not be attacked by a corrosion process alone, but, in combination with stress, is subject to localized attack. The mechanism is illustrated schematically in Fig. 15-8, and it may be seen that the corrosive attack is limited to the tip of a propagating crack. After crack growth, the fracture surfaces become passivated or covered by a protective film, which is ruptured when further slip occurs at the crack tip, producing fresh surface on which corrosive attack can occur. The process continues and the growth continues in a stepwise manner.

The crack path in SCC may be either transgranular or intergranular, depending on the particular material, corrosive environment, and temperature. If a material is particularly prone to grain-boundary attack—such as an austenitic stainless steel in which carbides have precipitated at grain boundaries, depleting the adjacent regions in chromium—widespread attack may result in an aggressive environment, for example, one containing chloride ions in solution.

In other cases, the cracking may be transgranular, with little or no crack branching, as shown in Fig. 15-9. All SCC fractures, however, are characterized by little ductility during the fracturing process.

The fracture-surface morphology produced can vary widely, an oxide coating often being present on the surface. Figure 15-10 shows what could easily be mistaken

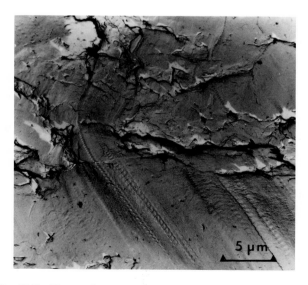

Fig. 15-7. Tire tracks on the fatigue fracture surface of AISI 4140 steel in the quenched and tempered condition (two-stage, plastic-carbon replica viewed in the TEM).

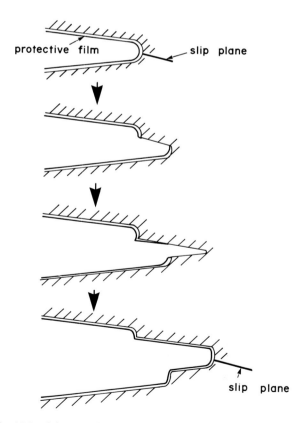

Fig. 15-8. Schematic illustration of the mechanism of stress-corrosion cracking, the crack propagating by rupture of the protective film by plastic deformation at the crack tip; subsequent corrosion, repassivation, and building up of stress repeat the cycle.

USE OF MICROSCOPY IN FAILURE ANALYSIS

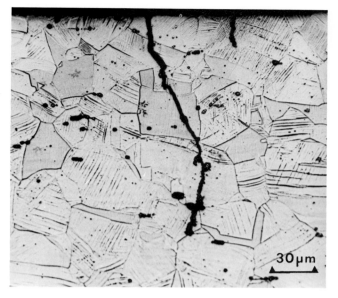

Fig. 15-9. Stress-corrosion cracking in an AISI 316 austenitic stainless steel (Fe–0.08%C–17%Cr–12%Ni–2%Mo). Etched in diluted aqua regia.

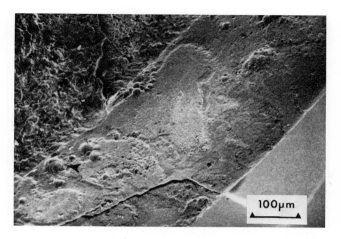

Fig. 15-11. Fracture surface of AISI 4140 steel bolt from a condenser after exposure and fracture in a wet H_2S-containing environment. The arrow indicates a pit in the treads from which a stress corrosion crack initiated. (SEM view.)

for a fatigue-fracture surface, with striations at right angles to the direction of crack growth and linear features, termed *tunnelling*, running in the crack growth direction.

The view shown in Fig. 15-11 is of the fracture surface at the root of the threads in an alloy steel bolt (AISI 4140) from a condenser exposed to a wet H_2S-containing environment. The rough, corroded nature of the fracture surface may be seen, together with pitting attack on the thread faces. In this case, cracks could be traced as originating at pits—a common initiation point for SCC.

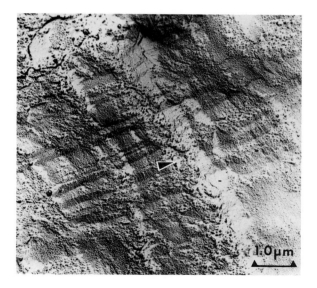

Fig. 15-10. Stress-corrosion cracking striations on the fracture surface of AISI 316L austenitic stainless steel (Fe–0.03%C–17%Cr–12%Ni–2%Mo) with linear features parallel to crack growth direction (arrowed), indicative of "tunnelling." (Two-stage, plastic-carbon replica viewed in the TEM.)

Corrosion Fatigue. Fatigue resistance and fatigue life are reduced in the presence of corrosive conditions. The fracture surface produced during corrosion fatigue sometimes shows no evidence of the fine striations normally associated with fatigue, such features being obscured by the action of the corrosive medium or by the presence of corrosion deposits. At other times, in aluminum alloys for example, more brittle striations, suggesting crystallographic fracture, are produced in these conditions.

Fretting. Fretting invoves relative oscillatory motion between contacting surfaces under vibration or a cyclic stress. The amplitude of the movement is of the order of 25 μm or less. The surface debris produced consists of small metallic platelets covered in oxide, and eventually the layer may consist entirely of oxide. Where the fretting action arises from cyclic stressing of one of the contacting components, fatigue cracks may result from the fretting.

Microscopy may be utilized to characterize the damage, look for the formation of cracks, and examine the nature of the debris. More detailed discussion is provided in the book by Waterhouse.[6]

High-Temperature Failures

Components that operate at high temperature are subject to deterioration and failure by oxidation or by continued deformation that will eventually lead to fracture. Such time-dependent deformation is termed *creep*. In some components, the limiting, or failure, condition is the build-up of excessive plastic deformation; in others, the failure may be by rupture, e.g., of a superheater tube. It is the latter type of failure that is of greatest concern here.

Creep Failures. At high temperatures and with low strain rates, as occurs under normal service conditions

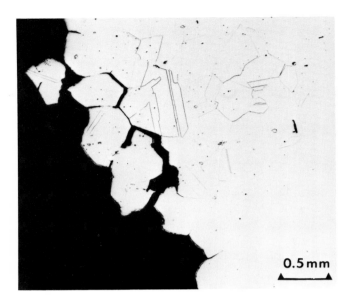

Fig. 15-12. Intergranular creep failure of a nickel-base superalloy gas-turbine blade (dilute aqua regia etch).

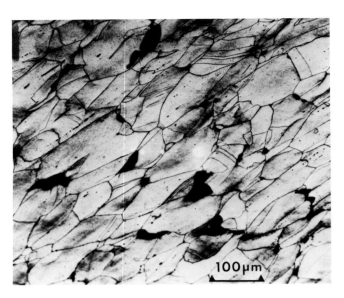

Fig. 15-14. Wedge-type cavities in AISI 316 austenitic stainless steel that failed from creep (dilute aqua regia etch).

for such components, polycrystalline metals that normally fracture in a transcrystalline ductile manner undergo fracture of an intergranular nature. Thus, the majority of creep failures in service are intergranular. A typical example is shown in Fig. 15-12.

The development of intergranular creep fracture depends on the nucleation, growth, and subsequent linking up of voids on the grain boundaries. Two types of void may be identified as of "r", or rounded, type and "w," or wedge, type, respectively. They are illustrated in Figs. 15-13 and 15-14, respectively.

Wedge cracks form at triple points as a result of grain-boundary sliding and may be promoted by decohesion at interfaces between grain-boundary precipitates and the matrix, as shown in Fig. 15-15.

Creep fractures occur by the growth and linking of the voids, and the morphology can vary widely. Figure 15-16 shows a fracture formed by the linking up of r-type voids, which may be contrasted with the section shown in Fig. 15-12, in which grain-boundary sliding may be identified. Figure 15-17 shows the fracture surface of the failure shown in cross section in Fig. 15-16.

In some cases, there may be precipitates identified on the fracture surfaces, as shown in Fig. 15-18, in which failure was initiated at the precipitate-matrix interfaces shown in Fig. 15-15.

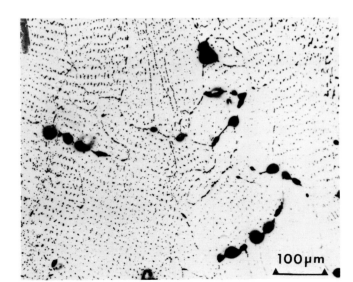

Fig. 15-13. Rounded-type voids in AISI 316L weld metal that failed from creep at high temperature (dilute aqua regia etch).

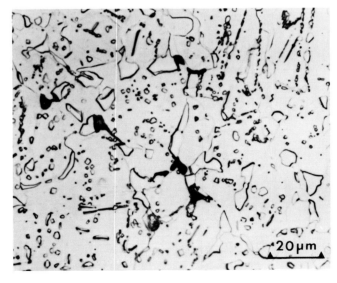

Fig. 15-15. Decohesion along particle/matrix interfaces during creep fracture of AISI 316 stainless steel (dilute aqua regia etch).

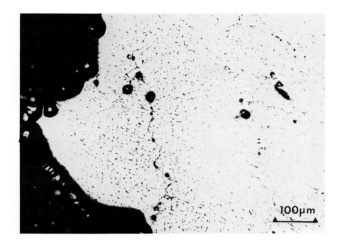

Fig. 15-16. Creep fracture in a weldment in austenitic stainless steel, showing linking of voids to cause fracture (dilute aqua regia etch).

In conducting an examination of a component that has failed by creep, it is important to determine if there have been changes in the original microstructure as a result of the long exposure to temperature and stress. For example, in an unstable structure, grain growth may have occurred or precipitation of an embrittling precipitate may have taken place. Also, if a steel component has been exposed to excessive temperature, decarburization may have occurred in an oxidizing environment, weakening the material; an example is shown in Fig. 15-19. Other effects that may sometimes be seen are carburization—and resulting embrittlement—in tubes used in petrochemical processing plants at high temperature, and the build-up of heavy deposits that can lead to local overheating because of reduced heat transfer.

In concluding this section on creep failures and their assessment using microscopy, a potentially important de-

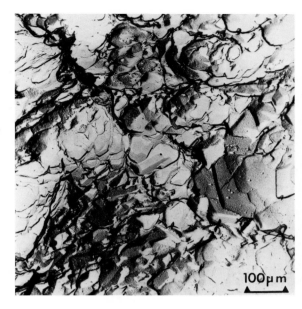

Fig. 15-18. Fracture surface of creep failure shown in cross section in Fig. 15-15, the precipitates being clearly seen on the fracture surface (two-stage, plastic-carbon replica viewed in the TEM).

velopment is the utilization of metallographic procedures for the examination of components that have been operating at high temperature for an extended period in order to assess the extent of the damage and the remaining safe life. In some cases, this can be done by replicating the outside of the tube or other component after it has been prepared by local polishing and etching. The replica can then be examined in the laboratory to determine the extent of void formation or fine cracking at the surface. This subject has recently been reviewed,[7] and further developments are to be expected.

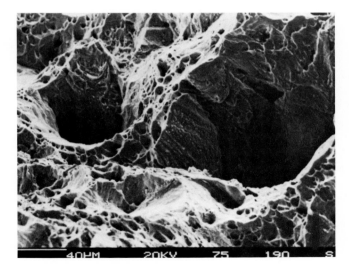

Fig. 15-17. Fracture surface of creep failure shown in Fig. 15-16, which occurred as a result of void growth and ductile rupture of ligaments between the voids (SEM view).

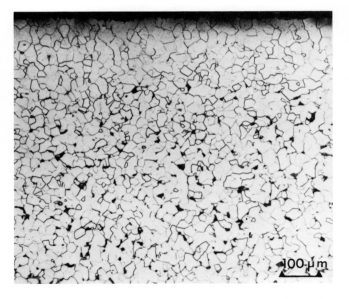

Fig. 15-19. Decarburized surface layer in boiler tube on the fire side (etched in nital).

Hydrogen-Related Failures

Hydrogen can reduce the fracture resistance of materials—in particular, of steels—and can also affect their plastic deformation. There is not a clear agreement on the mechanisms, but the metallographic features are not in any dispute.

Hydrogen-Assisted Cracking. This kind of cracking can occur in many components, for example, in pipeline steels in the presence of carbonate solutions or hydrogen sulfide, both of which contribute to reactions with release of hydrogen. Figure 15-20 shows an example of hydrogen-assisted cracking in a low-carbon steel.

The fractures produced are normally transgranular, but this result depends on the microstructure and the level of stress present. The crack shown in Fig. 15-20, for example, is largely intergranular. The similarity to SCC should be noted, and hydrogen may play a significant part in the occurrence of the latter.

The fracture surfaces may show cleavage facets, although small dimples have been found on apparently cleaved areas from specimens fractured in the presence of hydrogen.[8] In most cases, the fracture surfaces are clean and free from corrosion products. This is not always the case, however, as the liquid may contaminate and react with the fracture surfaces after the crack has developed.

Hydrogen Embrittlement. Embrittlement by hydrogen can occur either in service in hydrogen-containing environments or through its introduction during processing or fabrication. Hydrogen tends to accumulate at defects, producing a high internal pressure that may lead to rupture, particularly in alloys of high strength level. Hydrogen pickup can arise from moisture in the scrap used in steelmaking, the gas having become trapped during solidification; it may arise from pickling or in a plating operation involving cadmium or zinc; it may be introduced during welding if the electrodes have a moist coating or if the coating is not of the low-hydrogen type; or it may result from a hydrogen-containing service environment, as already noted.

Figure 15-21 shows schematically some of the features to be found in steel contaminated with hydrogen. Surface blistering is obvious, but microscopic examination of the cross section is required to show the occurrence of the internal cracking. Of particular concern is the stepwise cracking, which can occur through the build-up of hydrogen at parallel colonies of rolled-out MnS inclusions and the linking of cracks through the section.

Figure 15-22 shows an example of hydrogen cracking in the heat-affected zone (HAZ) of a weld in a high-strength steel plate, the electrode coatings not having been properly dried out. Such cracks, occurring in the HAZ, may not be visible on the surface, and they may grow gradually during service, potentially leading to failure.

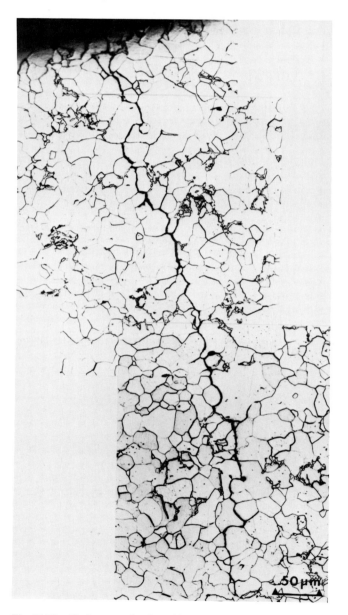

Fig. 15-20. Hydrogen-assisted cracking in a low-carbon pipeline steel; note the largely intergranular path with extensive branching (etched in nital).

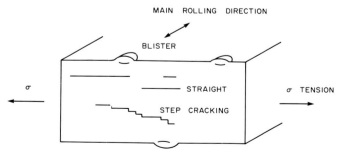

Fig. 15-21. Schematic view of features found in hydrogen-embrittled carbon steel.

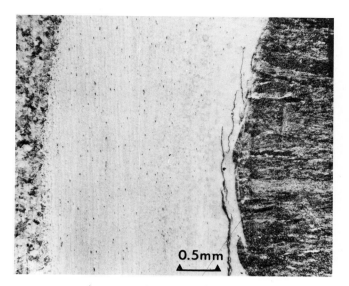

Fig. 15-22. Hydrogen cracking in the heat-affected zone of a high-alloy steel plate welded with improperly dried electrodes (etched in nital).

FAILURE ANALYSIS METHODOLOGY

To conclude this chapter, it is relevant to review the methodology of failure analysis, putting the use of microscopy into its proper context. The main stages in a failure analysis are as follows:

1. Collection of background information and relevant specifications and selection of samples for examination.
2. Visual examination of the failed part(s) and photographic recording.
3. Nondestructive examination of the part(s).
4. Mechanical testing as required, taking care not to destroy critical parts.
5. Selection and cleaning of specimens for more detailed examination.
6. Macroscopic examination.
7. Microscopic examination of fractures and surfaces.
8. Preparation of metallographic specimens and microscopic examination of them.
9. Chemical analyses as required.
10. Loading analysis of the component or structure.
11. Testing under service or related conditions, as required.
12. Analysis of evidence, determination of failure mechanisms, sequence, and cause, and preparation of report.

As may be seen, the use of microscopy is often a key factor in the failure investigation. In some cases, the mechanisms and cause of failure may be obvious from the visual examination alone, and it may be possible to move directly to an analysis of loading conditions and then prepare the report. In others, very detailed microscopic analysis may be required. It is very important to observe a logical approach, however, and not to destroy evidence before it has been thoroughly documented. Similarly, it is essential to record the exact locations from which specimens for metallographic and fractographic analysis are taken and to label the specimens appropriately.

CONCLUSION

The metallographer has a key role to perform in many failure analyses through application of microscopy techniques. The various possible failure mechanisms must be considered, and the metallographer must be familiar with the characteristic features of each. It is hoped that the limited number of illustrations and examples provided here will at least point the way to the proper use of microscopy in failure analysis. The interested reader is urged to consult more complete works of reference, some of which have been mentioned already.

REFERENCES

1. *Failure Analysis and Prevention.* Metals Handbook, 8th ed., vol. 10. Metals Park, OH: American Society for Metals, 1975.
2. McCall, James L., and French, P. M., eds. *Metallography in Failure Analysis.* New York: Plenum Press, 1978.
3. *Metallography, Structures and Phase Diagrams.* Metals Handbook, 8th ed., vol. 8. Metals Park, OH: American Society for Metals, 1973.
4. *Fractography and Atlas of Fractographs.* Metals Handbook, 8th ed., vol. 9. Metals Park, OH: American Society for Metals, 1974.
5. Le May, Iain. *Principles of Mechanical Metallurgy.* New York: Elsevier, 1981.
6. Waterhouse, R. B. *Fretting Corrosion.* Oxford: Pergamon Press, 1972.
7. Le May, I., and da Silveira, T. L. *Proceedings of the Second International Conference on Creep and Fracture of Engineering Materials and Structures,* pp. 1117–1133. Swansea: Pineridge Press, 1984.
8. Lynch, S. P. Hydrogen embrittlement and liquid-metal embrittlement in nickel single crystals. *Scripta Metall.* 13:1051–56 (1979).

16
MICROSCOPY AND THE STUDY OF WEAR

William A. Glaeser

Battelle–Columbus Division

As soon as man began to use tools, wear became a problem. The history of metallurgy, or man's continuing quest for tougher and harder materials, can be seen to have been motivated by the desire to have tools, especially if weapons may be categorized as tools, that will hold a sharp edge. To more peaceful ends, farmers once embedded sharp stones in wooden plow faces to increase their useful life. It has only been in the past 50 years, however, that man has recognized the economic impact of wear prevention and done more than just take wear for granted.

The first attempts to control wear involved mostly trial and error. In general, it was found that the harder the material, the more resistant it was to wear. Wear testing of a large number of engineering materials produced conflicting data and unpredictable results. Wear properties often proved to be influenced by environmental and operating conditions. At the same time, failure analysis of worn parts indicated that microstructure appeared to influence the wear process. Furthermore, a recognition of several wear processes (abrasive wear, adhesive wear, erosive wear) came into being.

Classification of wear modes necessitated close inspection of the microtopography of the wear scars produced. The wear processes were found to involve microscopic events such as micrometer-size scratches and the transfer of submicrometer-size particles. Microscopy has become as important to the diagnosis of causes of wear failures as it has to fractography. With the development of the scanning electron microscope (SEM), wear characterization has advanced considerably because the depth of focus of the SEM has made possible the scanning and photographing of rough microtopography generated by wear that, for the most part, would be out of focus in a light microscope.

Recognition and classification of wear modes require some experience in the interpretation of micrographs of worn surfaces. For instance, adhesive wear, or the transfer of material from one surface to another, is often difficult to distinguish from mild abrasion. Careful examination of the microscopic ridges and valleys formed by the high-point contacts is required to detect the differences caused by occasional adhesion points and the resulting scoring damage. In the abrasive wear of mining machinery, it is necessary to distinguish between gouging abrasion and low-stress abrasion when choosing a microstructure to control the wear rate. In a recent investigation, recognition of the difference between solid-particle erosion and liquid-drop erosion in an air compressor was necessary before the erosion problem could be solved.

Verification of wear test methods often requires characterization of the wear mode. This provides an excellent basis for determining how closely actual operating conditions have been simulated. It will also show whether, in accelerating the wear by increasing load or speed, or both, the wear mode has been changed and a different material response has occurred.

Research into the mechanisms of wear involves microscopy in all its forms. Not only are surface and subsurface microfeatures of worn materials being investigated after the fact, but also wear experiments are being carried out within SEMs and under optical systems to reveal microprocesses as they occur.[1,2]

This chapter will provide information on the processing of specimens for wear microscopy, the types of analysis used and why they are chosen, the types of wear microstructures, their interpretation, and photographic examples.

WEAR MODES

Examination of wear scars by microscopy is often used to identify the type of wear that has occurred. This is very much the same procedure used in identifying fracture modes by fractography. A number of wear modes are recognized, as described in the following sections.

Adhesive Wear

Adhesive wear occurs most often during dry sliding contact between metal surfaces. It is encountered in mechani-

Fig. 16-1. Example of adhesive wear: SEM micrograph of transfer of bronze to steel.

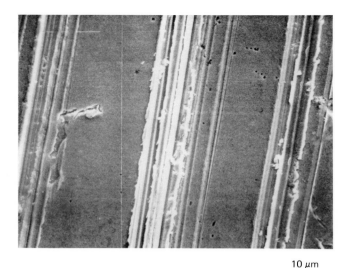

Fig. 16-2. Example of abrasive wear, cutting mode: SEM micrograph of AISI 4130 steel surface abraded by hard, angular alumina particles.

cal components such as those exposed to the vacuum of outer space (in space vehicles and satellites) and is the result of high points on contacting surfaces coming into sliding contact and cold welding. These microjunctions then shear off, causing transfer of material from one surface to the other and roughening of the surface in that zone. As this disturbance travels over the mating surface, it marks it with a scratch. The whole process of adhesion, transfer, and marking has characteristics that can be used in identifying the adhesive mode. A typical example of adhesive wear is shown in Fig. 16-1.

Abrasive Wear

Several modes of wear result from abrasion. Abrasion occurs when particles harder than the contacting surfaces are dragged across the surface and cut or plow the surface. These may be free particles (*three-body abrasion*) caught between rubbing surfaces, or they may be embedded in one surface and act like sandpaper (*two-body abrasion*). Whether it is two- or three-body abrasion, the process involves hard particles moving over one surface parallel to another surface and held in contact with the surface by the mating body. This causes characteristic microscratches in the surface, the geometry of which varies depending on the kind of abrasion.[3]

Cutting abrasion is the result of angular particles held with a facet at the right rake angle for the particle contact to act like a cutting tool. The resulting mark and debris are characteristic. A cutting abrasion mark and its resulting chiplike debris are shown in Fig. 16-2.

Plowing type abrasion, on the other hand, produces displacement of surface material without its removal. Wear striations are formed by mounding material up ahead of the contacting asperity and pushing it to the side as ridges. Some extrusion occurs as the ridges are flattened and occasional particles break off. An example of plowing wear is shown in Fig. 16-3.

High-stress abrasion and low-stress abrasion often occur in mining and construction machinery. *High-stress abrasion* can be found in jaw crushers used in rock crushing. In this process, the abrasive material is ground against opposing surfaces under contact stresses large enough to cause fracture of the abrasive grains. This produces severe wear as sharp cutting edges are formed continually. *Low-stress abrasion*, on the other hand, results when abrasive particles are held between sliding surfaces but with contact stresses low enough so that particle fracture does not occur (see Fig. 16-4). It occurs, for example, when a scoop is dragged across soil during a drag-line operation.

Gouging abrasion occurs when large hard pieces of mineral are forced into the surface of a part with very high contact stress. This causes the abrasive to gouge the surface and dig out relatively large pieces of metal (see Fig. 16-5).

Fretting Wear

Fretting wear is a special form of wear that occurs when two surfaces contact and move back and forth relative to one another with a very small amplitude. The amplitude of the motion can be as small as a few millionths of an inch. When the relative motion exceeds a few tenths of an inch, the wear process is regarded as conventional

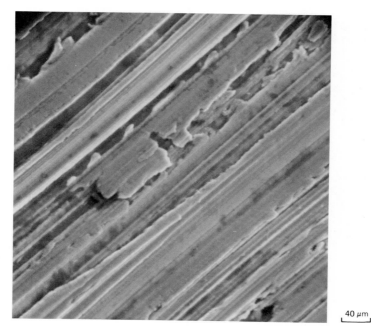

Fig. 16-3. Example of plowing abrasion.

wear. Fretting occurs at the roots of turbine-blade inserts, between the wheel hub and shaft in a pressed-on wheel, between vibrating fuel rods and fuel-rod spacers in nuclear reactors, and between heat exchanger tubes and tube support sheets when coolant flow around the tubes causes them to vibrate. It often occurs where surface attrition is not anticipated.

The fretting process in the presence of oxygen or reactive media and metal parts occurs as fretting corrosion. Fretting corrosion produces a reaction product, or an oxide.[4] The rust formed on steel parts during fretting is red in color and can be seen exuding from between the contact surfaces. The fretted surface is difficult to diagnose and takes careful microscopy to identify. An SEM micrograph of a fretted surface is shown in Fig. 16-6. Corrosion, however, is not a necessary element in the fretting process. High-point or asperity contact and adhesion also initiates the process and will produce a fine debris composed of the contacting materials if no chemical reactions occur. Thus, platinum and gold will fret and produce debris.

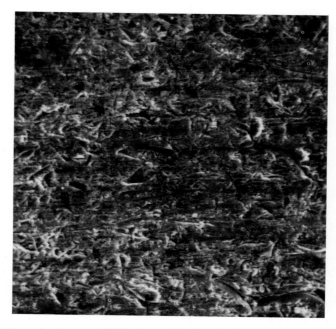

Fig. 16-4. Example of low-stress abrasive wear: SEM micrograph of steel surface subjected to the rubber-wheel abrasion test.

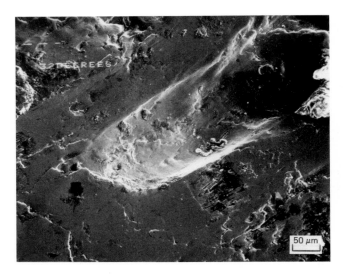

Fig. 16-5. Example of gouging abrasion: SEM micrograph of surface of wire from wire rope used in drag-line excavator.

Erosion

Erosion, cavitation, and liquid-drop impingement are all similar types of surface attrition caused by matter impacting a surface. Although these processes are similar, it is often essential to identify exactly which one has caused material removal. For instance, in a centrifugal gas compressor, the blades can be eroded by impingement of solid particles (ingested dust) or by impingement of water drop condensate. To remedy this situation, one must know whether to eliminate the operating conditions that produce the condensate, or to filter out the dust, or, if that is not possible, to coat the blades with an erosion-resistant material. In addition, if solid-particle erosion is diagnosed, one must know the angle of impingement before a suitable-erosion resistant coating can be chosen. The impingement angle of the impacting particles can be estimated

Fig. 16-6. Fretted surface of turbine blade root with a fretting fatigue failure; note fatigue crack at edge of fretted zone.

by the use of stereo pairs of photomicrographs in SEM analysis.

Impingement erosion can be caused by water drops as well as by solid particles. Liquid-drop erosion has some similarity to cavitation erosion because cavitation can occur as the drop impacts a surface. Water-drop erosion is a problem for helicoptor rotor blades in the presence of rain and for reentry vehicles passing through rain clouds.

Cavitation erosion occurs in mechanical parts submerged in a fluid. For instance, babbitt journal bearings are subject to cavitation damage owing to gas evolution in the oil film during vibration normal to the bearing surface. Cavitation damage is quite common in piping where fluid flow changes direction suddenly. The best known example of cavitation occurs on ship propellers or water-pump impellers. Solid-particle erosion occurs in turbines, slurry pipe lines, coal gasification plants, and helicoptor rotor blades. Examples of cavitation erosion, particle erosion, and liquid-drop impingement are shown in Figs. 16-7, 16-8, and 16-9.

Rolling Contact Fatigue

Rolling contact fatigue is a specific type of surface damage found in ball and roller bearings, cams, and gears. This damage takes the form of pits or spalls. The state of stress developed in the contact of rolling elements is cyclic and results in a fatigue process that causes surface and subsurface cracking. Phase changes can also occur in the steel microstructure as a result of cyclic contact stress. This behavior will be discussed in more detail. Since contact fatigue is often associated with inclusions in bearing steels, this type of surface damage is often investigated with the use of metallographic sections. Typical rolling-contact fatigue spalls are shown in Fig. 16-10.

MICROSCOPY

Wear surface analysis requires the use of both light microscopy and SEM to provide enough information for diagnosis. In some cases, high-resolution transmission electron microscopy (TEM) is also used. In wear failure analysis, the wear scar should be photographed using low-power magnifications (5 to 10×) for documentation and as future reference for SEM and metallographic analysis. If a lubricant is involved, the part should be photographed as received and after its surface has been cleaned. Examination with a hand magnifier (10× is the most useful size) is recommended.

Further examination can then be pursued with a stereoptician microscope and a metallographic microscope. The stereoptican is useful to about 60× magnification and provides a three-dimensional view of surface topography. The metallographic microscope is useful at 100× to 500× for studying details of surface damage. Light

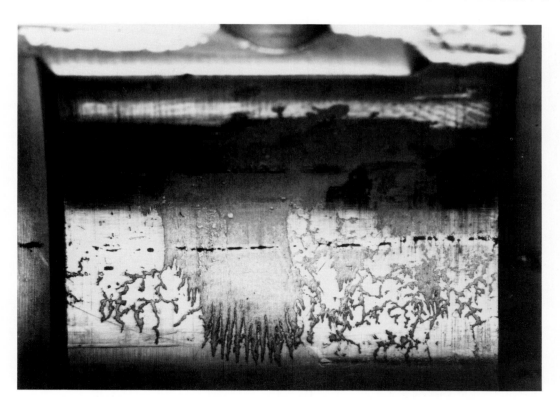

Fig. 16-7. Example of cavitation erosion: SEM micrograph of cavitated babbitt bearing (6×).

microscopy at higher magnifications produces diminishing returns if roughened surfaces are involved because of the extreme limitation of the depth of focus. Light microscopy allows one to work with relatively large pieces, reveals color differences, and permits surfaces to be probed with a needle or scribe while one observes through the microscope. Light microscopy should also be used for selecting areas for examination in the SEM; otherwise, much time can be lost searching over a surface in the SEM to find meaningful features. Comparison of SEM images with light photomicrographs of the same area is very helpful for interpretation of SEM data.

SEM analysis is much more time-consuming than light microscopy if effective analysis is required since large amounts of information of no relevance must be sorted out, leaving only the features that are significant to the wear mode. Before insertion of the part in the microscope, it is helpful to mark areas of interest with a felt-tip metal marking pen. Such marks will show up in the microscope as dark areas. An arrow pointing to an important feature can often save much searching time. In addition to marking the specimen, it is also helpful to make a sketch of it, showing its orientation on the microscope stage, its shape, and the location of important features. The use of stereo pairs is very effective in analysis of wear scars. In addition, montages of several micrographs to cover a large area is often useful.

Transfer material, debris, and reaction products can be identified by using EDS.* This feature is essential in wear analysis. The areas in which a given element or phase is located can be delineated with EDS by means of a dot-pattern mapping process. In this process, the element in question is outlined by a concentration of white dots on the photomicrograph, as shown in Fig. 16-11. The intensity of the dot concentration is related to the amount of material present. It must be kept in mind that the electron beam used in EDS penetrates the surface of the specimen to a depth of about 1 micrometer and X-rays come from a volume about 2 micrometers in diameter. Therefore, positive identification of very thin films less than a micrometer in thickness is not possible with EDS. EDS will also not detect elements with atomic numbers lower than 11 (Na). Therefore, one cannot identify hydrocarbon deposits and some lubricant reaction products on wear surfaces.

---

* Energy-dispersive spectroscopy for detection of X-rays characteristic of each element (atomic number 11 and higher).

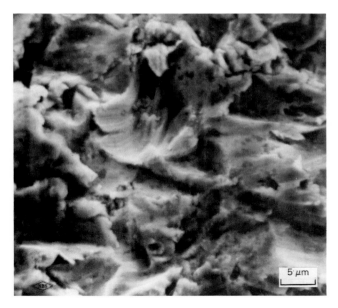

Fig. 16-8. Example of hard-particle erosion. (Courtesy, Dr. A. W. Ruff, National Bureau of Standards.)

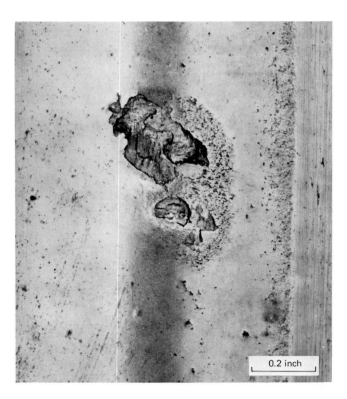

Fig. 16-10. Example of rolling-contact fatigue spalls: Micrograph of a ball bearing race (5×).

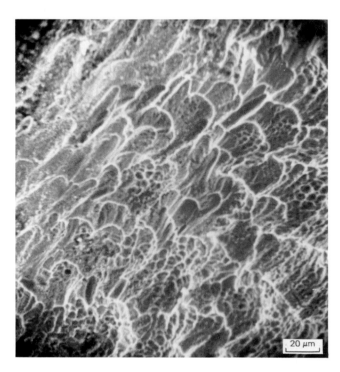

Fig. 16-9. Example of liquid-drop-impingement erosion: SEM micrograph of surface from boiler feed-water pump eroded by condensate.

Scanning Auger analysis is useful for identification of surface composition. An example of scanning Auger analysis of a hydrocarbon film on a bearing surface is shown in Fig. 16-12. Experience with interpretation of variations in the grey tones on the surface of an SEM micrograph makes it possible to detect hydrocarbon films, mineral intrusions, metal transfer, etc. Examples of such features are shown in Fig. 16-13.

Sectioning of a worn part and preparing a thin foil for investigation of the near-surface area by TEM has proven to be useful for identification of transfer material as well as of the subsurface microstructure produced by the wear process.[5] Suggested techniques for preparing wear-sample thin foils are given in the section on TEM later in this chapter. The TEM can also be used to examine fine debris in wear topography. A replica made of the wear scar is shadowed by conventional techniques and viewed by transmission electron microscopy. An example of this technique is shown in Fig. 16-14, which shows a TEM micrograph of a worn ball-bearing race surface. The finishing marks, seen easily in the SEM, are enlarged in the TEM, and, in addition, the very fine wear scratches have been resolved. The surface topography of wear scars, which appears somewhat exaggerated in the SEM, seems somewhat flattened by the TEM replication method.

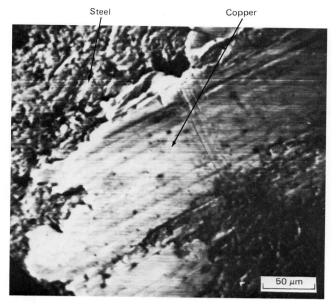

(a) SEM micrograph of transfer

Fig. 16-12. Scanning Auger micrograph of bronze-bearing surface showing iron transfer; dots show iron concentration to a depth of 5 nm.

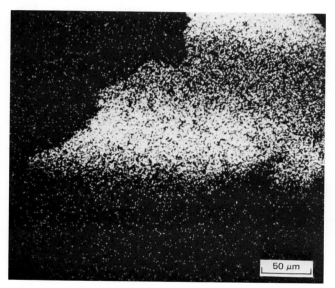

(b) EDS dot pattern scan for copper

Fig. 16-11. Energy-dispersive-spectroscopy (EDS) elemental dot pattern showing location of transferred copper on steel surface: (a) SEM micrograph of transfer area, and (b) EDS elemental dot scan for copper.

PREPARATION OF SPECIMENS

Before examining a wear specimen, it is necessary to process the surface to be examined. It is always best to look carefully at the wear surface prior to any surface alteration because there may be remnants of the failure process (decomposed lubricant, contaminants, etc.) that would be removed by any cleaning process. A photograph of the original as-received surface is a good record-keeping practice. Once the preliminary examination is completed

Fig. 16-13. Scanning Auger micrograph of hydrocarbon deposits.

and documented, measures should be taken to remove any obscuring contamination from the surface. This step is especially important for SEM analysis because loose debris and nonmetallic particles will fluoresce in the electron beam and wash out adjacent features. The presence of surface contamination can also reduce the resolution or sharpness of the image in the SEM, especially when high-magnification examination is desired. Residual oil, grease, or decomposed hydrocarbons in surface crevices will outgas in the SEM and reduce image quality. In short, the same practices used in quality SEM analysis should be practiced in wear analysis.

The simplest preparation method is ultrasonic cleaning in suitable solvents. This can be followed by ultrasonic cleaning in a detergent-water solution, followed by an alcohol rinse and drying with moving air. If it is necessary

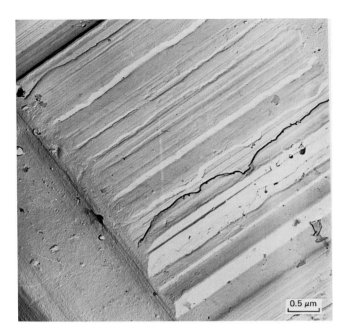

Fig. 16-14. High-magnification detail of ball-bearing race obtained by examination of shadowed replica with the TEM.

to preserve particles mechanically bonded or pressed into the surface, it must be cautioned that ultrasonic cleaning often removes these surface features. Those particles not removed by cleaning can be removed by surface replication if such is required. The easiest way to do so is to smear a thin layer of acetate cement over the surface, allow it to harden, and then remove it by pressing a piece of scotch tape over it and pulling the tape off with a smart snap. This procedure can also be used to obtain high-resolution surface replicas in the field. Replication can also be used to follow changes in surface microfeatures as wear progresses.[6] A Knoop hardness indentation in the surface can be used to measure minute amounts of wear.

If the surface to be examined is corroded or rusted so that the original wear scar is hidden, the corrosion product can be removed without damaging the microscopic details of the scar by an electrolytic cleaning method described in *Practical Metallography*[7] and reproduced in Appendix C.

An important wear feature may sometimes be obscured by a smeared surface layer. An example of this problem occurred in an analysis of wear mechanisms in tungsten-carbide rock drill bits. Sample tungsten-carbide inserts removed from a worn bit were examined with the SEM. Very little detail could be observed in the wear scars, and no evidence of the carbide grains could be seen. The specimen was then given a brief treatment in an acid etch, which removed a thin, smeared layer of cobalt from the surface and revealed the carbide phase and its condition after sliding contact. Although the amount of cobalt removal was not enough to penetrate between the carbide grains, the cobalt depletion between the grains was noted. This was confirmed in subsequent sectioning. Part of the wear mechanism was attributed to the extrusion of cobalt to the surface, causing loss of support of the surface carbides.

Better resolution with the SEM can usually be attained by coating the wear surface with gold or carbon. Nonconducting materials (minerals, polymers) must be coated with a conducting layer to prevent charging of the surface during SEM examination. The coating is put on by a physical vapor-deposition process and must be in electrical contact with the specimen holder. Plastic replicas for the SEM or TEM are also coated prior to insertion. The same vapor-deposition equipment used in transmission electron microscopy is used. Gold coating provides the best resolution, but carbon coating is satisfactory if elemental analysis using EDS is contemplated.

Surface replication may be required to examine the surface microtopography of large mechanical components in the field. This is a nondestructive procedure for analyzing a surface that cannot be physically placed within the microscope. Surface replication is also used in the wear analysis of features beyond the resolving capability of the SEM. The replica is taken, coated, and examined in the TEM. Replicas of large surface features can also be made with dental cement. This type of replica can be used for physical measurements (depths of gouges) and examined with light microscopes.

Replication can also be used to remove transfer material from a worn surface for metallographic analysis. In one process, developed by Tyzh-Chiang Sun of Ohio State University,[8] the wear scar is copper-plated, and the copper plate is then stripped off the surface. The plate pulls off the transfer material with it. The interface side of the stripped plate is then copper-plated so that the transfer particle is embedded in copper. The composite is then mounted edge-on in epoxy and sectioned and examined or thinned for TEM. Individual particles can be analyzed using high-resolution scanning transmission electron microscopy (STEM) by electron microdiffraction.

INTERPRETATION

Scratches or striations usually characterize a wear scar. Striations are not the features of the same name seen on a fatigue fracture surface but are rather a series of parallel microgrooves caused by high points plowing a softer surface. Scratches (random orientation) can be visible to the naked eye or so small that they require surface replication and TEM to resolve. The striations are present both in abrasive wear and adhesive wear. Even in rolling contact, microscopic surface striations are produced because the geometry of contact in bearings or gears produces mixed rolling and sliding. This fact helps in the interpretation of performance characteristics of ball or roller bearings. For instance, a circumferential band of

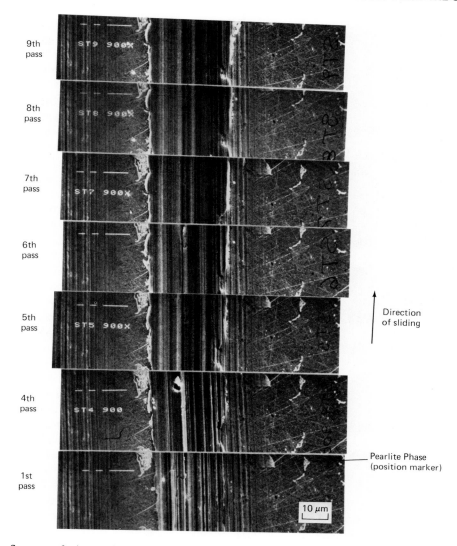

Fig. 16-15. Sequence of microscopic deformation events in a wear track from *in-situ* SEM wear experiment (mild steel).

fine scratches around each ball in a ball bearing indicates that the bearing has been operating under excessive thrust load. Examples of microscopic wear striations can be seen in Figs. 16-1 through 16-7.

Following the wear process in the SEM reveals the way in which surface striation morphology changes as wear progresses. A sequence taken in the SEM (real-time wear) is shown in Fig. 16-15.[1] Note that the area location can be indexed by the characteristic pearlite colonies. Initially, the striations were very narrow and sharply peaked. After several passes, however, the sharpness of the features diminished, and the striation ridges became more rounded. Note that many of the striations persisted, but the ridge contour was modified. This occurred because there was a minimum of transfer and the original asperities that produced the first sharp scratches have been worn and rounded, causing a change in contact conditions and a rounding of the resulting contact topography. This chain of events illustrates that it is important to consider how long a microscopic feature in a wear scar has persisted. If it represents a new condition developed in the last pass of the contacting surface, you are seeing only the final stage of the wear. This was found to be true for steel sliding against bronze in the *in-situ* SEM wear experiments shown in Fig. 16-15. The surface microstriations completely changed in appearance with each contact pass. This change is related to the effect of the transfer of bronze to the steel surface.

Adhesive wear usually produces striations with subtle differences in features that enable one to distinguish between adhesively generated scratches and abrasively produced scratches. The initiation of an adhesive wear striation is shown in Fig. 16-16. Initial contact by the asperity can be identified by the very short furrow before the lump (marked in the figure). The adhesive junction then became strong enough to induce plastic deformation, folding of surface layers, and shearing of some material. The result was a roughening of both surfaces in that local area and

270 APPLIED METALLOGRAPHY

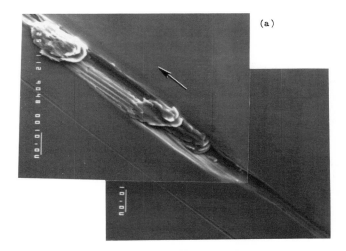

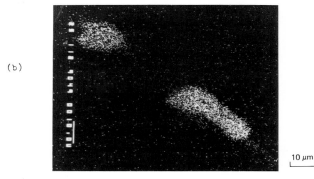

Fig. 16-16. SEM micrographs showing initiation of adhesive transfer: (a) SEM image of transfer of copper to iron, and (b) EDS elemental dot pattern for copper in preceding micrograph. (Courtesy of L. H. Chen, Ohio State University.)

work hardening of the roughened zone. As this area proceeded through further sliding contact, a furrow was plowed into the surface much like an abrasive wear scratch. Note, however, that downstream from the initial adhesion, there are transferred islands and surface tearing (shown by the arrows). The events shown in the photomicrographs occurred early in the wear process and at widely scattered sites. Considerable searching is necessary to find initiation or post-initiation features of this type of adhesive wear.

Flakelike features on a worn surface can originate from one of several processes. It is important to recognize that there can be more than one method by which a flake can be produced by wear. One possibility is by a contact-fatigue process—rolling contact fatigue, in fact, is the most commonly recognized process. Spalling from surface fracture, which may originate below the surface or at the surface and propagate under the surface, will produce a solid piece of the base material. Usually, this piece fragments as soon as it is released from the surface, and evidence of its presence is seen only as dents and scratches around the pit or spall from which it originated. This process is very rare in ductile materials like bronzes, aluminum alloys, and tin- or lead-base bearing materials.

Hardened steels, cemented carbides, white cast irons, and ceramics are more likely to produce surface-contact fatigue. Fatigue failure of aluminum and lead- and tin-base bearing materials does occur, but this is different from contact fatigue and is usually related to the cyclic loading of bearings in internal combustion engines (combustion and inertial forces transmitted through the piston rods).

Adhesive wear of ductile materials will produce flakelike surface deposits in the wear scar. An example is shown in Fig. 16-17. The structure of this transfer material often appears as particulate or as very fine grains. It can be composed of agglomerated fine particles or exist as a high-strain structure of extremely fine grains—the result of heavy smearing and working. A section through such a flake is shown in Fig. 16-18; note the particulate structure. Another source of flakes on the surface, especially if the flakes are very small, is extrusions from ridges during plowing of the surface. An extrusion is shown in Fig. 16-19. The above observations indicate how important it is to analyze the microstructure as well as the morphology of surface flakes before stating their probable origin.

Abrasive wear has been described as a scoring or machining of a surface. The micromachining marks from abrasion have characteristics that enable one to identify the type of abrasion. Three-body abrasion, or the abrasion by loose, hard particles trapped between two surfaces, produces a pattern much like that shown in Fig. 16-3. The scratch pattern is random, consisting of short, sometimes curved scratches and a series of dents caused by the particle rolling over the surface. The dents are often seen as a line like a line of periods (..........). Cutting abrasion produces relatively clean, straight scratches like those shown in Fig. 16-2. The presence of curled microchips in the debris helps to verify cutting abrasion. Gouging abrasion, often found in mining machinery, has irregular scratches with variable depths and widths. The edges of the scratches are heavily rolled over, and considerable flow and deformation is evident. An example of gouging

Fig. 16-17. Example of flakelike adhesive transfer.

MICROSCOPY AND THE STUDY OF WEAR

Fig. 16-18. TEM micrograph of transfer patch removed from worn surface. (Courtesy, Jarlen Don, Ohio State University.)

abrasion is shown in Fig. 16-5. Low-stress abrasion, as occurs in the ASTM rubber wheel abrasion test, shows the effect of rounded abrasive particles since the contact pressures are not high enough to fracture the abrading particles. An example of low-stress abrasion is shown in Fig. 16-4.

Other less prevalent wear modes have characteristic microtopography features readily distinguished in the SEM and with light microscopy. *Electrical discharge pitting* occurs in bearings and gears carrying electrical current. The current flow may be intentional, as in some electrical equipment, or accidental, as in machinery that generates static electricity. The pitting is caused by electrical discharge, and melting occurs around the edges of each pit. These effects can be seen in Fig. 16-20. *"Frosting"* occurs in ball bearings, cams, and gears when they slide or skid under heavy contact stress. The "frosted" area and the microspalling associated with the pitting is shown

Fig. 16-19. SEM micrograph of microextrusion occurring during the deformation process of wear (bronze surface).

Fig. 16-20. Example of electrical discharge pitting: Macrograph shows bearing races with "fluting" marks caused by combined vibration and electrical discharge.

in Fig. 16-21. These features can be seen with relatively low-power light microscopy. Corrosive wear produces a pitting type of attack accompanied by a removal of corrosion product and often a passivating film. The process, therefore, can be quite aggressive. An example of corrosive wear is shown in Fig. 16-22.

METALLOGRAPHY

Complete analysis of wear mechanisms requires detailed microtopography of the wear surface and also requires microstructural information from the near-surface regions under the wear scar. The kinds of information needed include the extent of subsurface deformation, subsurface cracking, internal corrosion, phase changes, oxide layers, microhardness, segregation, heat-induced structural changes, etc. Before sectioning the specimen, it is necessary to analyze the placement of the cut and the orientation of the section desired. This involves looking at the results of the surface investigation and giving some consideration to the geometry and dynamics of the wear process.

Sectioning parallel to the sliding direction will reveal microstructure alterations with directional aspects, that is, grain boundaries will be seen bending over and oriented in the direction of sliding. Measurement of the amount of tilt in these boundaries can be used to estimate the amount of strain in the near-surface region.[9] Dislocation cell structures, common in ductile materials subject to wear, will often be elongated in the direction of sliding. Grain-boundary bending is shown in Fig. 16-23.

272 APPLIED METALLOGRAPHY

Fig. 16-21. Micrograph showing details of "frosted" area from ball bearing race (6×).

Evidence of flow around hard inclusions and the elongation of soft inclusions can also be found in this way. Surface laps and subsurface cracks may likewise be revealed as part of the wear process. An example of subsurface cracking in a carburized chain link is shown in Fig. 16-24. An example of subsurface flow patterns in an unhardened chain link is shown in Fig. 16-25. This photomicrograph shows the effect of oscillating sliding motion

Fig. 16-22. Example of corrosive wear.

Fig. 16-23. Metallographic section through wear scar showing near-surface deformation delineated by grain-boundary bending (section was cut parallel to direction of sliding contact).

on the formation of a hardened lump in the center of the contact.

Sectioning across the sliding direction will reveal the profile of abrasion striations and the details of the structure in the heavily deformed ridges. Hard transfer material will often embed in the softer substrate it encounters. This embedded transfer can be detected in sections taken across the sliding direction, as shown in Fig. 16-26. Any lateral flow resulting from the contact process will show up in this type of metallographic section.

Before sectioning, the worn part should be nickel-plated to preserve the surface topography. Good plating practice requires that surfaces be carefully cleaned before plating, especially of wear samples because they are often contaminated by oil or grease. Conventional metallographic grinding and polishing techniques can destroy surface features essential to diagnosing wear processes.[10] Preservation of the specimen edge containing the wear surface is impor-

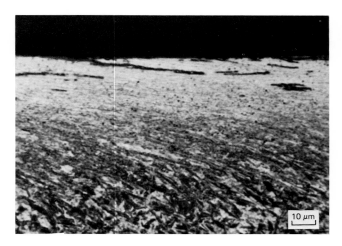

Fig. 16-24. Metallographic section from carburized chain link showing flow and subsurface cracking.

Fig. 16-25. Metallographic section through unhardened steel chain link showing subsurface flow pattern resulting from oscillatory contact sliding from mating link.

tant; otherwise, in brittle materials, grinding particles will cause shearing or extrusion of the edges or cracking. Nickel plating is essential to preserve these near-surface features. The grinding direction should always be away from the edge and toward the interior of the specimen.

A sectioned specimen should be examined in the light microscope prior to etching. Some delicate edge features in the section can be destroyed by heavy etching. Near-surface microstructures can be so fine that it cannot be resolved by light microscopy. If a heavy etch is used to bring out the structure in high relief, the microsection can be examined in the SEM. This technique has been used in the study of near-surface microstructures produced by laser surface melting and alloying. An example is shown in Fig. 16-27.

When investigating a surface exposed to erosion, it is helpful to make a metallographic section. If the section

Fig. 16-26. TEM micrograph of section through wear scar showing embedding of hard wear debris in soft counter surface. (Courtesy, Tyzh-Chiang Sun, Ohio State University).

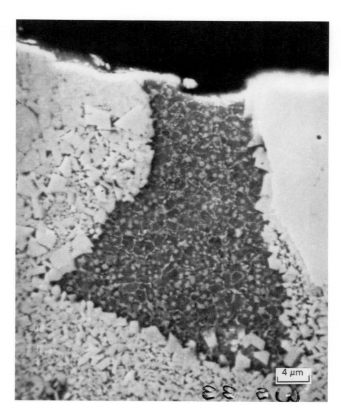

Fig. 16-27. Micrograph of section through laser-melted tungsten carbide part (etched prior to examination).

is made parallel to the direction of impingement, the extent of surface extrusion can be observed. Subsurface cracking from impact can also be detected. In soft metals like aluminum, considerable mixing of surface material with subsurface material can often be found. An example is shown in Fig. 16-28.

Taper sectioning is often used in the study of very fine surface features and shallow, near-surface microstruc-

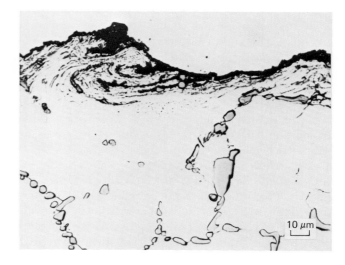

Fig. 16-28. Metallographic section of worn aluminum alloy showing mixing of surface debris in highly deformed near-surface region.

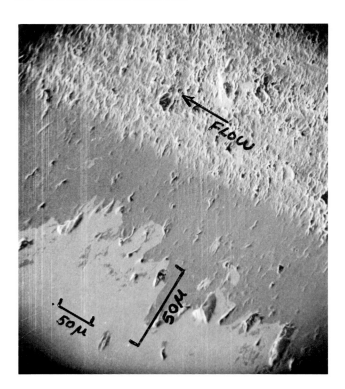

Fig. 16-29. Taper section of plasma-sprayed steel part after wear testing; surface features can be matched to subsurface structure.

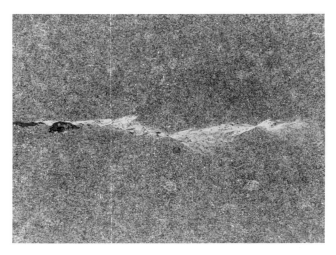

Fig. 16-30. Metallographic section of ball bearing race showing a "butterfly" defect associated with hard inclusion.

tural changes. The process is described in the Appendix. An example of a taper section is shown in Fig. 16-29.

Many subsurface features can be found in ball or roller bearings after use, especially in heavily loaded bearings. Detailed metallographic descriptions of rolling contact fatigue has been documented.[10-13] Among the features that can be observed are dark-etching zones, white-etching zones, "butterfly wings," and cracking. Most of these features can be observed with light microscopy using common etchants. These features vary in appearance depending on etching time and the etchant used. They are found below the rolling-element contact surface at the zone of maximum shear stress. Effective identification of the composition of these phase changes in thin foils has been done with SEM and TEM. The changes in etching characteristics are related to a decay process of the martensite and carbides in the original heat-treated structure. The dark-etching area is composed of partly decayed martensite and thin bands of ferrite. The white-etching area appears later in the decay process and is composed of highly strained ferrite with no carbides (dislocation cell structure). Carbide disks on the order of 1-micrometer thick are part of the decay process of carbides. The white etching "wings" are found around hard, nonmetallic inclusions and are composed of cell-structure ferrite decomposition products from the original martensite. The hard oxide inclusions produce a stress concentration in the material surrounding them and thus localize the matrix decay process. White-etching areas are highly strained zones composed of dislocation cells so small that they cannot be resolved by light microscopy. An example of a butterfly is shown in Fig. 16-30.

White-etching zones can also be found in near-surface regions of cast irons and steels that have been subjected to high-contact-stress rubbing contact (wire rope, rails, drag line buckets, grinding balls, jaw crushers, brake disks, etc.). There is some controversy as to the composition of these surface white-etching layers. They are much harder than the substrate metal and may be high-strain cell structures or nitrided zones. They can often be seen with the naked eye after the surface is etched with nital without any other preparation.

STEREO-PAIR ANALYSIS

In electron microscopy, it is often very effective to use stereographic observations. It is useful not only for determining which topographic features are up and which down, the steepness of depressions, and the depth of pits or cracks; looking at a three-dimensional view of a complex surface microtopography will also help interpret surface features. Significant aspects of a surface structure often become obvious when viewed in three dimensions. Contour measurements can be made of abrasion marks—heights of ridges, depths of valleys—using a simple parallax measuring system.[14] The feature to be measured is first photographed in the SEM with the specimen oriented so that the distances to be measured are horizontal. Two photographs are made at two different stage tilt angles. It is best to make the angle between tilts relatively large. Ten degrees is a good angle to choose because the mea-

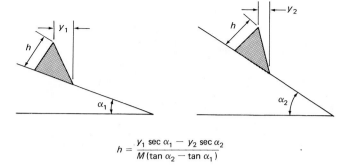

Fig. 16-31. Diagram of geometric features in stereo-pair analysis.

surement differences will be larger and easier to handle in the calculations. Using the sketch in Fig. 16-31, measurements y_1 and y_2 are made using the two micrographs. The following equation is then used to calculate the height, h:

$$h = \frac{y_1 \sec \alpha_1 - y_2 \sec \alpha_2}{M(\tan \alpha_2 - \tan \alpha_1)} \quad (16\text{-}1)$$

Where M is the magnification, α_1 is the stage angle on the first micrograph, and α_2 is the stage angle on the second micrograph.

A more sophisticated system has been developed at SKF Research Labs in King of Prussia, Pa.[15] This involves the control of the electron beam in the SEM and digitization of the images. The computer will calculate the profile of any feature selected.

TRANSMISSION ELECTRON MICROSCOPY

The use of surface replication and transmission electron microscopy for detailed analysis of surface features has been discussed. More recently, the TEM has been used to study the defect structure in near-surface regions and transfer-layer structure associated with wear of metals.[16-19] Foil specimens extracted from sections made at wear scars have revealed a dislocation cell structure similar to that found in highly strained regions resulting from the cold deformation of metals. Although the concept for producing specimens that will transmit electrons is the same as that used in dislocation studies, techniques for preparation of the specimens have been modified.

The process for the preparation of a thin-section specimen from the extreme surface of a wear specimen is lengthy and tedious and, in addition, has a low probability of success. A number of attempts must usually be made before a satisfactory foil is obtained. The ideal approach is to make thin foils of sections taken in three orientations: parallel to the direction of sliding, normal to the direction of sliding, and parallel to the surface. Doing so provides a three-dimensional picture of the near-surface defect structure. This approach, however, can be expensive as well as very lengthy, especially if the metal alloy has several phases, inclusions, and carbides in it.

The easiest method involves making a thin section parallel to the surface and eliminates the problem of preserving the boundary between the outer surface and the substrate—one of the difficult aspects of thin section preparation for wear analysis. The parallel section can be ion milled to a few micrometers thickness and then electrolytically or chemically treated until a region that will transmit electrons had been achieved. The required foil thickness for transmission of electrons will, of course, depend on the voltage of the microscope to be used and the atomic number of the metal specimen. With a parallel section specimen, the dislocation cell structure can be revealed.

Sections made perpendicular to the wear surface will provide more information about the near-surface structure, including the structure of any transfer layers, any variation of dislocation cell morphology with depth, and the thickness of the heavily disturbed zone. The near-surface zones found in association with wear are shown in Fig. 16-32. An example of cell structure and transfer layer structure is shown in Fig. 16-33 (see also Fig. 16-23). These thin foil sections can be analyzed with high-resolution STEM systems with EDS, electron microdiffraction, and use of Kikuchi line analysis.

WEAR DEBRIS

Analysis of wear debris is just as important as microanalysis of wear scar surfaces and near-surface metallography. Both the chemical composition and morphology of wear debris provide important information about the wear process. Collection of wear debris in circulating oil systems makes it possible to diagnose impending failure processes at work in engines.[20]

There are many techniques for the collection of wear debris. If a liquid lubricant bathes the wear zone, it will wash away loose wear debris and carry it to a filter where the debris can be extracted. Some debris tends to stick to the wear surface (i.e., it is mechanically bonded) and

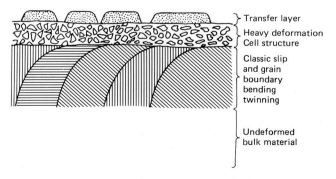

Fig. 16-32. Diagram of near-surface zones of deformation associated with wear process; these features can be seen with careful sectioning and thinning of sections for TEM.

Fig. 16-33. TEM micrograph of thinned section through wear scar showing cell structure and fine structure in transfer layer. OFHC copper. (Courtesy, Prof. David Rigney, Ohio State University.)

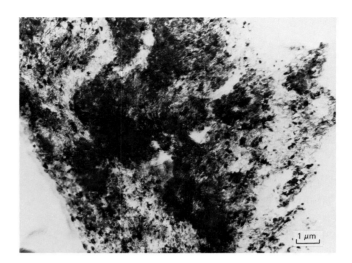

Fig. 16-34. Wear debris as observed in scanning transmission electron microscope (STEM); debris was collected from lubricated bronze sample.

can be removed by surface replication. The debris comes off with the plastic replica. In dry systems, the debris can be brushed off the moving surface so that it falls into a collector. Magnetic devices can be used to collect ferromagnetic debris.

If the debris is suspended in a fluid, either in the liquid lubricant from the system or in a solvent wash, the solid particles can be extracted by one of several methods. Ferrography does this by making the liquid flow over a glass plate with a magnet under it. Any particle attracted by the magnet will be extracted from the liquid. The particles deposited on the plate in this way are fixed to the plate and observed by microscopy. This method separates the particles by size. A number of techniques have been developed for use with this system of debris analysis.[20]

Another method involves centrifuging the suspension, removing the concentrated debris, suspending it in a solvent, and washing it over a fine filter with a flat, smooth surface. A nuclear pore filter works very well for this purpose since the debris is not incorporated in a tangle of fibers. The filter is dried and then carbon coated. The composite is cut into wafers the size of the grids used in electron microscopes. Removal of the filter material from the wafers by dissolution in ether results in a carbon-debris composite that can be observed in transmission in a TEM or STEM. Examples of debris microscopy are shown in Fig. 16-34 and Fig. 16-35.

Wear debris size and morphology can be used to determine the process of removal or the wear process. For instance, cutting abrasion produces curled chips like microscopic versions of machining chips. Flakes are associated with adhesive wear or near-surface fracture. Some wear flakes will contain imprints of the surface they have been pressed against during a transfer process. Very small rounded particles (submicrometer size) are often the result of chemical action—oxidation or corrosion. Perfectly spherical particles have often been reported. There has been some attempt to associate them with rolling contact conditions.[21] Careful examination of these perfect spheres, however, usually shows that they have solidified from a molten state. The dendritic structure shown in Fig. 16-35 is a clue. This type of particle is ubiquitous and has several origins. A few examples of possible sources for spherical particles includes grinding, welding, casting, and the burn-up of meteorites as they enter the atmosphere. There are some less than perfect spheres that have been found to develop from being rolled between two surfaces; these particles are usually more cigar-shaped or oblate.

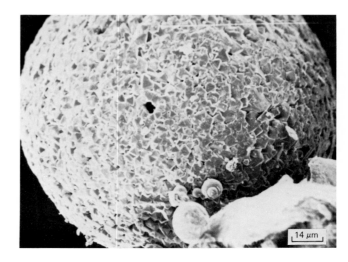

Fig. 16-35. SEM micrograph of spherical particle found in wear debris sample.

SUMMARY

The microscopic features resulting from surface contact (sliding, rolling, impact) are useful for diagnosing wear modes and for the study of basic mechanisms of wear. The photomicrographs reproduced in this chapter do not include all possible wear morphologies. For further study, it is recommended that the journal, *Wear,* and the proceedings of the International Conferences on the *Wear of Materials* (ASME) be consulted; their pages contain a wealth of examples of wear microscopy.

As pointed out in this chapter, subsurface modification to existing microstructures is just as important as surface topographical features in defining wear processes. Equally important is the morphology of the debris produced by the wear.

Experience is by far the most effective component in the interpretation of wear microfeatures. Understanding and translation of complex scratch patterns are achieved by a process of learning (pattern recognition). It is thus important to spend whatever time is required to examine in detail many examples of wear microstructure, especially examples of well-documented diagnosed wear modes. This experience should be supplemented with tribology theory and principles, rudimentary understanding of machine dynamics, and surface chemistry.

In the future, a steady increase in the use of surface-science techniques—such as scanning Auger, secondary ion mass spectrometry (SIMS), photoelectron spectroscopy or electron spectroscopy for chemical analysis (ESCA), Raman spectroscopy, and ellipsometry—will supplement wear microscopy. These powerful techniques are moving from pure research areas to the solution of practical problems. They promise to answer many questions concerning the mechanisms of wear that have to date proved extremely refractory. Publication of this sort of multidisciplinary investigation can be expected to increase.

REFERENCES

1. Glaeser, W. A. Wear experiments in the scanning electron microscope. *Wear* 73:371–378 (1981).
2. Calabrese, S. J.; Ling, F. F.; and Murray, S. F. Dynamic wear tests in the SEM. *ASLE Transactions* 26(4):455–465 (Oct. 1983).
3. Huffington, J. D. Abrasion groove sizes and shapes in relation to the mechanism of abrasion. *Wear* 49:327–337 (1978).
4. Glaeser, W. A. Erosion-corrosion, cavitation and fretting. In *NACE Handbook 1, Forms of Corrosion Recognition and Prevention,* pp. 71–78, 1982.
5. Heilman, P.; Don, J.; Sun, T. C.; Glaeser, W. A.; and Rigney, D. A. Sliding wear and transfer. In *Wear of Materials 1983,* pp. 414–425. ASME 1983.
6. Glaeser, W. A. Wear measurement technique using surface replication. *Wear* 40:135–137 (1976).
7. Yuzawich, P. M., and Hughes, C. W. An improved technique for removal of oxide scale from fractured surfaces of ferrous materials. *Practical Metallography* 15:184–195 (1978).
8. Sun, T. Unlubricated friction and wear in dispersion hardened copper systems. Ph.D. dissertation, The Ohio State University, Columbus, OH, 1983.
9. Dautzenberg, J. H., and Zaat, J. H. Quantitative determination of deformation by sliding wear. *Wear* 23:9–19 (1973).
10. Torrance, A. A. The metallography of worn surfaces and some theories of wear. *Wear* 50:169–182 (1978).
11. Swahn, H.; Becker, P. C.; and Vingsbo, O. Electron-microscope studies of carbide decay during contact fatigue in ball bearings. *Metal Science* 35–39 (Jan. 1976).
12. Osterlund, R., and Vingsbo, O. Phase changes in fatigued ball bearings. *Metallurgical Transactions* 11A:701–706 (May 1980).
13. Becker, P. C.; Swahn, H.; and Vingsbo, O. Structural changes in ball bearing steel caused by rolling contact fatigue. *Mechanique Materiaux Electricite* No. 320–321:8–13 (Aug./Sept. 1976).
14. Boyde, A. Quantitative photogrammetric analysis and qualitative stereoscopic analysis of SEM images. *Journal of Microscopy* 98:452–471 (Aug. 1973).
15. Aggarwal, B. B., and Wher, R. F. Final technical report on development of a 3-D microscale topography system to Office of Naval Research, Contract No. n00014–80–C–0937 (1983).
16. Ruff, A. W.; Ives, L. K.; and Glaeser, W. A. Characterization of wear surfaces and wear debris. In *Fundamentals of Friction and Wear of Materials,* Proceedings 1980 ASM Materials Science Seminar, pp. 235–289. Metals Park, OH: American Society for Metals, 1980.
17. Ohmae, N. Transmission electron microscope study of the interrelationship between friction and deformation of copper single crystals. In *Fundamentals of Tribology,* pp. 201–220. Cambridge, MA: MIT Press, 1980.
18. Hsu, K. L.; Ahn, T. M.; and Rigney, D. A. Friction, wear and microstructure of unlubricated austenitic stainless steel. In *Wear of Materials 1979,* pp. 12–26. New York: ASME, 1979.
19. Sun, T. Technique for preparation of wear debris for transmission electron microscopy. *Wear* 79:385–388 (1982).
20. Seifert, W. W., and Wescott, V. C. A method for the study of wear particles in lubricating oil. *Wear* 21:27–42 (1972).
21. Christensen, C. On the origin of spherical particles found on fatigue surfaces and ferrograms. *Wear* 53:189–193 (1979).
22. Godfry, D. Diagnosis of wear mechanisms. In *Wear Control Handbook,* edited by Peterson, M., and Winer, W., pp. 302–309. New York: ASME, 1980.

APPENDIX
Special Techniques Used in Wear Microscopy

ELECTROCHEMICAL REMOVAL OF RUST[7]

Surfaces encrusted with rust can be cleaned sufficiently to reveal microscopic detail of wear scars hidden under the oxide layer. The removal process should not damage or alter these features and involves an electrochemical system that immerses the specimen in an electrolyte. The electrolyte is composed of a solution of Endox 214 (Enthone, Inc.) in cold water (8-oz Endox in 1000-ml water). A small amount of photoflo is added to the solution (the solution contains sodium cyanide and should therefore be used with caution under a hood). The specimen is made the cathode in an electrolytic cell and preferably should be set up to rotate. The anode consists of carbon or platinum. An electric potential is applied and adjusted to obtain a current density of 250 mA/cm^2. The current is allowed to flow for about 1 minute while the specimen is rotated. After this first minute, the specimen surface is inspected. If not sufficiently cleaned, it should be given another minute of treatment. Complete removal of the rust may take several 1-minute cycles. Final cleaning should be done by ultrasonic cleaning in Alconox, followed by a rinse in methanol.

TAPER SECTIONING

Taper sectioning is used to magnify and reveal near-surface details of wear scars. This method allows study of fine surface details by light microscopy that would be missed or unresolvable in normal sections. Taper sectioning preserves much of the near-surface material that would be removed in normal sectioning during the grinding and polishing (rounding of the edges). Before taper sectioning, the wear surface should be cleaned and nickel-plated. A normal section should be made first so that certain features that might be distorted by the taper section can be identified and understood.

Two approaches to the taper-sectioning process are shown in Figs. C-1 and C-2. Figure C-1 shows the result of mounting the sample so that it can be ground at an angle of 5 degrees. The grinding direction should be away from the wear surface and toward the interior. Figure C-2 shows a method for producing a taper section in a cylindrial specimen (e.g., a bearing shaft).

The taper section can be used for measuring near-surface microhardness, elemental composition, and depths or heights of surface microfeatures in profile. Direct measurements from the taper section can be converted to actual dimensions by using the following equation (see Figs. C-1 and C-2):

$$h = l \sin \theta$$

where:

h = vertical thickness
l = measured thickness on photograph
θ = taper angle

A taper section can be used to determine the effect of surface disturbances on local subsurface structure. If a taper section is made without plating the wear surface and the ground and polished surface is etched, the specimen can be broken out of the mount and set up in the SEM so that it can be viewed edge-on, with both the wear surface and the near-surface section observable simultaneously. An example of surface–subsurface viewing is shown in Chap. 16 in Fig. 16-29.

SURFACE SPOT TESTS[22]

Metal particles are often found embedded in wear scars. Determination of the chemical content of these particles is often helpful in diagnosing their origin. Spot chemical tests can be used to identify iron, manganese sulfide, and aluminum quickly.

Iron Test

Placing a drop of a saturated solution of copper sulfate slightly acidified with sulfuric acid on a surface will turn

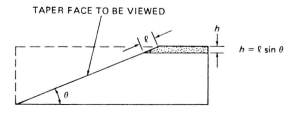

Fig. C-1. Diagram of taper section of rectangular specimen.

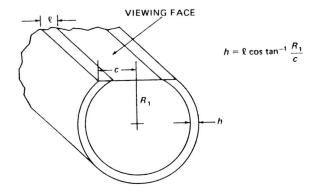

Fig. C-2. Diagram of method for obtaining taper section of cylindrical specimen.

the drop pink in color if iron is present. A filter paper can be soaked in the solution and pressed against the metal surface in question (e.g., a babbitt bearing). Spots on the paper will show the distribution of iron particles.

Sulfide Test

Photographic paper soaked with 2-percent sulfuric acid and pressed against a metal surface will show the location of iron or manganese sulfide inclusions. Two minutes of contact will reveal the sulfides by the appearance of black spots. Excessive amounts of iron sulfide indicates possible corrosive wear.

Aluminum Test

Concentrated sodium hydroxide placed on the metal particle on a steel surface will react with any aluminum present to produce a black spot. Appearance of hydrogen bubbles will locate the source of the aluminum.

17
MICROSCOPY AND THE STUDY OF CORROSION

W. E. White
Petro-Canada Inc.

The study of corrosion, whether in research or in the analyses of corrosion-related engineering problems, is always challenging and often difficult. Frequently, it only takes small environmental changes to precipitate catastrophic electrochemical or oxidation reactions leading to early systems or component failures.

Corrosion may be defined generally as the degradation of metals and alloys from interaction with their environments.[1-3] This broad definition encompasses classical electrochemical (or aqueous) corrosion and environmentally assisted fracture as well as material degradation from high-temperature oxidation. Corrosion is thermodynamically predictable,[4-6] but the courses (or paths) of corrosion may be as varied as the microstructures of the metals supporting the interfacial electrodic reactions and the environments interacting with them.[4-7] Often, seemingly minor microstructural or environmental variations can dramatically alter rates (kinetics) and forms of corrosive attack adversely. Consequently, microstructural, fractographic, chemical, X-ray diffraction, or, more generally, metallographic tools and techniques of any kind can be usefully applied to gain a better understanding of corrosion and corrosion control.[8-30]

The application of metallography to the study of corrosion forms a part of the subject matter of this chapter. The illustrations used to demonstrate the importance of microstructural studies in corrosion science and engineering practice are primarily based on the author's experiences. The techniques described and applications of them are by no means exhaustive. More comprehensive coverage of much of the material presented in this text may be found in the literature,[9-37] and the reader is encouraged to make use of this extensive resource.

FORMS AND MECHANISMS OF CORROSION

There are four basic elements required in an electrochemical cell: a conducting metal (electron conductor), a conducting electrolyte (ion conductor), anodes, and cathodes. Remove any one of them and electrochemical (aqueous) corrosion will cease.[2-4]

Corrosion, localized or otherwise, is often a consequence of dissimilar conditions either within the environment or within the metal itself. Differential corrosion cells may have their origins in localized differences in temperature (differential temperatures cells), differences in oxygen concentration (differential aeration cells), differences in hydrogen ion activity (differential pH cells), differences in flow of fluids (differential velocity cells), differences in materials in contact (galvanic or bi-metallic corrosion cells), stress gradients (differential stress cells), or from microstructural dissimilarities even within particular types of metals or alloys. In fact, residual stresses within materials as a result of thermo-mechanical processing are often sufficient to initiate corrosion-related failures; also, weld-decay or intergranular corrosion may sometimes be a unique form of galvanic or bi-metal corrosion. Moreover, specific, dissimilar conditions may act singly or conjointly to promote particular forms of corrosion.

Fontana and Greene[31] classified corrosion on the basis of the forms in which it manifests itself and defined and described eight of these. Recognition of the different forms of corrosion is not always sufficient, however, to make it possible to state the root cause(s) for the occurence of a particular form.

The forms listed by Fontana and Greene are as follows: (1) uniform or general corrosion (see, for example, Fig. 17-1), (2) galvanic or bi-metallic corrosion (Fig. 17-2), (3) crevice corrosion (Fig. 17-3), (4) pitting corrosion (Fig. 17-4), (5) intergranular corrosion, (6) selective leaching (e.g., dezincification of brass), (7) erosion-corrosion (Fig. 17-5), and (8) stress-corrosion (Fig. 17-6). Each of these may have had its origins in a variety of sources, including, as will be shown subsequently, microstructural variables. The use of microscopy is often paramount in determining the many factors contributory to corrosion. The following section will discuss its role by using specific practical examples taken from laboratory and field studies.

282 APPLIED METALLOGRAPHY

Fig. 17-1. General corrosion of a carbon-steel "Quench string" taken from a heavy oil production well: Corrosion from reservoir fluids in the well bore caused perforation of the steel tubing in several areas.

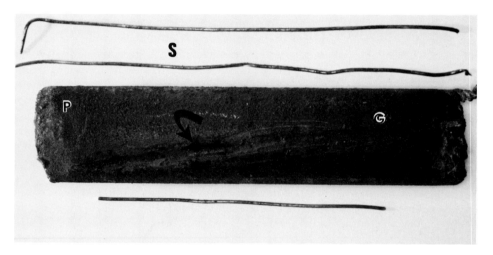

Fig. 17-2. Galvanic or bi-metallic corrosion between the stainless-steel capillary tubing (S) and the carbon-steel pipe (P) caused the formation of the spiral groove (G) and, ultimately perforation (arrows) and loss of pressure.

Fig. 17-3. Crevice corrosion on the head of a surgical stainless-steel screw used to fixate a bone plate to the femur to facilitate bone union.

MICROSCOPY AND THE STUDY OF CORROSION 283

Fig. 17-4. Pitting corrosion on carbon-steel tubing used in oil production.

Fig. 17-5. Corrosion erosion of carbon steel tubing used in oil production: The flow of well fluids caused extensive wall thining leading to longitudinal splitting, as shown.

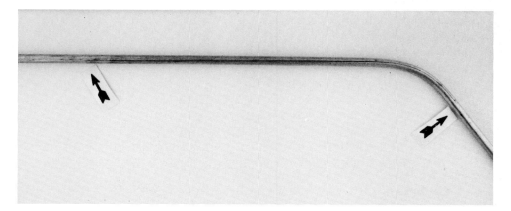

Fig. 17-6. Stress-corrosion cracking of stainless-steel, high-pressure tubing.

INVESTIGATIVE METHODS IN CORROSION SCIENCE AND ENGINEERING PRACTICE

Before presenting practical examples of applied microscopy in corrosion studies, it may be useful to list specific techniques that, individually or collectively, yield information diagnostic of factors contributing to corrosion. Visual inspection, low-power optical stereo microscopy, metrology, chemical spot testing, X-ray diffraction analysis (XRD), chemical analysis, mechanical and physical testing, potentiostatic/kinetic testing, optical microscopy, electron microscopy (including SEM, TEM, and STEM) and Auger scanning microscopy are the tools and techniques, among others, that may be usefully applied in corrosion research, engineering practice, and failure analysis. The science and theory on which these methods of analyses are based are well-covered in the open literature,[14–16,32–44] and much of this subject has also been described elsewhere in this book. Proper preparation and preservation of corrosion samples for microscopy is extremely important, and specialized techniques may often be required, as has been amply demonstrated elsewhere in this book and in the open literature.[33–35,38–41]

Because the emphasis in this chapter is on applications of microscopy, a brief listing of important information obtainable from applying the methods previously mentioned will be given before presenting specific examples.

Visual examination and low-power microscopy enables one to make general characterization of damage from corrosion (whether on laboratory or field samples) by observing the morphology, color, texture, and so forth of the corrosion products; some examples have been shown in Figs. 17-1 to 17-6, inclusive.

Metrology is concerned with physical measurements (dimensions) of component sizes (pipe diameters and wall thicknesses, for example), weight losses (from corrosion coupons and laboratory test samples), corrosion pit depths, crack lengths, the degree of wear (erosion), and measurements of other features relevant to the particular corrosion problems at hand. Microscopy is of fundamental importance in grain size measurements, inclusion counts, quantitative measurements in fractography, pit depth measurements (whether directly by use of a measuring microscope or from pit profiles on metallographic samples), and other measurements of a microscopic nature that are usually classified under metrology.

Chemical spot tests, X-ray diffraction analysis, energy-dispersive X-ray analysis (EDXA), Auger microscopy, and other allied techniques facilitate qualitative or quantitative determination of the types of corrosion products and other deposits.

Mechanical tests (hardness, tension, fatigue, impact, fracture toughness) and chemical analyses of test materials will yield data that can be used to check materials and components conformance (or otherwise) to relevant component standards and materials specifications.

Optical and electron microscopy is used, not only to characterize microstructures generally, but also to examine any corrosion or corrosion-related damage in terms of the supporting microstructures. Also, the specific morphologies of corrosion products imaged through the use of microscopy may be partially diagnostic of their types and origins.[22,23,30]

The potentiostat has gained general acceptance in recent years as a useful tool for corrosion monitoring and testing. Automation and computer control of potentiostatic instrumentation have greatly facilitated the efficient collection of corrosion data over extremely short periods of time in very complex systems. The development of this technology has fostered a variety of accelerated electrochemical test methods that provide information on general corrosion rates, pitting and crevice corrosion rates, active-passive material tendencies, pitting susceptibilities, and pitting potentials. In general, they will permit corrosion behavior to be characterized over the full practical redox potential range. The use of the potentiostat for the development of etchants is described elsewhere in this book. Although it is beyond the scope of this chapter to describe these and allied test methods (i.e., galvanostatic, a-c impedance, slow-strain-rate SCC tests, among others) in detail, a brief introduction to the potentiostat and its application to anodic, cathodic, and cyclic polarization testing will be given.[6,32,42–44]

A schematic diagram of the basic components of the potentiostat and polarization cell is shown in Fig. 17-7,[6] and the system may be seen to contain a three-electrode configuration in the test cell. In potentiostatic (or kinetic) testing, the electrical potential of the test specimen or working electrode (WE) is measured against a reference electrode (RE), and the response to controlled changes in this potential by the potentiostatic circuitry is measured and recorded as a flow of current between the specimen and an auxiliary or counterelectrode (AE), the latter usually being made from a cathode-efficient material (carbon

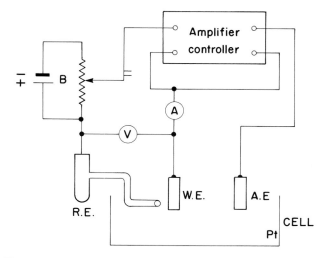

Fig. 17-7. Basic circuit for a potentiostat and electrochemical cell.[6]

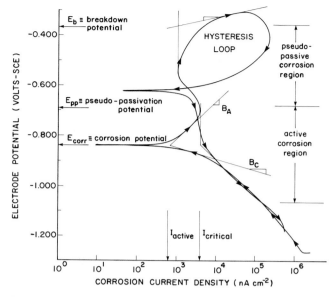

Fig. 17-8. Forward and reverse (arrows), cathodic and anodic polarization of a line pipe steel exposed to water containing H_2S at 25°C.

or platinum, for example). One of many ways in which to present the data obtained is shown in Figs. 17-8 and 17-9.

Figure 17-8 displays the results from a cathodic and anodic forward and reverse polarization scan, on which are defined several significant potentials, corrosion rates, and regions of fundamental importance to corrosion studies. The data given in this figure were obtained from tests on CSA Z245-category I grade of line pipe steel that was exposed to a simulated sour gas condensate, the partially passivating sulfide films being removed easily at higher anodic overpotentials (i.e., $> E_b$).

An anodic cyclic polarization graph obtained from tests on a surgical grade of stainless steel (AISI 316L) that was exposed to Hank's physiological solution appears in Fig. 17-9 and shows the breakdown potentials, pitting potentials, and pitting susceptibility ranges for this steel as influenced by variations in chloride ion concentrations.

The data illustrated in Figs. 17-8 and 17-9 were obtained from laboratory research, much of which has been reported in the literature.[45-48] These results will be used in the discussions that follow; more comprehensive coverage of the theory of these techniques is available elsewhere.[49,50]

APPLIED MICROSCOPY—CASE STUDIES

Microscopy is more than just a tool to characterize the microstructure of materials, whether one suffering from field failure or a laboratory sample, although such characterization is extremely important. In fact, laboratory corrosion studies reported in the literature for specific types of materials often ignore the metallurgical state of these materials altogether. Adequate documentation of whatever metallurgy may be involved should always be of great importance in laboratory corrosion research, especially for those heat-treatable alloys that vary widely in microstructure. Nevertheless, microscopy in corrosion studies has much broader implications than the mere characterization of microstructures.

PITTING CORROSION—LABORATORY STUDIES

A typical set of polarization curves for weld metal and parent metal obtained from research to investigate the susceptibility of spiral-welded pipe to corrosion in ice or permafrost in the Canadian arctic are shown in Fig. 17-10.[45,51]

Test samples from the weld zone [which included weld metal, heat-affected-zone (HAZ) metal, and parent metal] were polished, lightly etched, and then placed in the polarization test cell and anodically polarized to a potential a few millivolts more noble than the corrosion potential, E_{corr}. Samples were removed after exposures of 5, 10, 20, and 60 minutes and examined using optical and scanning electron microscopy (SEM). It was observed that initial corrosion was in the form of pitting attack along the interfaces between nonmetallic inclusions and the steel's matrices as well as along coarse-grained ferrite grain boundaries in the overheated zone of the HAZ.[51] The morphologies and progression of pitting corrosion were followed with microscopy.

The micrographs shown in Figs. 17-11 through 17-13 inclusive illustrate the pearlite, ferrite, and inclusion morphologies of parent and HAZ metal prior to testing. The

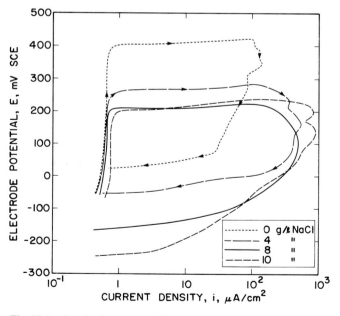

Fig. 17-9. Results from potentiokinetic cyclic experiments using surgical grade type 316L stainless steel in Hanks' physiological solution at 37°C with variable concentrations of NaCl showing hysteresis loops characteristic of susceptibility to crevice corrosion and/or pitting attack. Note: Arrows indicate scan directions.

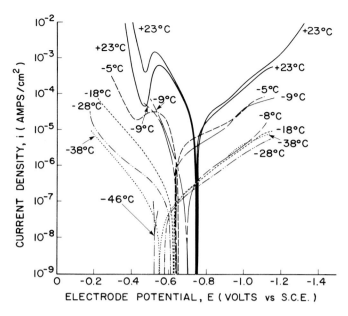

Fig. 17-10. Experimental polarization curves for a carbon-steel spiral welded pipe exposed to simulated waters at the temperatures indicated characteristic of soil-water chemistries in the Canadian arctic.

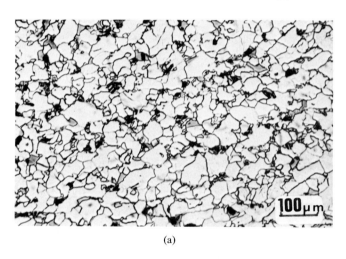

(a)

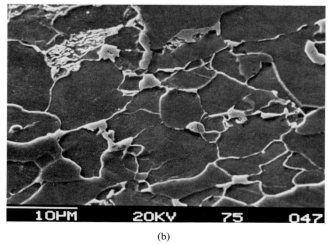

(b)

Fig. 17-11. Optical (a) and SEM (b) micrographs of parent metal showing fine-grained ferrite and pearlite (2-percent nital etch).

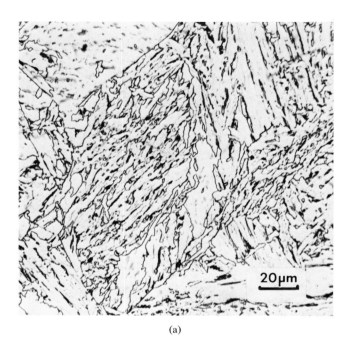

(a)

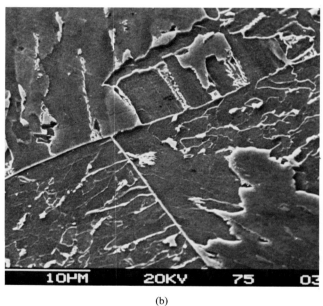

(b)

Fig. 17-12. Optical (a) and SEM (b) micrographs showing coarse-grained microstructure in heat-affected zone (2-percent nital etch).

parent metal contained normal, fine-grained ferrite and pearlite, whereas the HAZ had a coarse microstructure with some degenerated pearlite. The inclusions were of mixed, multiphase character. Energy-dispersive X-ray analysis (EDAX) was used to assess the character of the multi-phase inclusions qualitatively [see Fig. 17-13(b)].

The progression of pitting corrosion after exposures of 5 minutes and 20 minutes is illustrated in Figs. 17-14 and 17-15, respectively. The interfacial corrosive attack in the early stages of pitting is clearly shown, and progression of the pitting was observed to be primarily a consequence of matrix dissolution.

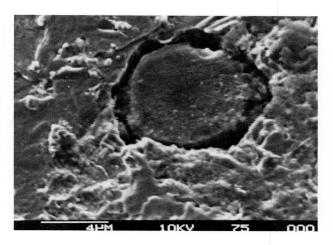

Fig. 17-15. SEM micrograph showing more extensive dissolution of metal at inclusion/matrix interface after 20 minutes of exposure. Note also some corrosion-product formation on matrix.

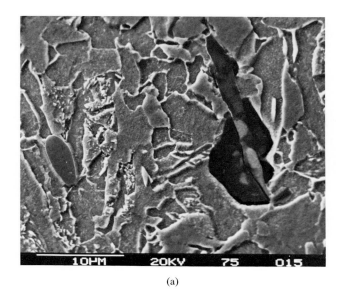
(a)

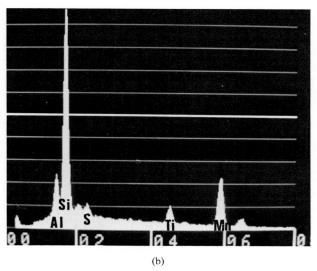
(b)

Fig. 17-13. SEM micrograph (a) showing multi-phase inclusions in line pipe steel and the corresponding energy-dispersive X-ray analysis (b) of the inclusion (M).

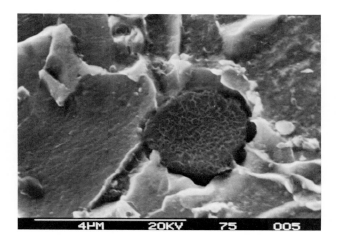

Fig. 17-14. SEM micrograph showing pitting corrosion initiating at inclusion/matrix interface after 5 minutes of exposure.

PITTING CORROSION—FIELD STUDIES

Corrosion monitoring of an injection well in a fireflood, enhanced-oil-recovery (EOR) pilot plant involved the use of corrosion coupons positioned on mandrels at various depths in the injection tubing. Microscopic examination of the corrosion coupons after exposure to the injection fluids revealed that the morphologies (and rates) of pitting were influenced considerably by the anisotropy of the microstructures of the coupons.

The "banded" nature of the ferrite and pearlite is revealed clearly in the micrographs of Fig. 17-16. Subsequently, optical and SEM images of the corrosion-pit profiles (shown in Figs. 17-17 and 17-18) showed that pits transverse to the "fiber" of the microstructure were shallow and broad-based, whereas those pits that developed along the fiber of the microstructure were deeply penetrating. The SEM was also useful for pit-depth measurements and to examine the morphology and density of the observed pitting attack (Fig. 17-19). EDXA was used to detect the presence of chlorides and sulfides at the base of the exposed pits, which, in conjunction with data obtained from other tests, was used as partial explanation of the origin of pitting corrosion.

A final example to illustrate the application of microscopy to pitting-corrosion studies involved examination of an AISI type 316 stainless-steel, Sperry-Sun tubing used for "downhole" pressure measurements in an observation well. The good depth-of-field and resolution of the SEM to record the morphology of pitting attack and the use of EDXA to detect chlorides, among other constituents, at the base of the corrosion pits are clearly illustrated in Fig. 17-20. The "faceted" texture of the corrosion pits was indicative of active acid corrosion of the stainless steel as the pits developed.

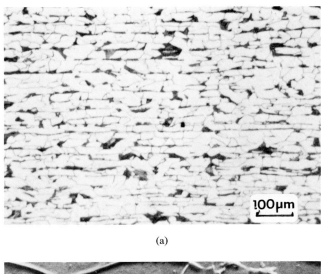

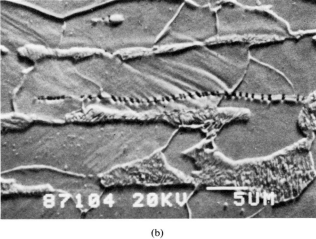

Fig. 17-16. Optical (a) and SEM (b) micrographs illustrating a "banded" microstructure of ferrite and pearlite (2-percent nital etch).

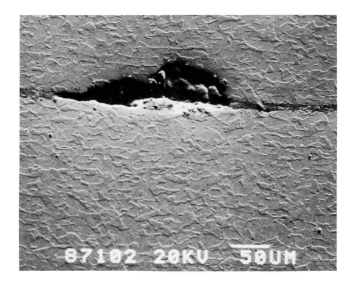

Fig. 17-17. SEM micrograph of corrosion pit in "profile" illustrating shallow, broad-based pitting transverse to "Banding."

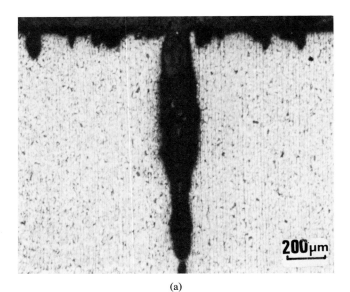

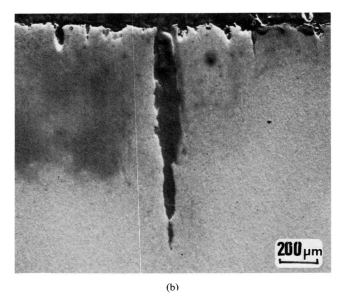

Fig. 17-18. Optical (a) and SEM (b) micrographs illustrating the deep-penetrating nature of corrosion pits growing parallel to the "banded" microstructure.

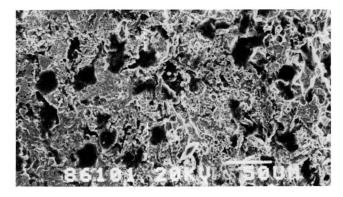

Fig. 17-19. SEM image of densely packed corrosion pits on a C1018 carbon-steel corrosion coupon.

MICROSCOPY AND THE STUDY OF CORROSION

CORROSION–EROSION

Microscopy was usefully employed to study the corrosion–erosion damage of the tubing sample in Fig. 17-5. Extensive wall thinning from fluid flow through the tubing was observed to be nonuniform (see Fig. 17-21). It was also observed that considerable variation in grain size, not only in the weld zone of the seam weld, but generally throughout the tubing, contributed significantly to the nonuniform reductions in wall thicknesses (see Fig. 17-22). Coarse-grained, predominantly ferrite regions of classical "Widmanstätten" morphology were generally coincident with the thinnest sections of the tubing walls.

The SEM was used to study flow damage in greater depth. The flow lines were observed to be alternately composed of etched regions, essentially free of corrosion products, and lines of corrosion-product "build-up" (see Fig. 17-23). The texture and character of corrosion products as revealed by electron microscopy are often partially diagnostic of their type (i.e., oxides, carbonates, sulfides, and so forth).[22,23,30]

STRESS-CORROSION, EMBRITTLEMENT, AND FRACTURE

Microscopy is an invaluable aid in the study of environmentally assisted fracture of metallic components, in which the conjoint action of corrosive environments and residual or externally applied loads (stresses) promote failure in a shorter time than do isolated and separate effects of either the environment or the imposed stresses. The literature contains numerous examples of microscopy applied to laboratory research on stress corrosion[8,10,27] and in failure analysis.[12,20,21,27] This section will not necessarily distinguish between the often subtle differences be-

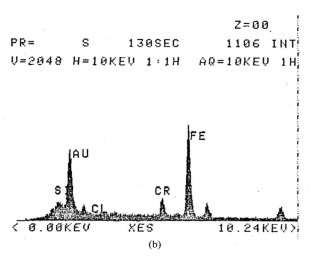

Fig. 17-20. (a) Pitting corrosion of AISI type 316 capillary tube used for "downhole" pressure measurements in oil production, and (b) EDAX analysis in pitted region showing presence of chlorides (Cl⁻).

Fig. 17-21. Tubing sample of Fig. 17-5 viewed "end-on" and showing extensive wall-thickness reductions from erosion-corrosion.

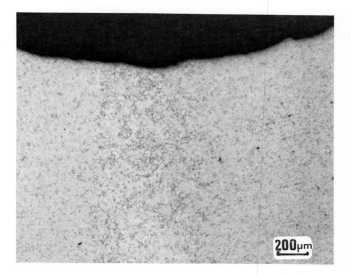

Fig. 17-22. Optical micrograph of "thinned" region in "profile" showing that areas of greatest wall thickness reductions were coincident with coarse-grained microstructures (2-percent nital etch).

(a)

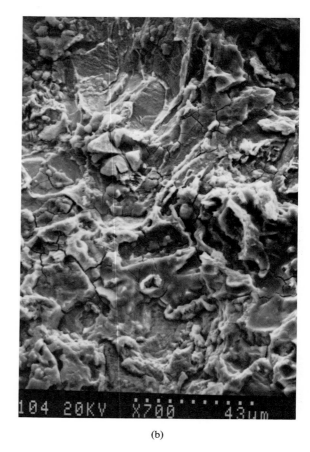

(b)

(c)

tween stress-corrosion cracking, sulfide-stress cracking, chloride-stress cracking, hydrogen-assisted cracking, and so forth. Rather, it will discuss and illustrate the use of microscopy to detect features diagnostic of particular modes of stress-corrosion failures.

During the course of research and failure analyses concerned with corrosion and fracture of metallic orthopedic implants,[47,48,52,53] microscopy was extensively used to assess the damage. In some of these studies, once the polarization data were obtained (as shown in Fig. 17-9), specimens of the stainless steels (both stressed and unstressed) were polarized into the pitting corrosion susceptibility range, and the damage was assessed using microscopy. Microscopy was also used extensively in implant failure analyses.[52,53]

It has often been observed that anodic stress-corrosion cracks are initiated from corrosion pits that develop from passivation-break-down of stainless steel in the presence of Cl^- and that crack propagation is assisted by active metal dissolution at the crack tips, as may be seen from the optical micrographs in Fig. 17-24. Note the corrosion

Fig. 17-23. SEM examination of corrosion–erosion showing (a) longitudinal lines of corrosion products alternating with corrosion-product-free regions. The micrograph in (b) illustrates the "etched" texture of the corrosion-product-free regions, and the micrograph in (c) shows the texture of the corrosion products more clearly.

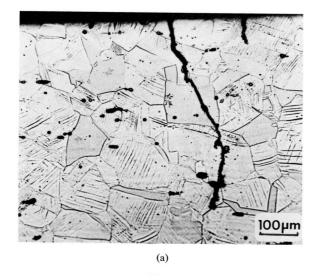

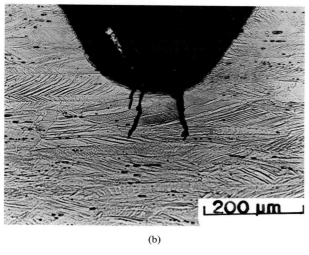

Fig. 17-24. Optical micrographs illustrating anodic stress-corrosion cracking in a stainless-steel fixation plate (a) and at the roots of threads in a bone-plate screw (b).

pits at the surface and the blunted crack tips from metal dissolution. The character of these cracks is significantly different from that which develops by a cathodic stress-corrosion cracking mechanism, the latter often being intergranular, with sharp, penetrating crack tips.[8,12,18]

Anodic stress-corrosion crack propagation is often assisted by hydrogen cracking. Although the crack-tip regions are anodic and actively corrode, microcathodes are formed in the regions of the crack tips to support reduction reactions. Monatomic hydrogen, being produced from hydrogen-ion reduction, will ingress into the metal lattice or penetrate grain boundaries, and secondary intergranular cracks or hydrogen fissures will develop. Evidence for these various phenomena associated with stress-corrosion failures can be found by means of scanning electron microfractography or transmission electron microfractography.

The TEM fractographs shown in Fig. 17-25, which

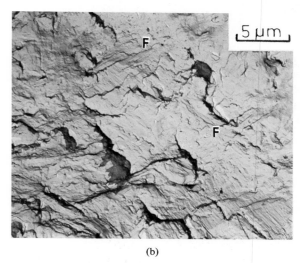

Fig. 17-25. TEM microfractographs showing in (a) striations associated with transgranular anodic stress-corrosion cracking cycles (C) and intergranular hydrogen-assisted cracking (G). The micrograph in (b) shows hydrogen fissures (F).

were obtained from replicas of fracture surfaces, illustrate anodic stress-corrosion crack striations produced as a result of the dissolution/no-dissolution cycles of crack propagation. The intergranular facet torn from the metal as shown in Fig. 17-25(a) was a consequence of secondary hydrogen intergranular cracking. Hydrogen fissures may be seen running parallel to the crack propagation direction in Fig. 17-25(b).

The SEM fractograph of Fig. 17-26 illustrates crack arrest lines often observed on stress-corrosion fractures, these being analogous to beach lines in fatigue.

The fracture of an orthopedic fixation screw is illustrated in Fig. 17-27, and the utility in applying electron microscopy to the analyses of such fractures is demonstrated. It is clearly shown that crack initiation was environmentally assisted, whereas fast fracture was simply a consequence of overload ductile crack growth.

292 APPLIED METALLOGRAPHY

Fig. 17-26. SEM microfractograph at low magnification showing crack "arrest" lines often typical of SCC in metallic orthopedic implant failures. The crack propagation direction is shown by the arrow.

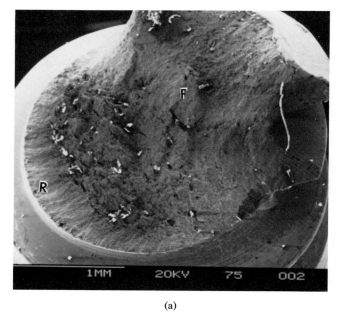

(a)

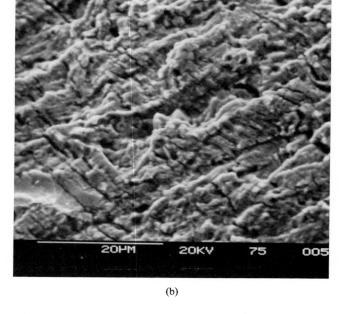

(b)

Fig. 17-27. SEM microfractographs of a metallic orthopedic fixation screw (a) showing that crack initiation (R) was induced by stress-corrosion, as illustrated in (b), and that fast fracture (F) was induced by dutile overload crack growth mechanisms, as shown by the characteristic dimple topography in (c).

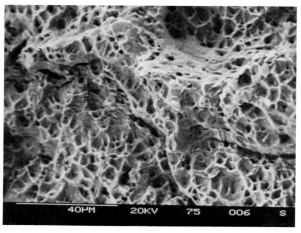

(c)

Fig. 17-28. Optical (a) and SEM (b) fractographs showing brittle fracture of a threaded connection and confirming that the cracks originated at corrosion pits.

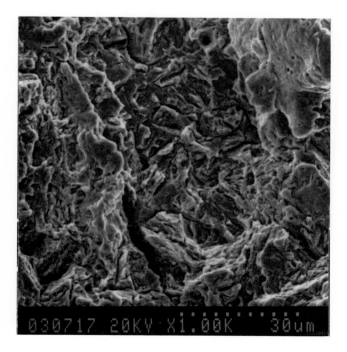

Fig. 17-29. SEM microfractograph showing secondary cracking associated with hydrogen-assisted crack growth and embrittlement.

The fracture shown in Fig. 17-28(a) developed from the failure of a drilling tool at a threaded connection, the tool being made from AISI Type 4340 steel that had been quenched and tempered to $\approx$ HRC 40 hardness. The optical fractographic features (radial lines emanating from the crack origins) are typical of brittle overload failure. Several crack initiation points were observed, and scanning electron microscopy revealed, as shown in Fig. 17-28(b), that corrosion pits were associated with each of them. The secondary cracks shown in the SEM fractograph of Fig. 17-29 were partially diagnostic of hydrogen embrittlement, the material having been embrittled from pickling and coating the threads during manufacture. Impact tests done on material remote from the threaded ends and on specimens machined from the threaded ends confirmed the brittle nature of the threaded connections. Fractographic examination revealed classic ductile dimple topography in the former case and quasi-cleavage fracture in the latter, as may be seen in Fig. 17-30.

As a final, rather simple example of the importance of microscopy in corrosion-related investigations, the stress-corrosion cracking seen in the stainless-steel, high-pressure tubing shown in Fig. 17-6 was observed to be associated with a longitudinal seam weld. Fracture occurred along the fusion line, as may be seen in the optical micrographs in Fig. 17-31. As a closing comment, specifications required that the tubing be seamless!

294 APPLIED METALLOGRAPHY

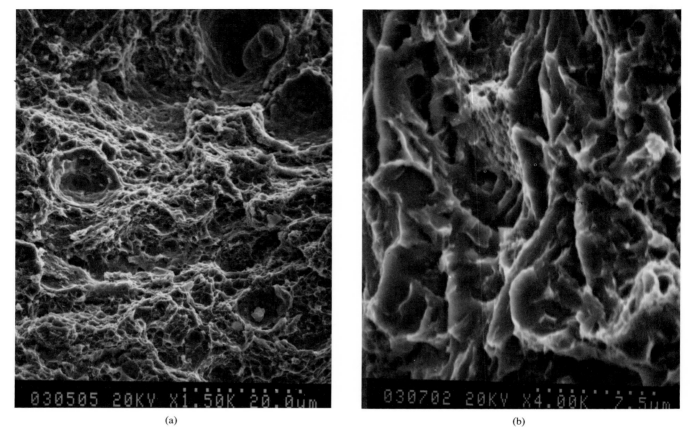

Fig. 17-30. SEM fractographs illustrating (a) dimple-fracture topography characteristic of ductile crack growth, and (b) quasi-cleavage in tempered martensite characteristic of brittle fracture.

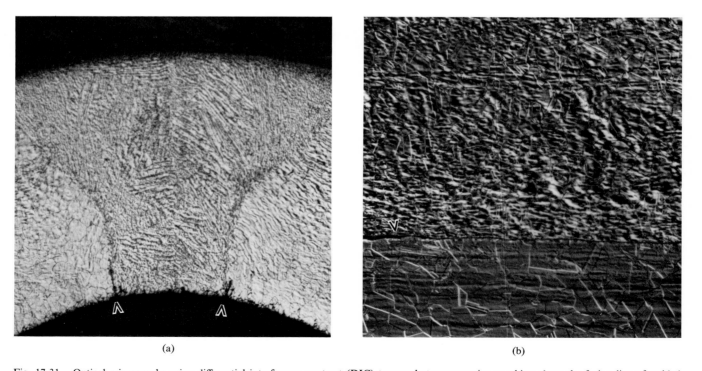

Fig. 17-31. Optical micrographs using differential interference contrast (DIC) to reveal stress-corrosion cracking along the fusion line of welded stainless-steel pressure tubing. (a) transverse view, and (b) longitudinal view (diluted aqua-regia etch).

SUMMARY

Microscopy, in part, encompasses the processes and techniques of using microscopes as investigative tools in science and engineering practice. Through the use of practical, illustrative examples, the importance, power, and utility of microscopy as an aide to the study of corrosion has been demonstrated. The material presented, however, is far from exhaustive, and more comprehensive coverage is available in the literature cited. The serious student and researcher is encouraged to make ample use of this vast resource.

REFERENCES

1. Uhlig, H. H., ed. *Corrosion Handbook.* New York: John Wiley and Sons, 1948.
2. Brasunas, A. de S., ed. *Basic Corrosion.* National Assoc. of Corr. Eng., 1978.
3. Shreir, L. L. *Corrosion,* vol. 1. London: Newnes-Butterworth, 1976.
4. Bockris, J. O'M., and Reddy, A. K. N. *Modern Electrochemistry,* vols. 1 and 2. New York: Plenum Press, 1977.
5. Pourbaix, M. *Lectures on Electrochemical Corrosion.* New York: Plenum Press, 1973.
6. Shreir, L. L., *Corrosion,* vol. 2. London: Newnes-Butterworth, 1976.
7. Crow, D. R. *Principles and Applications of Electrochemistry,* London: Chapman and Hall, 1974.
8. Burns, D. S. Laboratory test for evaluating alloys for H_2S service. In *H_2S Corrosion in Oil and Gas Production—A Compilation of Classic Papers,* edited by R. N. Tuttle and R. D. Kane, pp. 275–282. NACE, 1981.
9. Briant, C. L., and S. K. Banerji. Intergranular failure in steel: The role of grain boundary composition. *Int. Met. Rev.* 23, No. 4:164–199 (1978).
10. Bernstein, I. M., and Thompson, A. W. Effect of metallurgical variables on environmental fracture of steels. *Int. Met. Rev.* 21, No. 4:269–287 (1976).
11. Hoeppner, D. W. Environmental effects in fretting fatigue. In *Fretting Fatigue,* edited by R. B. Waterhouse, pp. 143–158. Essex, England: Applied Science Publishers, 1981.
12. Forsyth, P. J. E. Occurrence of fretting fatigue failures in practice. Ibid, pp. 99–126.
13. Guttmann, V., and Mertz, M., eds. *Corrosion and Mechanical Stress at High Temperatures.* London: Applied Science Publishers, Ltd., 1981.
14. Andrews, K. W.; Dyson, D. J.; and Keown, S. R. *Interpretation of Electron Diffraction Patterns,* London: Hilgers and Watts Ltd., 1967.
15. Goldstein, J. I., and Yakowitz, H., eds. *Practical Scanning Electron Microscopy.* New York: Plenum Press, 1977.
16. Loretto, M. H., and Smallman, R. E. *Defect Analysis in Electron Microscopy.* London: Chapman and Hall, 1975.
17. Louthan, Jr., M. R., et al. Hydrogen embrittlement in metals. In *Hydrogen Damage,* edited by C. D. Beachem, pp. 289–300. ASM, 1977.
18. Nielson, N. A. Observations and thoughts on stress corrosion mechanisms. Ibid, pp. 219–254.
19. Scheidl, H. Transmission electron microscope and microfractographic studies on steel X20 Cr13 for turbine blades. *Praktische Metallographie* 9:503–516 (1973).
20. Effertz, P. H.; Schuller, H. J.; and Wiume, D. High temperature corrosion damage. *Praktische Metallographie* 9:409–426 (1976).
21. Colangelo, V. Metallographic analysis of corrosion failures. In *Metallography in Failure Analysis,* edited by J. L. McCall and P. M. French. New York: Plenum Press, 1977.
22. Northwood, D. O.; White, W. E.; and Vander Voort, G. F., eds. *Corrosion, Microstructure and Metallography.* Microstructural Science, vol. 12. ASM, 1985.
23. Miley, D. V., and Smolik, G. R. SEM examinations of oxidation characteristics of alloy 800H. In *Microstructural Science,* vol. 10, edited by W. E. White, J. H. Richardson, and J. L. McCall, pp. 43–50. New York: American Elsevier 1982.
24. Ganesan, P.; Mehrotra, P. K.; and Sargent, G. A. A microstructural study of the erosion-corrosion behavior of type 304 and 310 stainless steels. Ibid, pp. 91–102.
25. Jonard-Guerin, F., and Frade, G. Study of the dezincification of the Aℓ-Zn alloys with 11% Zn by transmission electron microscopy. *Metallography* 8:489–508 (1975).
26. Agaruala, V. S., and Murty, Y. V. A controlled-potential corrosion study of Aℓ-4.5 cu alloy in 3.5% NaCℓ solution. *Metallography* 10:451–461 (1977).
27. Shei, S. A., and Kim, C. D. A microstructural study of the sulfide stress cracking resistance of a Cr–Mo–V–B steel. *Corrosion* 41:12–18 (1985).
28. Jeffrey, Y., and Mui, P. Corrosion mechanisms of metals and alloys in the silicon-hydrogen-chlorosilane system. *Corrosion* 41:63–68 (1985).
29. Roques, Y., et al. Pitting of Zircaloy-4 in chloride containing aqueous methanolic solution. *Corrosion* 40:561–566 (1984).
30. Moriya, M., and Ives, M. B. The structure of anodic films formed on nickel and nickel-13 w/o molybdenum alloy in pH 2.8 sodium sulfate solution. *Corrosion* 40:62–72 (1984).
31. Fontana, M. G., and Grenne, N. D. *Corrosion Engineering.* New York: McGraw-Hill, 1967.
32. Edeleneau, C. The potentiostat as a metallographic tool. *J.I.S.I.* 185:482–488 (1957).
33. *Atlas of Microstructures of Industrial Alloys.* Metals Handbook, 8th ed., vol. 7. American Society for Metals, 1972.
34. *Metallography, Structures and Phase Diagrams,* Metals Handbook, 8th ed., vol. 8. American Society for Metals, 1973.
35. *Fractography and Atlas of Fractographs,* Metals Handbook, 8th ed., vol. 9. American Society for Metals, 1974.
36. Easterly, K. E. Recent Developments in quantitative electron microscopy. *Int. Met. Rev.* 22:1–24 (1977).
37. Beeston, B. E. P.; Horne, R. W.; and Markham, R. Part II: Electron diffraction and optical diffraction techniques. In *Practical Methods in Electron Microscopy,* edited by A. M. Glauert. New York: American Elsevier, 1972.
38. *Metallographic Preparation for Corrosion Studies,* Metal Digest, vol. 22, No. 2. Lake Bluff, IL: Buehler Ltd., 1983.
39. *Metallographic Sample Preparation,* Metal Digest, vol. 20, No. 2. Lake Bluff, IL: Buehler Ltd., 1981.
40. Goodhew, P. J. Specimen preparation in materials science. In *Practical Methods in Electron Microscopy,* edited by A. M. Glauert. New York: American Elsevier, 1972.
41. Cihal, V. and Prazak, M. Corrosion and metallographic study of stainless steels using potentiostatic techniques. *J.I.S.I.* 193:360–367 (1959).
42. ASTM Annual Standard G3-74, Part 10: Conventions applicable to electrochemical measurements in corrosion testing. American Society for Testing and Materials, Philadelphia, PA (1980).
43. ASTM Annual Standard G5-78, Part 10: Standard reference method for making potentiostatic and potentiodynamic anodic polarization measurements. American Society for Testing and Materials, Philadelphia, PA (1980).
44. ASTM Annual Standard G61-78, Part 10: Practice for conducting cyclic potentiodynamic polarization measurements for localized corrosion. American Society for Testing and Materials, Philadelphia, PA (1980).
45. King, R. J. Corrosion of Steel Weldments in Permafrost. Ph.D. Thesis, The University of Calgary, Calgary, Canada, 1983.
46. White, W. E.; King, R. J.; and Coulson, K. E. W. Preliminary

observations of corrosion on carbon steel in permafrost. *Corrosion* 39:346–353 (1983).
47. Ogundele, G. I. Corrosion of metallic orthopaedic implants. *M.Sc. Thesis,* The University of Calgary, Calgary, Canada 1980.
48. Ogundele, G. I., and W. E. White, Polarization studies on surgical grade stainless steel in Hank's physiological solution. In *Corrosion and Degradation of Implant Materials,* pp. 233–248. Philadelphia: ASTM STP 859, 1985.
49. Baboian, R., and Haynes, G. S. Cyclic polarization measurements, experimental procedure and evaluation of test data. In *Electrochemical Corrosion Testing,* pp. 274–282. Philadelphia: ASTM STP 727, 1981.
50. Sarkar, N. K. Application of potentiokinetic hysteresis technique to characterize the chloride corrosion in high copper dental amalgams. Ibid, pp. 283–289.
51. White, W. E., and King, R. J. Pitting corrosion of carbon steel weldments in permafrost. In *Microstructural Science,* vol. 11, pp. 435–456. New York: American Elsevier, 1982.
52. White, W. E., and Le May, I. Optical and electron fractographic studies of fracture in orthopaedic implants. In *Microstructural Science,* vol. 3, Part B, pp. 911–930. New York: American Elsevier, 1975.
53. White, W. E.; Postlethwaite, J.; and Le May, I. On the fracture of orthopaedic implants. In *Microstructural Science,* vol. 4, pp. 145–158. New York: American Elsevier, 1976.

INDEX

Abrasive wear, 262–264
Adhesive wear, 261–262
Additives, 211–212, 214–224
Alpha-beta titanium alloys, 243–245
Alpha titanium alloys, 241–243
Aluminum bronze, 2–3
Aluminum-coated sheet steel, 42–43
Area point count, 90–92, 98
Artifacts, 75–76, 238–239
Atomic number contrast, 140–146

Backscattered electrons, 140–147, 153–162, 178, 213, 215, 225
Beach marks, 253
Beam diffraction, convergent, 181–186, 191
Beta titanium alloys, 245–247
Bismuth-treated steel, 214–215
Boron-treated steels, 212, 223–224
Brass (alpha-beta), 6, 8
Bright-field imaging, 176–178
Brittle fracture, 251–253

Calcium-treated steels, 222–224
Carbides:
    optical characteristics, 48–50
    sintered; see Sintered carbides
    tint etching, 11–12
Carbon tool steel, etching, 11–12
Channeling contrast, 146
Chemical etch compositions, 4–7, 78, 163–165
Chemical polishing, of stainless steels, 76–77
Chemical spot tests, 279–280
Cleaning specimens, 268, 279
Cleavage, 252
Cliff-Lorimer factor, 187, 189
Contrast, image:
    light microscopy, 139
    scanning electron microscopy, 139–170
    visual, 140
Convergent beam diffraction, 181–186, 191
Corrosion failure analysis, 281–296
    analysis methods, 284–285
    crevice corrosion, 281–282
    corrosion–erosion, 289–290
    corrosion fatigue, 255
    forms and mechanisms, 281–283
    galvanic corrosion, 281–282
    general corrosion, 281–282
    hydrogen-assisted cracking, 258
    hydrogen embrittlement, 258, 293–294
    pitting corrosion, 285–289
    potentiostatic methods, 284–286
    stress-corrosion cracking, 80, 82–85, 254–255, 289–294
Corrosion mechanisms, 281–283

Cracking, stress–corrosion, 80, 82–85, 254–255, 289–294
Creep failures, 255–257
Crevice corrosion, 281–282
Crystallographic contrast, 146–147
Cutting tools, 232

Dark-field imaging, 178
Deep etching, 104–105, 163–165
Delta ferrite, 5, 15, 30, 32, 55–58, 60, 86
Depth of focus, 151
Differentiation ratio, 27, 31
Diffraction, beam; see Beam diffraction
Domains, magnetic, 55–56
Ductile fracture, 251

Edge preservation, 272–273
Electrical discharge pitting, 271
Electrolytic etching, 14–16, 79–80
Electrolytic inclusion extraction, 224–228
Electrolytic polishing:
    stainless steel, 75–76, 78–79
    titanium, 238
Electrons, backscattered; see Backscattered electrons
Electron backscatter coefficient, 141
Electron channeling contrast, 146
Electron energy loss, 179–181, 189, 192–194
Embrittlement, hydrogen, 258, 293–294
Energy-dispersive spectroscopy:
    corrosion studies, 286–287, 289
    wear studies, 265–266
Environmentally assisted fracture, 80, 82–85, 254–255, 258, 293–294
Erosion, 264–266
Etchants:
    composition of, 4–6, 34–35, 78–79
    deep etching, 163–165
    selective, 4–6
    for stainless steels, 78–79
    tint; see Tint etchants
Etching:
    chemical, 6–7, 76–79, 163–165
    deep, 160–165
    electrolytic, 14–16, 79–80
    magnetic; see Magnetic etching
    potentiostatic; see Potentiostatic etching
    selective, 3–18
    tint, 7–14
Extraction of inclusions, 224–228

Failure analysis, 251–259
    corrosion, 281–296
    failure mechanisms; see Failure mechanisms
    methodology, 259
    wear, 261–280

298 INDEX

Failure mechanisms:
  abrasive wear, 262–264
  adhesive wear, 261–262
  brittle fracture, 251–253
  corrosion–erosion, 289–290
  corrosion fatigue, 255
  creep, 255–257
  ductile fracture, 251–252
  erosion, 264–266
  fatigue, 253–254
  fretting, 255, 262–264
  high-temperature, 255–257
  hydrogen-assisted cracking, 258
  hydrogen embrittlement, 258–259, 293–294
  overload, 251
  pitting, 285–289
  rolling contact fatigue, 264, 266
  stress-corrosion cracking, 80, 82–85, 254–255, 289–294
Fatigue, 253–254, 255, 264, 266
Fatigue striations, 117–119, 253–254
Feature count, 91, 92, 98
Feature-specific measurements, 91, 93
Ferrofluid, 53–55
Field measurements, 91
Focus, depth of, 150
Fractals, 102, 111
Fractography; see Quantitative fractography
Fracture:
  brittle, 251–253
  ductile, 251–252
  environmentally assisted, 80, 82–85, 254–255, 258, 293–294
Free-machining steels, 211–236
  additives, 211–212
  bismuth-treated, 214–215
  boron-treated, 223–224
  calcium-treated, 222–224
  compositions, 216
  extractions, 224–228
  leaded, 211, 213–215, 220–221
  metallographic preparation, 212–213
  microstructure, 213–224
  quantitative metallography, 224–230
  resulfurized, 214–220
  selenium-treated, 220–221
  tellurium-treated, 220–221
Fretting, 255, 262–264
Fusion zone, 197

Galvanic corrosion, 281–282
Gas-contrasting method, 63–70
  apparatus, 64–65
  contrast development, 67–69
  examples, 66–67; see also color insert pages
  theory, 65–69
Grain size, 97

Half-cell potentials, 22–24
Heat-affected zone, 197, 207–208
Heat tinting, 5–6, 16–18
High-temperature failures, 255–257
HOLZ lines, 182–184, 186–187
Hydride formation in Ti, 238
Hydrogen-assisted cracking, 258
Hydrogen embrittlement, 258, 293–294

Identification, of phases:
  heat tinting, 16–18
  morphology, 1
  polarized light, 2–3, 17, 237–238
  quantity, 1
  reflectance, 1–2
  relief, 3
  selective etching, 3–18, 86
  tint etching, 7–14
Image analysis, 93–94, 101, 106, 126–127, 165–167
Image contrast:
  light microscopy, 139
  scanning electron microscopy, 139–150
  visual, 140
Inclusions; see Metallic inclusions and Nonmetallic inclusions
Inclusion extraction, 224–228
Integral mean curvature, 92, 98
Interdendritic eutectic, 200–201
Interference layer method, 41–51
  apparatus, 45–46
  applications, 49–51
  deposition layers, 46–49
  optical characteristics, 41–42, 46–49
  reflectivity, 41–44, 50
  specimen preparation, 43–45
  technique, 45–46
  theory, 42–43
  see also color insert pages
Ion beam thinning, 238–239

Klemm's reagent, 4, 6, 9–11, 146–147
Knoop hardness, 124–125, 126–127

Leaded steels, 211, 213–215, 220–221
  lead print, 213
  metallographic preparation, 212–213
  microstructure, 213–214
  SEM examination, 213–215
  solubility of lead, 213
Lead print, 213
Length in unit volume, 92, 98
Light microscopy, 139
Light-section microscope, 109
Line intercept count, 90–92, 97

Machinability, 211–236
  additives, 211–212, 214–224
  bismuth-treated steels, 214–215
  boron-treated steels, 212, 223–224
  built-up edge, 231–232
  calcium-treated steels, 222–224
  cutting tools, 232
  direct observation of machining, 231–232
  leaded steels, 211, 213–215, 220–221
  MnS shape effects, 216–220, 229–230
  MnS size effects, 216–220, 229
  MnS types, 216, 218, 221
  oxides, 221–223
  quick stop tests, 230–231
  resulfurized steels, 214–220
  selenides, 220–221
  tellurides, 220–221
  tool life, 230
Magnetic contrast, 148
Magnetic domains, 55–56
Magnetic etching, 53–61
  examples, 55–60
  history, 53
  technique, 53–55

Magnification:
  scanning electron microscopy, 151
  selection in stereology, 95
  stereoscopic images, 102–103
Manganese sulfides, 13–14, 211, 214–220, 228–230
  deformability, 218–220
  etching, 13–14
  identification line, 230
  Mn/S ratio, 216
  resulfurized steels, 214–220
  size and shape effects, 216–220, 228–230
  sulfur print, 213–214, 280
  types, 216, 218, 221
  X-ray maps, 148–149, 217, 225
Martensite, strain-induced, 56, 59, 75
Mean free path, 93
Mean lineal intercept, 92–93, 97, 98
Metallic inclusions:
  bismuth, 214, 215
  lead, 211, 213–215, 220–221
Metallography, quantitative; *see* Quantitative stereology
Microdiffraction, 181–184
Microhardness; *see* Microindentation hardness
Microindentation hardness, 123–138
  examples, 128–130
  fixturing, 132, 136–137
  hardness conversions, 128
  high-temperature tests, 132, 135
  inaccuracy and errors, 125–127
  indenter types, 124–125
  Knoop hardness, 124–125, 126–127
  methods, 123–128
  projected area hardness, 124–125
  scratch tests, 127–128
  specimen preparation, 132, 135–136
  standards, 125–126
  taper sections, 135–136
  tester selection, 129–132
Microscopic techniques
  corrosion, 284–285
  wear, 264–268, 271–274
Microscopy; *see* Scanning electron microscopy, Scanning transmission electron microscopy, Transmission electron microscopy
MnS characteristics; *see* Manganese sulfides
Murakami's reagent, 12, 15, 17, 86
  compositions, 5–6, 78

Nonmetallic inclusions:
  calcium-containing oxides, 222–224
  manganese sulfides, 13–14, 211, 214–220, 228–230
  oxides, 221–223
  selenides, 220–221
  tellurides, 220–221
Nonmetallic inclusion extraction, 224–228

Overload failures, 251
Oxides, 221–223
Oxygen enrichment, 5, 13–14

Phase identification:
  convergent beam diffraction, 182–186, 191
  heat tinting, 16–18
  morphology, 1
  polarized light, 2–3, 17, 139, 217, 237–238
  quantity, 1
  reflectance, 1–2
  relief, 3

selective etching, 3–18
tint etching, 7–14
Photogrammetric method, 103–104
Pitting, electrical discharge, 271
Pitting corrosion, 285–289
Point count, 90–92, 96–97
Polarization curves; *see* Potentiostatic etching
Polarized light, 2–3, 17, 139, 217, 237–238
Potentiostatic etching, 21–39
  etch compositions, 34–35
  examples, 34, 36–38
  film formation, 31–33
  orientation effects, 28
  polarization curves, 25–26, 28, 30, 32, 36, 285–286
  potentiostat, 25, 284–286
  procedure, 33–34
  theory, 21–24, 27–33
Profile measurements, 101–104, 108–117
Projected area Knoop hardness, 124–125
Projected images, 101, 105–108

Quantitative fractography, 101–122
  deep surface etching, 104–105
  fatigue striations, 117–119
  fractals, 102, 111
  geometrical relationships, 104
  photogrammetric method, 103–104
  profile measurements, 101–104, 108–117
  projected images, 101, 105–108
  roughness parameters, 102, 110–115
  stereological relationships, 101–102, 105–108, 119–121
  stereoscopic methods, 101–103
Quantitative metallography; *see* Quantitative Stereology
Quantitative stereology, 89–99, 101–102, 105–108, 119–121, 228–230
  area point count, 90–92, 98
  counting measurements, 90–92
  equipment, 90, 93, 95
  examples, 96–98, 117–121
  extractions, 224–228
  feature count, 91, 98
  field selection, 95–96
  fractography, 101–122
  grain size, 97
  integral mean curvature, 92, 98
  length, 92, 98
  light microscopy, 88–99
  line intercept count, 90–92, 97
  magnification, 95
  mean free path, 93
  mean lineal intercept, 92, 97–98
  meaning, 92–93
  number of fields, 95
  point count, 90–92, 96
  planning measurements, 94–96
  scanning electron microscopy, 165–167
  size and shape, 93, 229–230
  statistical analysis, 95–98
  surface area per volume, 92, 97
  technique, 89–90
  volume fractions, 92, 96–97, 165–167, 229
Quasi-cleavage, 252–253
Quick-stop tests, 230–231

Reflectance, 1–2
Reflectivity, 41–44, 50
Relief polishing, 3
Replicas, 162, 165

Resolution:
    scanning electron microscopy, 150
    scanning transmission electron microscopy, 174–176
Resulfurized steels, 214–220
Retained austenite, 16, 41–43
Rolling-contact fatigue, 264, 266
Roughness parameters, 102, 110–117

Scanning electron microscopy, 139–170
    atomic number contrast, 140–146
    backscattered electrons, 140–147, 153–162, 178, 213, 215, 225
    channeling patterns, 146
    contrast mechanisms, 139–150
    crystallographic contrast, 146–147
    deep etching, 104–105, 160–165
    depth of focus, 151
    etching fractures, 162
    fatigue striations, 117–119, 253–254
    fractography, 101, 105–109
    image modes, 140
    magnetic contrast, 148
    microscope variables, 150–154
    quantitative metallography, 165–167
    replicas, 162, 165
    resolution, 150
    specimen preparation, 154–165
    stereo-parts, 102–104
    topographic contrast, 140
    two-surface analysis, 160
    wear analysis, 264–271
    X-ray contrast, 148–150
Scanning transmission electron microscopy, 171–210
    backscattered electron imaging, 178
    bright field imaging, 176–178
    dark-field imaging, 178
    electron energy loss, 179–181, 189, 192–194
    HOLZ lines, 181–184, 186–187
    imaging modes, 176–181
    instrument, 171–173
    microdiffraction, 181–186, 191
    resolution, 174–176
    secondary electron imaging, 178
    specimen preparation, 172–174, 193–194
    specimen thickness, 193–194
    wear debris analysis, 276
    X-ray analysis, 184, 186–190, 192
    X-ray imaging, 179
Scratch hardness, 127–128
Selected area diffraction, 181–182
Selective etching, 3–18
Selenium-treated steels, 220–221
SEM examination, 213–215
Sensitization, 79–81
Sigma phase:
    etching, 85–86
    polarized light response, 2
    potentiostatic etching, 30, 32–33
Sintered carbides:
    etching, 18
    gas contrasting, 66–67
    heat tinting, 18
    relief polishing, 3
Specimen preparation:
    free-machining steels, 212–213
    interference layer method, 43–45
    microindentation hardness, 132, 135–136
    scanning electron microscopy, 154–165
    scanning transmission electron microscopy, 172–174, 193–194
    stainless steels, 44–45, 75–77
    titanium alloys, 237–238
    wear analysis, 267–268, 271–274
Spectroscopy; see Energy-dispersive spectroscopy
Spot tests, chemical, 279–280
Stain etchants; see tint etchants
Stainless steels, 71–88
    alloy compositions, 72
    etching, 5, 14–16, 76–80, 86
    heat tinting, 16–17
    magnetic etching, 55–60
    potentiostatic etching, 32–37
    specimen preparation, 44–45, 75–77
    thin-foil preparation, 86–87
    types, 71–75
Steels:
    calcium-treated, 222–224
    carbon tool, 11–12
    free-machining; see Free-machining steels
    leaded, 211, 213–215, 220–221
    resulfurized, 214–220
    selenium-treated, 220–221
    stainless; see Stainless steels
    tellurium-treated, 220–221
Stereology; see Quantitative stereology
Stereoscopic methods, 101, 102–103, 274–275
Strain-induced martensite, 56, 59, 75
Stress-corrosion cracking, 80, 82–85, 254–255, 289–292
Striations, 117–119, 253–254
Sulfur print, 213–214, 280
Surface area per unit volume, 92, 97
Sulfides, 211, 214–220
    optical characteristics, 48, 50–51
    tint etching, 13–14
    types, 216, 218, 221
    X-ray maps, 148–149, 217, 225
    see also Manganese sulfides

Taper sections, 135–136, 273–274, 279–280
Tellurium-treated steels, 220–221
Thin-foil preparation:
    ion beam thinning, 238–239
    stainless steels, 86–87
    titanium alloys, 238–239
    wear specimens, 275–276
Tint etchants, 4–5, 7–14, 146–147, 156–159
Tinting, heat, 5–6, 16–18
Titanium alloys, 237–249
    alpha and near-alpha alloys, 241–243
    alpha-beta alloys, 243–245
    allotropic transformation, 237, 240
    beta alloys, 245–247
    creep properties, 241–243, 245, 247
    cryogenic properties, 243, 247
    etching, 5–6, 18
    forging, 241–242, 245
    formability, 245
    fracture toughness, 243–245
    heat tinting, 17–18
    heat treatment, 239–241, 246
    metallographic preparation, 237–238
    microstructure, 239–247
    properties, 241–247
    stress-corrosion cracking, 242

superplasticity, 245
thin-foil preparation, 238–239
weldability, 243
Titanium aluminide, 241–242
Topographic contrast, 140
Transmission electron microscopy:
fracture replication, 101, 110
thin-foil preparation, 86–87, 238–239, 275–276
wear analysis techniques, 266, 275
wear surface replication, 268
Triple line length, 98
Two-surface analysis, 160

Volume fractions, 92, 96–97, 165–167, 229

Wear, 261–280
abrasive wear, 262–264
adhesive wear, 261–262
cleaning specimen, 268, 279
erosion, 264–266
fretting, 262–264
interpretation, 268–271
microscopic techniques, 264–268, 271–275
rolling contact fatigue, 264, 266
specimen preparation, 267–268, 271–274
spot tests, 279–280
stereo-pairs, 274–275
taper sections, 273–274, 279–280
wear debris, 275–276
wear modes, 261–264
Weld metal defects, 201–203
Welding metallography, 197–210
defects, 201–203
dissimilar metal welds, 208–209
fusion zone, 197
grain structure, 199–200
heat-affected zone, 197, 207–208
interdendritic eutectic, 200–201
multi-pass welds, 197, 206–207
single-pass welds, 197
solid-state transformations, 203–207
thermal cycle, 198

X-ray analysis, 184, 186–190, 192
X-ray contrast, 148–150
X-ray imaging, 179
X-ray mapping, 148–149, 179, 217, 225, 267, 270